International Conference on Materials Technology and Energy (ICMTE 2017)

IOP Conference Series: Materials Science and Engineering
Volume 217

Miri, Malaysia
20 – 21 April 2017

ISBN: 978-1-5108-4418-6
ISSN: 1757-8981

Printed from e-media with permission by:

Curran Associates, Inc.
57 Morehouse Lane
Red Hook, NY 12571

Some format issues inherent in the e-media version may also appear in this print version.

Printed by Curran Associates, Inc. (2017)

Institute of Physics
Dirac House, Temple Back
Bristol BS1 6BE UK

Phone: 44 1 17 929 7481
Fax: 44 1 17 920 0979

techtracking@iop.org

Additional copies of this publication are available from:

Curran Associates, Inc.
57 Morehouse Lane
Red Hook, NY 12571 USA
Phone: 845-758-0400
Fax: 845-758-2633
Email: curran@proceedings.com
Web: www.proceedings.com

Table of contents

Volume 217

International Conference on Materials Technology and Energy
20–21 April 2017, Curtin University, Malaysia

Accepted papers received: 28 June 2017
Published online: 7 July 2017

Preface

Papers

IOP Conf. Series: Materials Science and Engineering **217** (2017) 011001 doi:10.1088/1757-899X/217/1/011001

Foreword

It is a great pleasure to welcome you to the International Conference on Materials Technology and Energy (ICMTE) 2017.

ICMTE 2017 is an international conference aimed at bringing together academics and practitioners in the field of materials technology and energy. The conference is jointly organized by the Institution of Engineers Malaysia (IEM-Miri Branch), Institute of Materials Malaysia (IMM-Miri Chapter), and Curtin University Malaysia.

ICMTE 2017 has attracted more than 50 contributions around the world to bring forth the latest technology and research on materials technology and energy. Keynote speakers are reputable and eminent academic and industrial practitioners in their respective field. This conference will provide platform to discuss recent technical advancements, innovation, and best practices in relation to materials technology, and energy. This is an excellent opportunity for networking and exchanging of ideas as well as for discussing future challenges.

On behalf of the organizing committee, I would like to thank our sponsors, Sarawak Energy Berhad and Sarawak Convention Bureau. We appreciate your support for ICMTE 2017 and in research and development, and your foresight in nurturing cutting edge research into industrial applications. ICMTE 2017 would not be possible without the dedicated work and efforts of the organizing committee, who worked tirelessly in all aspects of the conference organization. I thank you for your hard work and commitment towards making ICMTE 2017a success.

Selamat Datai (Welcome) and enjoy the conference.

Dr Vincent Lee Chieng Chen
Organizing Chairperson

IOP Conf. Series: Materials Science and Engineering **217** (2017) 011002 doi:10.1088/1757-899X/217/1/011002

Organising Committee

Advisors

Ir Paul Chew Lik Ing
Ir Professor Lau Hieng Ho
Ir Associate Professor Edwin Jong Nyon Tchan
Professor Beena Giridharan

Chairperson

Associate Professor Vincent Lee Chieng Chen

Co-chairperson

Dr Yam Ke San

Secretariat

Ir Meheron Selowara Joo
Shirley Liew

Treasurer

Lee Chun Khiong
Ir Sean Wong

Committee

Dr Stephanie Chan Yen San
Dr Henry Foo Chee Yew
Ir Dr Wong Wei Kitt
Ir Desmond Chin Teck Eng (CPD and Proceedings)
Ir Tian Foo Howe (Sponsorship)
Bernard Maxmillan (Sponsorship)
Chang Chee Tsung (Publicity)

Peer review statement

All papers published in this volume of *IOP Conference Series: Materials Science and Engineering* have been peer reviewed through processes administered by the proceedings Editors. Reviews were conducted by expert referees to the professional and scientific standards expected of a proceedings journal published by IOP Publishing.

The effects of opening areas on solar chimney performance

L S Ling[1], M M Rahman[1,*], C M Chu[1], M S bin Misaran[1] and F M Tamiri[1]

[1]Material and Mineral Research Unit (MMRU), Universiti Malaysia Sabah, Kota Kinabalu, Sabah, Malaysia.

E-mail: mizanur@ums.edu.my

Abstract. To enhance natural ventilation at day time, solar chimney is one of the suitable options for topical country like Malaysia. Solar chimney creates air flow due to stack effect caused by temperature difference between ambient and inside wall. In the solar chimney, solar energy is harvested by the inner wall that cause temperature rise compare to ambient. Therefore, the efficiency of the solar chimney depends on the availability of solar energy as well as the solar intensity. In addition, it is very hard to get good ventilation at night time by using a solar chimney. To overcome this problem one of the suitable valid option is to integrate solar chimney with turbine ventilator. A new type of solar chimney is designed and fluid flow analyzed with the computational fluid dynamics (CFD) software. The aim of CFD and theoretical study are to investigate the effect of opening areas on modified solar chimney performance. The inlet and outlet area of solar chimney are varied from $0.0224m^2$ to $0.6m^2$ and $0.1m^2$ to $0.14m^2$ respectively based on the changes of inclination angle and gap between inner and outer wall. In the CFD study the constant heat flux is considered as $500W/m^2$. CFD result shows that there is no significant relation between opening areas and the air flow rate through solar chimney but the ratio between inlet and outlet is significant on flow performance. If the area ratio between inlet and outlet are equal to two or larger, the performance of the solar chimney is better than the solar chimney with ratio lesser than two. The solar chimney performance does not effect if the area ratio between inlet and outlet varies from 1 to 2. This result will be useful for design and verification of actual solar chimney performance.

1. Introduction

Malaysia is situated at the equatorial region therefore received high solar radiation intensity as well as high relative humidity. This hot and humid climate has posed difficulties in achieving indoor thermal comfort. As a result, the household sector in Malaysia has consumed high electricity mainly for cooling [1].

Solar chimney seems to be a good potential energy conservation strategy to reduce the room temperature. Significant numbers of studies have been done to increase its capability of reducing heat from the room as well as enhanced the rate of natural ventilation [2]. Moreover, solar chimney received energy mainly from the sun which is abundant in Malaysia. The daily average solar radiation received in Malaysia is about 4000 – $5000Wh/m^2$ and average sunshine duration of 4 – 8 hours per day [3].

The design of solar chimney is very simple which consists of glazing, air gap, absorber surface, and insulator. In the solar chimney, energy is harvested through the absorber surface and heating up the air presents between the glazing and the absorber. As a result, the air become lighter and pushed up. The hot and light air is delivered out from the chimney due to buoyancy force that is known as stack effect. The hot air delivered outside and relatively cold air enters from the room to fill the gap. The process continues as long as the energy is available in the absorber [4]. In order to ensure suitable and reliable ventilation during overcast condition as well as night time, solar chimney integrated with turbine ventilator is one of the good options. This option is suitable as availability of solar and/or wind energy at a reasonable intensity during a large part of any day [5]. Furthermore, this integration system is expecting to further enhance the natural ventilation performance compare to individual system.

In this study, a new type solar chimney is introduced that can be easily integrated with the turbine ventilator. In order to optimize the performance of solar chimney, it is important to understand the effect of different solar chimney's parameters. The predominant parameters are chimney aspect ratio (stack height/air gap depth), ventilation height (height between inlet and outlet), thermal characteristics of the absorber, and chimney inclination angle [6]. In addition, opening areas of solar chimney have direct and great influence on the air flow rate. The opening areas change the resistance to the flow at inlet and outlet of the chimney [7-8]. Although opening areas have impacts on the performance of solar chimney, but not much investigations have reported on it. Therefore, the aim of this study is to investigate the effect of opening areas on the modified solar chimney performance.

2. Description of analysed solar chimney

The new type solar chimney consists of outer part and inner part. As shown in figure 1, two sides of the outer part are covered by transparent glasses that act as glazing. The glazing allows solar radiation to pass through and the energy is received by the absorber at the inner part.

According to figure 2, two sides of the inner part are absorber and the other two sides are perforated plates. The perforated plates allow air to flow through the chimney. By assembling outer part and inner part together, the new type solar chimney along with its section view is shown in figure 3.

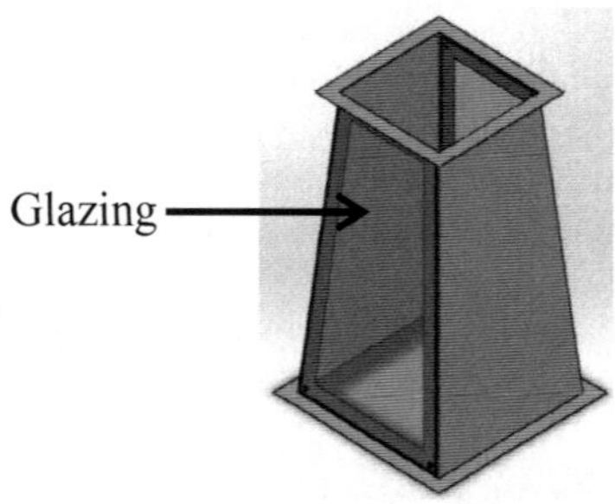

Figure 1. Outer part of solar chimney.

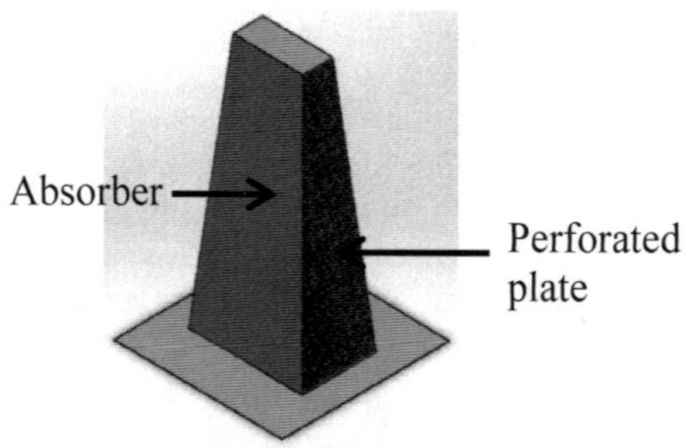

Figure 2. Inner part of solar chimney.

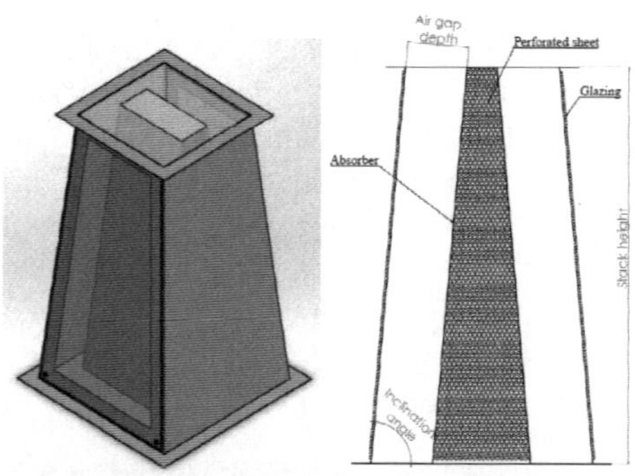

Figure 3. Solar chimney and its section view.

There are 16 different configurations to investigate the effect of opening areas on the performance of a solar chimney as shown in table 1. The inlet and outlet area of solar chimney are varied from $0.0224m^2$ to $0.6m^2$ and 0.1m2 to 0.14m2 respectively based on the changes of inclination angle and gap between inner and outer wall.

Table 1. Opening areas of the 16 configurations of the solar chimney.

Outlet area (m^2)	Inlet area (m^2)	$Ratio = \frac{Outlet}{Inlet}$
0.104	0.056	0.54
	0.169	1.62
	0.347	3.34
	0.597	5.74
0.115	0.045	0.39
	0.151	1.31
	0.322	2.80
	0.565	4.90
0.126	0.034	0.27
	0.133	1.05
	0.297	2.35
	0.532	4.21
0.138	0.022	0.16
	0.115	0.83
	0.272	1.97
	0.500	3.63

3. Modelling the solar chimney

Two approaches are taken to investigate the effect of opening areas on the performance of solar chimney. One of the approaches is via computational fluid dynamics (CFD) software named SolidWorks. Another approach is using the mathematical equations that had been developed by the Xu and Liu [9].

3.1. Computational Fluid Dynamics (CFD)

The following simplifying assumptions are made in modelling the solar chimney with CFD software named SolidWorks:

- Air is considered as flowing fluid
- The air follows the ideal gas law
- Wind effect is not included
- The entire system is at steady state condition
- Frictional losses is negligible due to the very low order of flow velocity
- Air leakage is negligible
- The solar chimney walls are considered as adiabatic wall
- The absorber is considered as pure black body

In this study, the solar chimney modeling has been done under constant heat flux. Assume that absorber acts as heating source and supplied heat energy at constant rate $500W/m^2$. The initial temperature of the whole model has been set to be the same as the ambient temperature at 301K including the solar chimney surface temperature. At the inlet and outlet, the pressure has been considered as ambient which is 101kPa.

The flow in the solar chimney has been simulated at steady state condition. Half of the chimney is considered as computational domain since other part of the chimney is symmetric. The details about the number of mesh have shown in table 2. Under the similar heating and initial condition setting, different configurations are compared by determining the air flow rate that can be achieved by each configuration.

Table 2. Number of mesh cells for different configurations of solar chimney.

Outlet area (m^2)	Inlet area (m^2)	Number of cells X	Y	Z
0.104	0.056	14	56	28
	0.169	16	46	32
	0.347	16	38	32
	0.597	16	32	32
0.115	0.045	14	56	28
	0.151	16	46	32
	0.322	16	38	32
	0.565	16	32	32
0.126	0.034	14	56	28
	0.133	16	46	32
	0.297	16	38	32
	0.532	16	32	32
0.138	0.022	14	56	28
	0.115	16	46	32
	0.272	16	38	32
	0.500	16	32	32

3.2. Theoretical calculation

According to Xu and Liu [9], the theoretical mass flow rate ($\dot{m}$), volumetric flow rate$\left(\dot{V}\right)$, and air velocity (v) inside the solar chimney can be estimated through the following equations.

$$(\dot{m}) = c_d \rho_{f1} A_0 \left[\frac{2gL_s sin\theta\left(T_f - T_r\right)}{(1 + A_r^2)T_r}\right]^{\frac{1}{2}} \tag{1}$$

$$\left(\dot{V}\right) = \frac{\dot{m}}{\rho_{f1}} = c_d A_0 \left[\frac{2gL_s sin\theta\left(T_f - T_r\right)}{(1 + A_r^2)T_r}\right]^{\frac{1}{2}} \tag{2}$$

$$v = \frac{\dot{m}}{\rho_{f1} A_0} = c_d \left[\frac{2g\, L_s sin\theta\left(T_f - T_r\right)}{(1 + A_r^2)T_r}\right]^{\frac{1}{2}} \tag{3}$$

where,

c_d = mean coefficient of discharge of air channel inlet
ρ_{f1}= density of air in the flow channel
A_0= cross sectional area of outlet
A_r= ratio of cross sectional area of outlet over cross sectional area of inlet
T_f= mean temperature of air in the channel
T_r= room temperature

L_s= stack height
g = acceleration due to gravity
θ= angle of inclination with the horizontal surface

4. Results and discussions

The experimental set up is shown in figure 4, where the absorber is heated by using a heater under constant heat flux (500W/m^2). A hot wire anemometer is placed below the solar chimney and at the center of the inlet opening area to measure the inlet air velocity after 1 hour heating.

Figure 5 shows that the simulation result is giving similar trend compare to experimental result. Therefore, similar simulation setting can be carried out to investigate the effect of opening areas on the performance of solar chimney. The variations in the volume flow rate with respect to opening areas are shown in figure 6 to figure 9.

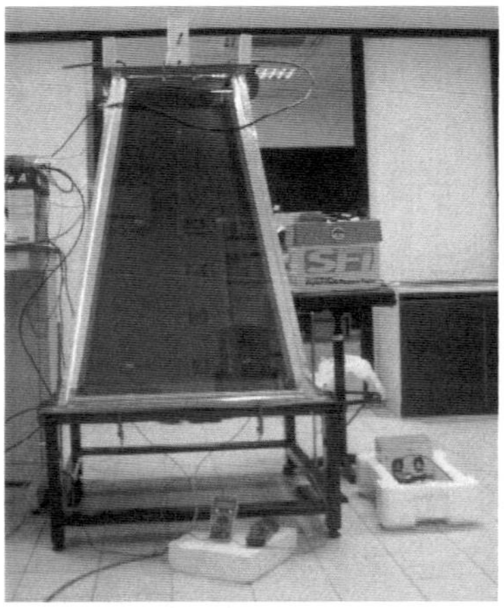

Figure 4. Experimental set up to investigate solar chimney performance.

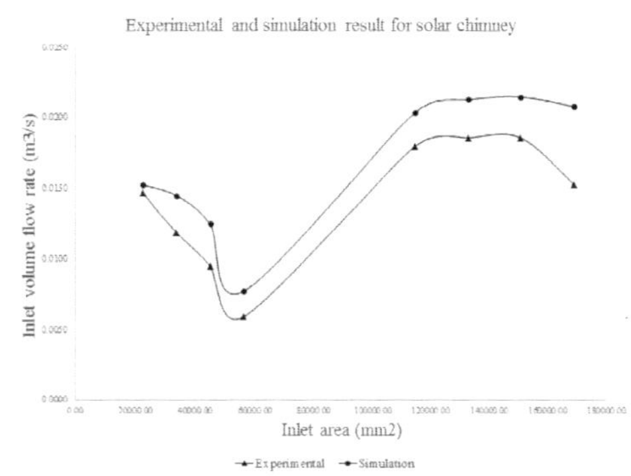

Figure 5. Experimental and simulation result for solar chimney.

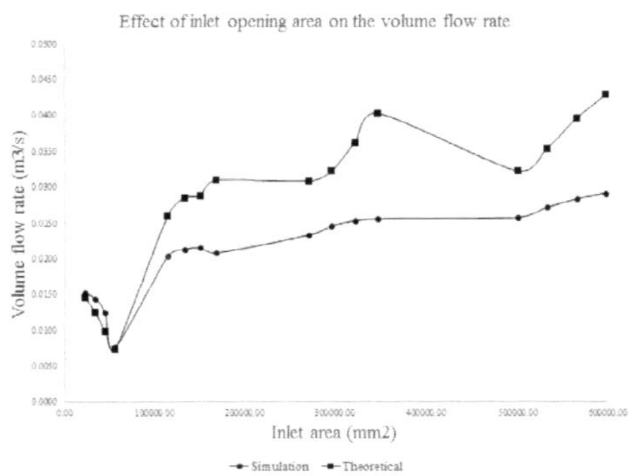

Figure 6. Effect of inlet opening area on volume flow rate.

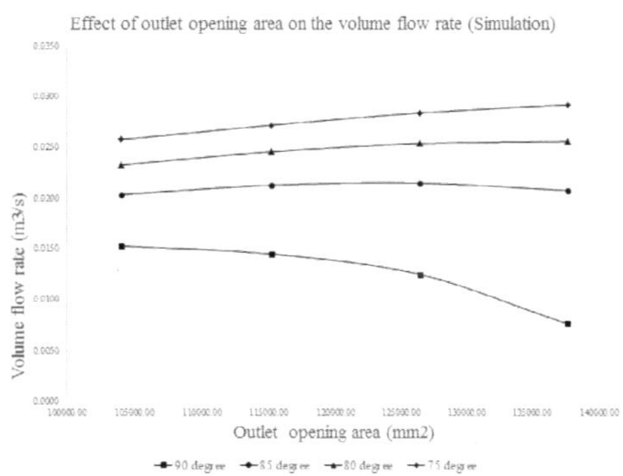

Figure 7. Effect of outlet opening area on volume flow rate (simulation).

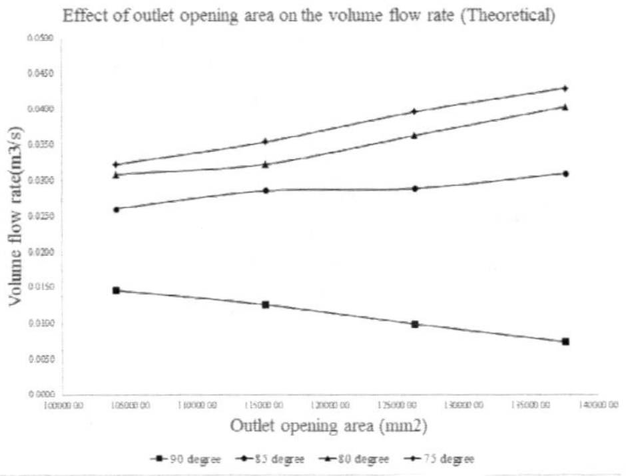

Figure 8. Effect of outlet opening area on volume flow rate (theoretical).

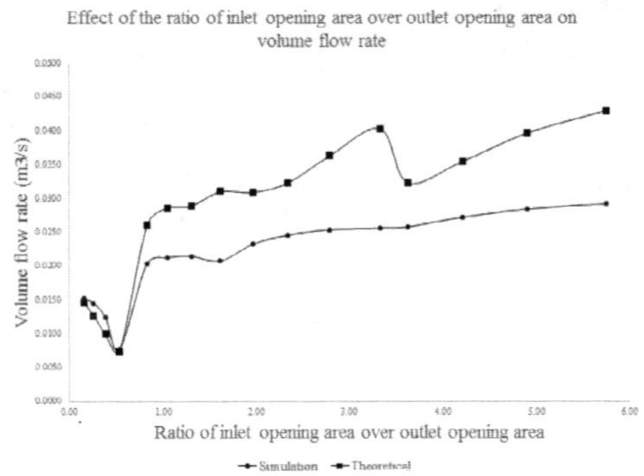

Figure 9. Effect of the ratio of inlet opening area over outlet opening area on volume flow rate.

The volumetric flow rate is lower when the inlet opening area is bigger (Figure 6). It is also found that the volumetric flow rate is higher when the inlet opening area is bigger starting from certain inlet opening area size (0.056m^2). From the Figure 6, it can be concluded that bigger inlet opening area is not necessary to have better flow rate than smaller inlet opening area. Similarly, bigger outlet opening area is also not necessary has better flow rate than smaller outlet opening area. This is shown in Figure 7 and 8 that significant flow rate is established at bigger outlet opening area for 75°, 80°, and 85°, but it is not significant when chimney angle is 90°.

Therefore, it is very hard to conclude the relationship between inlet as well as outlet opening area on the volume flow rate of solar chimney. On the other hand, some important correlation between opening area and performance of solar chimney is found in Figure 9. From Figure 9, the ratio of inlet opening area over outlet opening area equal to two or larger is performing better than the solar chimney with the ratio lesser than two. Moreover, the volume flow rate achieve by the solar chimney does not differ much when the ratio is between 1 and 2.

5. Conclusion

Solar chimney is workable only when the solar energy is available. In order to prolong the duration either cloudy day or night time and further enhanced the natural ventilation induced, a new type solar chimney is introduced that ease its integration with the turbine ventilator. In this study, the effect of opening areas on the performance of solar chimney has been studied. One of the findings is the inlet opening area should be at least double the outlet opening area. This is due to bigger outlet opening area than inlet opening area is performing badly. Another finding is the performance of solar chimney does not differ much when the area ratio between inlet and outlet varies from 1 and 2. All those information are useful when deciding the suitable opening areas of solar chimney.

References

[1] Ariffin A S, Rao S P and Inangda S N 2002 *Development of Passive Solar Design and Technology in Tropical Climates* 23 – 34. Universiti Sains Malaysia, Pulau Pinang.
[2] Zhai X Q, Song Z P and Wang R Z 2011 *Renewable and Sustainable Energy Reviews* **15** 3757 – 67.
[3] Mekhilef S, Safari A, Mustaffa W E S, Saidur R, Omar R and Younis M A A 2012 *Renewable and Sustainable Energy Reviews* **16** 386 – 396.
[4] Tan A Y K and Wong N H 2014 *Solar Energy* **102** 116 – 125.
[5] Shun S and Ahmed N A 2008 *Renewable Energy* **33** 1392 – 1397.
[6] Khanal R and Lei C 2011 *Energy and Buildings* **43** 1811 – 1819.

[7] Khedari J, Mansirisub W, Chaima S, Pratinthong N and Hirunlabh J 2000 *Energy and Buildings* **31** 171 – 178.
[8] Maerefat M and Haghighi A P 2010 *Renewable Energy* **35** 2316 – 2324.
[9] Xu J L and Liu W H 2013 *Energy and Buildings* **66** 467 – 469.

Acknowledgement
The author would like to express his gratitude to the Ministry of Higher Education of Malaysia and Universiti Malaysia Sabah for their financial assistance and facilities support through Fundamental Research Grant (FRG0429-TK-1/2015). In addition, the authors would like to express sincere thanks to the evaluators of ICMTE 2017 for valuable comments on this manuscript.

Experimentally Identify the Effective Plume Chimney over a Natural Draft Chimney Model

M M Rahman[1,], C M Chu [1], A M Tahir[1], M A bin Ismail[1], M S bin Misran[1] and L S Ling[1]

[1]Thermal and Environmental Research Group, Material and Mineral Research Unit (MMRU), Faculty of Engineering, Universiti Malaysia Sabah, 88400, Kota Kinabalu, Sabah, Malaysia

E-mail: mizanur@ums.edu.my

Abstract. The demands of energy are in increasing order due to rapid industrialization and urbanization. The researchers and scientists are working hard to improve the performance of the industry so that the energy consumption can be reduced significantly. Industries like power plant, timber processing plant, oil refinery, etc. performance mainly depend on the cooling tower chimney's performance, either natural draft or forced draft. Chimney is used to create sufficient draft, so that air can flow through it. Cold inflow or flow reversal at chimney exit is one of the main identified problems that may alter the overall plant performance. The presence Effective Plume Chimney (EPC) is an indication of cold inflow free operation of natural draft chimney. Different mathematical model equations are used to estimate the EPC height over the heat exchanger or hot surface. In this paper, it is aim to identify the EPC experimentally. In order to do that, horizontal temperature profiling is done at the exit of the chimneys of face area 0.56m^2, 1.00m^2 and 2.25m^2. A wire mesh screen is installed at chimneys exit to ensure cold inflow chimney operation. It is found that EPC exists in all modified chimney models and the heights of EPC varied from 1 cm to 9 cm. The mathematical models indicate that the estimated heights of EPC varied from 1 cm to 2.3 cm. Smoke test is also conducted to ensure the existence of EPC and cold inflow free option of chimney. Smoke test results confirmed the presence of EPC and cold inflow free operation of chimney. The performance of the cold inflow free chimney is increased by 50% to 90% than normal chimney.

1. Introduction

Chimneys are widely used in the industries like power plant, chemical processing plant, timber processing plant etc. to create sufficient draft. As a result, hot smoke or gases or air can escape to the atmosphere faster than the natural processes [1-3]. Two types of chimneys; natural and force draft are commonly used in the industries. In the forced draft chimney, a fan or driver is placed either at the bottom or above the tube bundle. The fan generates sufficient air flow that remove unwanted heat, smoke and dust particle from the system. In the natural draft chimney, the temperature difference between process sides and the ambient generates buoyancy force or stack effect. As a result, air flow through the chimney and remove heat waste from the system. This process is continuous until the buoyancy or stack effect is present in the chimney. There are no mechanical appliances used in the natural draft chimney. Therefore, it has more advantage than forced draft in terms of operational safety and reliability [4].

The identified problems in the natural draft chimney that may alter its performance are cross wind, flow reversal and hot air recirculation [5]. The installation of side wall can eliminate the effects of cross wind and hot air recirculation significantly. The flow reversal or cold inflow at exit point weaken the performance of the chimney by reducing the pressure draft in chimney [6-7]. Cold inflow occurs when inappropriate chimney's height is used in the system or system draft is overcome by its depressurization [8, 9]. Very limited experimental investigations have been carried out to understand the presence, behaviour and effects of flow reversal [6-11]. Most of the related researchers measured temperature around the chimney exit or reduction air inlet temperature as a tool to determine the cold inflow. The presence of effective plume above the chimney can also be used as an indication of cold inflow free natural draft chimney. An effective plume chimney is a free boundary flow of fluid from constant heat source and it is driven by buoyancy force.

2. Effective Plume-Chimney Height (EPCH)

The effective plume chimney is formed above the hot surface due to the buoyancy force. Therefore, in the effective plume, the air velocity decreased gradually and momentum increased continuously. In the effective plume chimney, the centreline temperature remains the same as source exit air temperature. The temperature drops due to mixing and loss of buoyancy [12]. Zinoubi et al. in the year 2005, has conducted an experimental study on plume thermos-siphon interaction. In this study the horizontal and vertical temperature profile inside a cylinder is determined. The average temperature reduced along the plume vertical axis due to penetration of cold air from the side of the model. At higher level, the temperature profile is noticeably uniform which indicates the establishment of the turbulence and no penetration of cold air inflow [13]. Therefore, the Effective Plume Chimney Height (EPCH) is the extreme height of the plume chimney where centreline temperature remains the same as source exit air temperature. It would act as a solid hardware chimney but invisible in nature. Doyle and Benkly's (1973) empirical formula is the first known predictive method on the EPCH of forced and induced draft air-cooled heat exchanger.

$$h_{EPCH} = 0.8\, F^{1/2} \tag{1}$$

and

$$F = 2.45\left(\frac{\Delta T}{T_a}\right) v_F D_h^2 \tag{2}$$

$$v_F = 1.08\left(\frac{\Delta T}{400}\right)^{3/8}\left(\frac{h}{6}\right)^{1/4}\left(\frac{6}{N_R}\right)^{1/2} \tag{3}$$

where,
A_F = Face area of tube bundle (m^2)
D_h = Equivalent hydraulic diameter (m)
N_R = Total tube rows number
ΔT = Process side and inlet temperature difference (K)
F = Ratio Buoyant force and ambient density (m^4s^{-3})
v_F = Velocity at inlet (ms^{-1})

This empirical formula does not consider the significant effect of density difference. A mathematical relation is suggested by Chu in the year 2002 to estimate the EPCH for turbulence flow as follows.

$$h_0 = 10.1\frac{L_B^{0.5}\beta^{0.8}\Delta T^{0.8}\eta^{0.3}\rho_o^{0.7}}{(\Delta\rho)g_n^{0.2}} \tag{4}$$

where,
h_0 = effective plume chimney height(m)
L_B= plate breadth (m)
β = coefficient of thermal expansion (k^{-1})
η = dynamic viscosity of fluid (Ns m^{-2})
ρ = density of fluid (kg m^{-3})
Equation (4) is a buoyancy-pressure drop balance equation solved for natural draft flow together with the heat transfer equation, yielding the airflow rate and the exit air temperature. This equation is suitable when the pressure and temperature change are linear. These mathematical formulas can be used to estimate the EPCH above the hot surface for quiescent ambient condition. [5, 14-19].

3. Experimental Procedure

Actual natural draft chimney is very large in size and shape. Therefore, it is difficult to conduct intensive research on an actual dimension natural draft chimney. In order to conduct research on natural draft chimney, different researchers used a different dimensional model that makes it difficult to compare experimental results with actual natural draft chimney. In this study, three models with chimney face areas 0.56 m^2, 1.00 m^2 and 2.25 m^2 with 0.3 m to 1.2 m solid chimney has been used. A wire mesh screen has been installed at the exit of the chimney. The experiments have been carried out with and without wire mesh screen for each heat load approximately 1.0kW, 1.5kW, 2.0kW and 2.5kW. An array of seven k-type thermocouples have been located above the natural draft chimney's exit to determine the centreline temperature. Five k-type thermocouples are also placed at the exit of the chimney to determine the exit temperature of the chimney.

4. Experimental Results and discussion

The exit air temperature has been measured for different solid chimney heights, face areas and heat loads. The temperature above the ambient value is considered as exit temperature rise. Figure 1 shows the exit air temperature rise at different heat loads. For the same heat load, it is found that wire mesh screen significantly increases the chimney's exit air temperature by reducing the penetration of cold air at exit point of the chimney. When the exit air temperature is low at a low heat load in the natural draft chimney, the cold air penetrated the chimney and mixed with hot air, resulting immediate temperature drops.

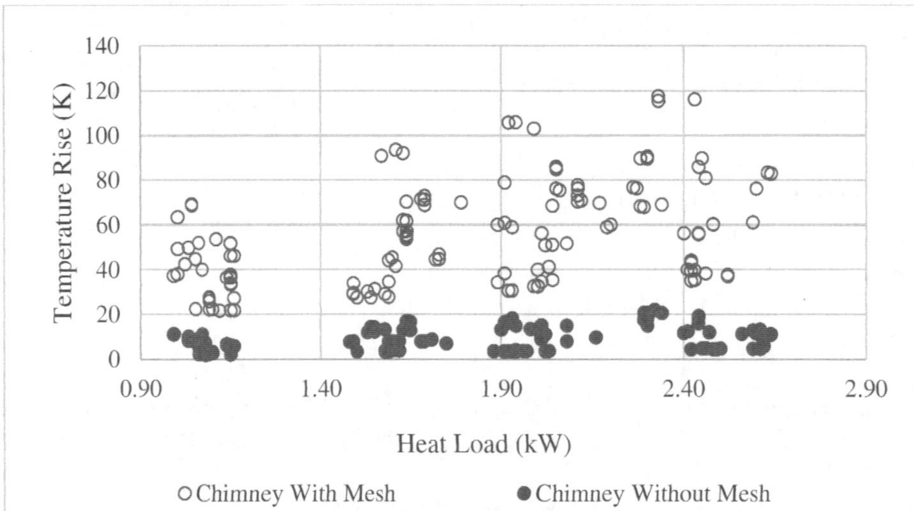

Figure 1. Exit air temperature rise for different heat load

The temperature rise is also compared with theoretical value of temperature rise and presented in the Figure 2 and Figure 3. In Figure 2, the relations are significant for the face area of $0.56m^2$, $1.00m^2$ and 2.56 m^2 whereas in Figure 3 the relations are insignificant due to fluctuation of exit air temperature which is caused by cold inflow. The cold air is mixed with the exit air and dramatically reduces the exit air temperature. Therefore, the ratio of the calculated value and measured value fluctuate a lot. The data also indicates that wire mesh screen at the natural draft chimney model may reduce the effect of cold inflow or help cold inflow free operation of natural draft chimney.

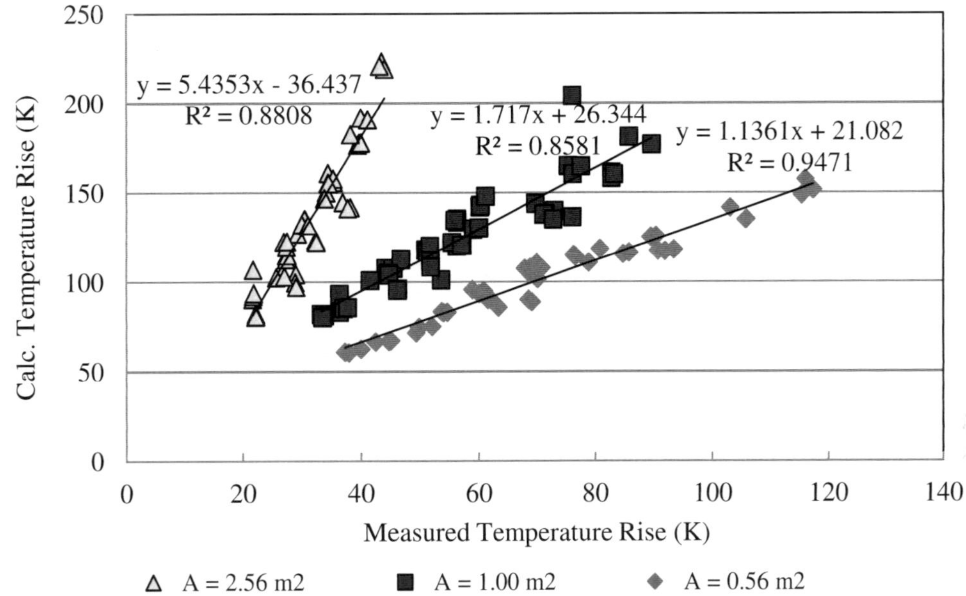

Figure 2. Relation between calculated and actual temperature rise in natural draft chimney with wire mesh screen.

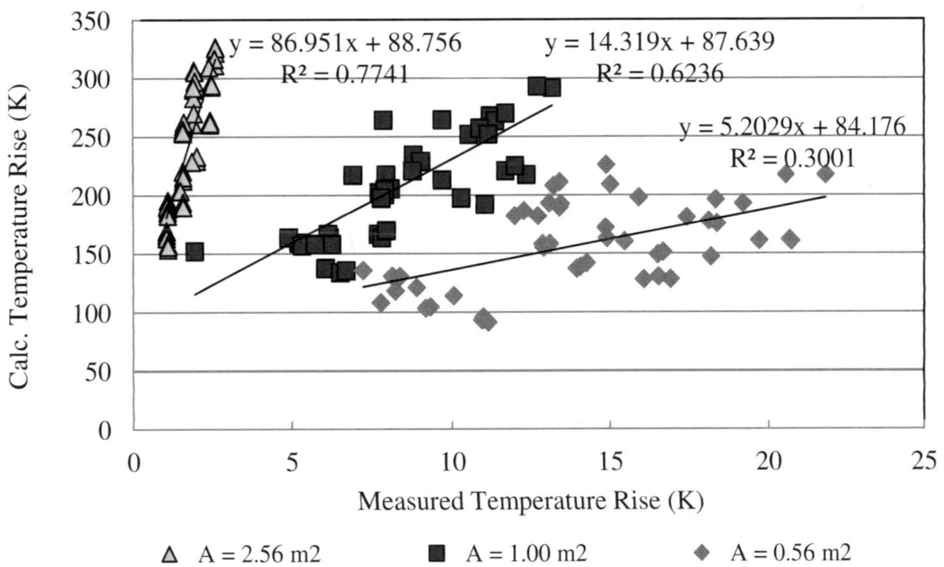

Figure 3. Relation between calculated and actual temperature rise in natural draft chimney without wire mesh screen.

In order to identify the effective plume chimney height, it is calculated from the total pressure draft and is presented against T rise ratio in Figure 4. It is found that in the model chimney, the heights of the effective plume chimney varied from 0.14 m to 0.59 m. Since the pressure draft depends on the presence of wire mesh screen and height of the solid wall chimney as well as loss in the heating system, therefore calculated effective plume chimney heights is not accurate. Hence, heights of the effective plume chimney are also calculated from the Doyle's and Chu's empirical correlation and presented in the Figure 5.

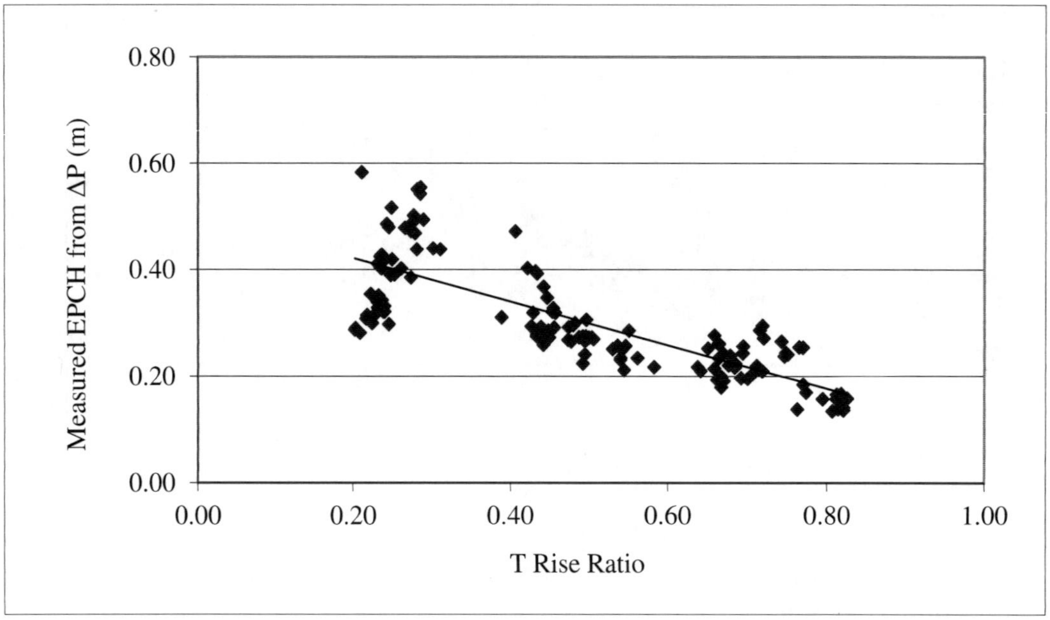

Figure 4. Relation between measure EPCH and T raise ratio

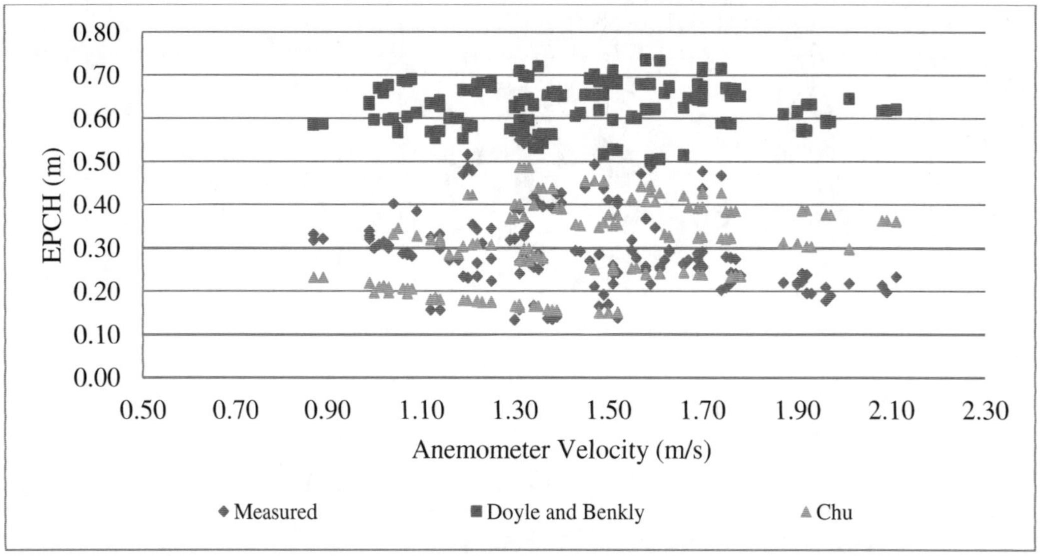

Figure 5. Compare EPCH from different model and measured value

The measured values have found to be significantly higher than the estimated values from Chu's empirical model but lower than Doyle's and Benkly's model. This is because estimated effective plume chimney heights from Chu's empirical model depends on bundle breadth and bundle area while Doyle's and Benkly's model equation is developed for forced draft cooling tower. It is also influenced by the face velocity. This is also supported by Chu 2006 [5]. Therefore, the effective plume chimney is investigated above the wire mesh screen with the ratio exit air temperature and solid chimney temperature at vertical axis and presented in the Figure 6, Figure 7 and Figure 8.

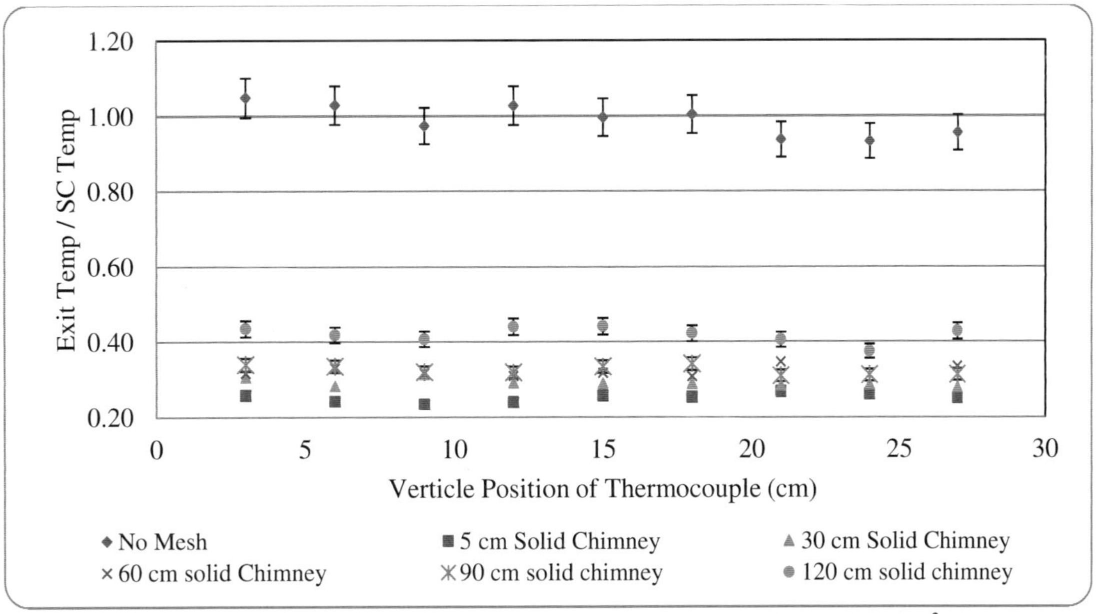

Figure 6. Exit temperature/ solid chimney temperature of model 0.56 m^2

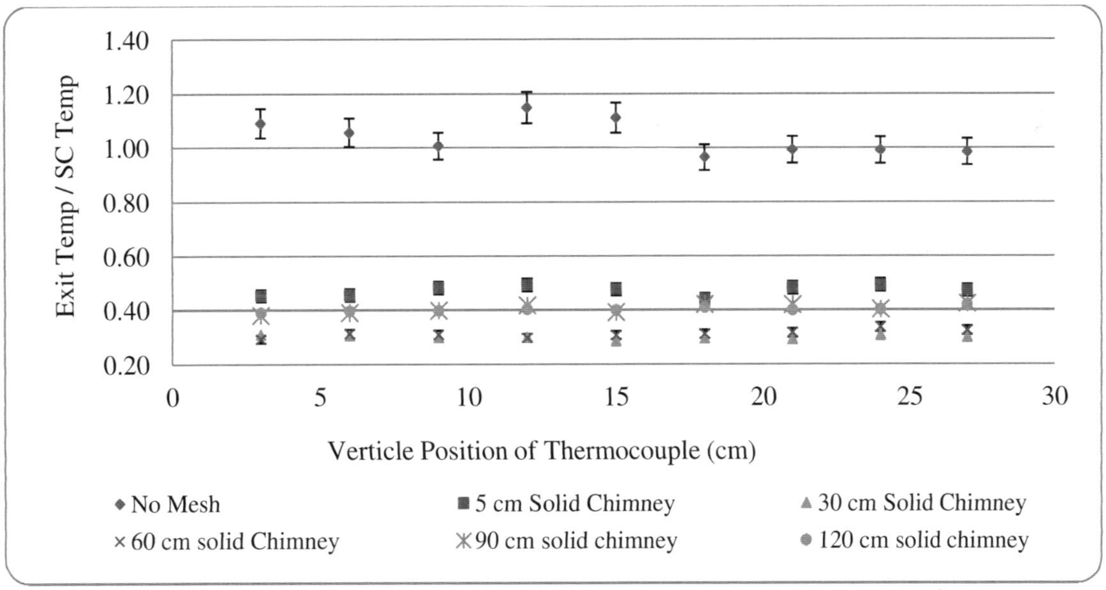

Figure 7. Exit temperature/ solid chimney temperature of model 1.00 m^2

The figures showed that without wire mesh at exit of the chimney, the exit air temperature is higher than the solid chimney temperature some times because the ratio is more than 1. This is only possible when cold air penetrates in the model chimney. In chimney model with wire mesh screen these ratio are always below the unit value that indicate exit air temperature is higher than solid chimney temperature. That is an indication of cold inflow free operation of chimney model.

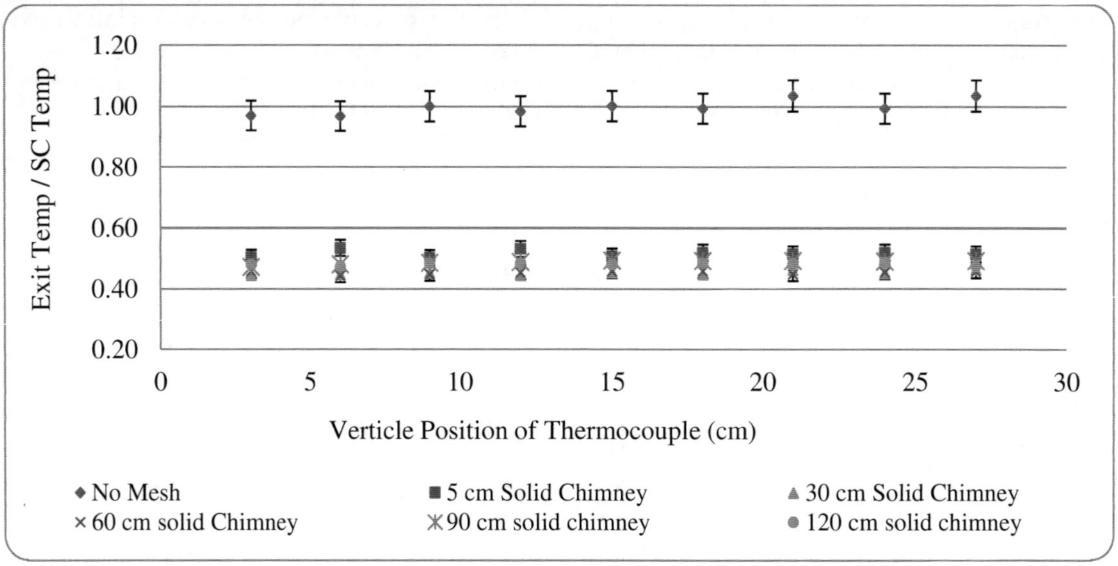

Figure 8. Exit temperature/ solid chimney temperature of model 2.56 m^2

According to the definition of effective plume chimney after the chimney exit the temperature in vertical direction is constant. The temperature is constant until the effective plume present above the chimney exit. Figure 9 and Figure 10 show the vertical temperature profile above the solid wall temperature.

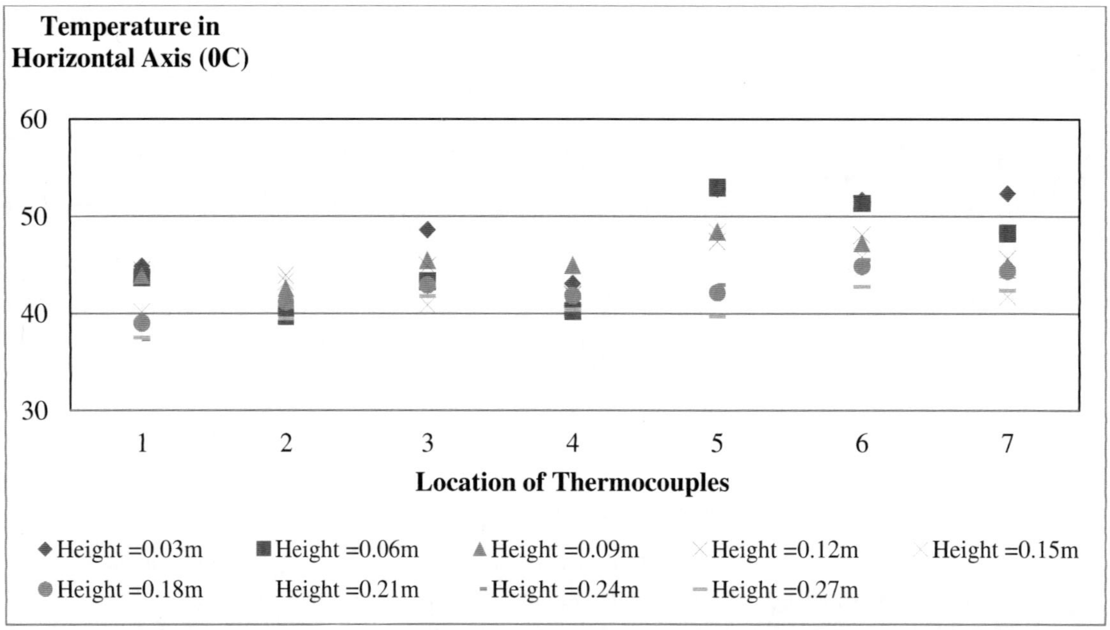

Figure 9. Vertical temperature profile above the solid wall chimney without wire mesh screen

In Figure 9, it is found that in 0.03 m height, the temperature values fluctuate from 45°C to 52^{0}C. It indicates the absence of plume chimney above the solid wall chimney where there is no wire mesh. In Figure 10, the temperatures are fluctuating at side but it is found to be almost linear at middle. The side thermocouples are disturbed by the surrounding air whereas plume is existing at the middle. It also found that wire mesh significantly protects the penetration of cold inflow.

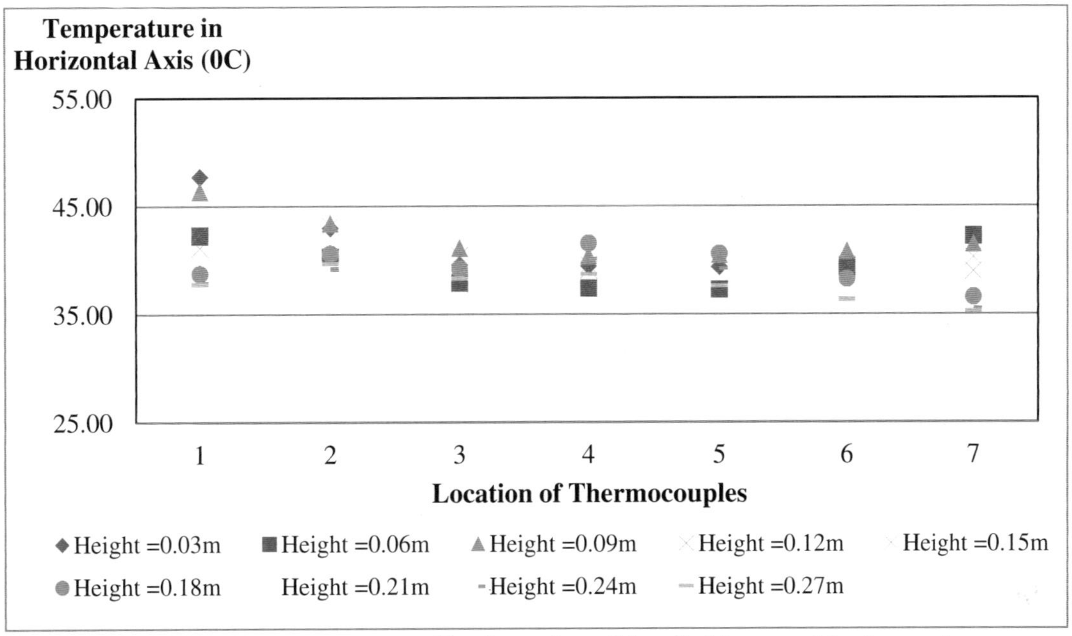

Figure 10. Vertical temperature profile above the solid wall chimney with wire mesh screen

5. Conclusion

The natural draft chimney performance is highly influenced by the flow reversal problems. The exit dimension, temperature distribution and draft loss in the natural draft chimney are responsible for the presence of cold inflow. The wire mesh screen significantly eliminates the cold air penetration problem and helps to establish an effective plume chimney about the solid wall chimney. Experimental results show that effective plume chimney exists in the natural draft chimney when it operates without the effects of cold inflow. The mathematical correlations could be used to state the presence of effective plume chimney but results have found significant error when compared with experimental values.

References

[1] Arce J, Jiménez M J, Guzmán J D, Heras M R, Alvarez G and Xamán J 2009 *Renewable Energy* **34** 2928-34
[2] Fisher T S, Torrance K E and Sikka K K 1997 *IEEE Transactions on Components, Packaging, and Manufacturing Technology: Part A* **20** 111-119
[3] Rahman M M and Chu C M 2007 *21st Symp. of Malaysian Chemical Engineers.* (Selangor: University Putra Malaysia)
[4] Chu C M 2002 *Heat Transfer Engineering* **23** 3-12
[5] Chu C M 2006 *Heat Transfer Engineering* **27** 81-85
[6] Andreozzi A, Buonomo B and Manca O 2009 *Int. J. of Thermal Sciences* **48** 475-487
[7] Zahedi P, Javadi S M, Yousefi K, and Pakdel A 2013 *Int. J. of Materials, Mechanics and Manufacturing* **1** 143-147
[8] Tongbai P, Chitsomboon T 2014 *J. of Power and Energy Engineering* **2** 22-29
[9] Nagda N L, Koontz M D, Billick I H, Leslie N P and Behrens D W 1996 *J. of the Air & Waste Management Association* **46** 838-846

[10] Fisher T S and Torrance K E 1999 *J. of Heat Transfer* **121** 603-609
[11] Thrasher W W, Fisher T S and Torrance K E 2000 *J. of Electron Package* **122** 350–355
[12] Henderson B 1983 *Applied Mathematics Modeling* **7** 395–397
[13] Zinoubi J, Maad B R and Belghith A 2005 *Applied Thermal Engineering* **25** 533-544
[14] Al-Waked R 2010 *Int. J. of Thermal Sciences* **49** 218-224
[15] Bender T J, Bergstrom D J and Rezkallah K S 1996 *J. of Wind Engineering and Industrial Aerodynamics* **64** 61-72
[16] Doyle P T and Benkly G J 1973 *Hydrocarbon Processing* **52** 81-86
[17] Jörg O and Scorer R S 1967 *Atmospheric Environment* **1** 645–646
[18] Zhai Z and Fu S 2006 *Applied Thermal Engineering* **26** 1008-17
[19] Wei Q D, Zhang B Y, Liu K Q, Du X D and Meng X Z 1995 *Journal of Wind Engineering and Industrial Aerodynamics* **54-55** 633-64

Acknowledgments
The authors would like to express their gratitude to the Ministry of Higher Education of Malaysia and Universiti Malaysia Sabah for the financial assistance and facilities supports through Fundamental Research Grant (FRG0429-TK-1/2015). In addition, the authors would like to express sincere thanks to the evaluators of ICMTE 2017 for valuable comments on this manuscript.

Fruit based Dye Sensitized Solar Cells

M C Ung[1,*], C S Sipaut[1], J Dayou[2], K S Liow[1], J Kulip[3], R F Mansa[1]

[1]Energy and Materials Research Group, Faculty of Engineering, Universiti Malaysia Sabah, Jalan UMS, 88400, Kota Kinabalu, Sabah, Malaysia.

[2]Vibration and Sound Research Group, Faculty of Science and Natural Resources, Universiti Malaysia Sabah, Jalan UMS, 88400, Kota Kinabalu, Sabah, Malaysia

[3]Biology Tropical and Conservation Institute, Universiti Malaysia UMS, 88400,, Jalan Maktab Sabah, 88100 Kota Kinabalu, Sabah, Malaysia

*Corresponding author: meeching_88@hotmail.com

Abstract. Dye Sensitized Solar Cell (DSSC) was first discovered in 1991 by O'regan and Gratzel. This new type of solar cell was reported to have lower production cost with efficiency as high as 12% which is comparable to conventional silicon solar cell. Initially, it uses ruthenium dye as light sensitizer for the operation. However, DSSC with ruthenium dyes are facing environment friendly issues due to the toxic chemicals and costly purification in processing ruthenium dye. Regardless of the poor performance in DSSC, natural dyes which are easy to prepare, cheap and environmental friendly still appear to be an alternative as dye sensitizer. In this study, dye sensitized solar cells (DSSCs) were fabricated using anthocyanin source dyes extracted from several local fruits. All the extracts absorb a wide range of the visible light and ultraviolet spectrum. Therefore, all of the natural dyes show light absorption properties which is important for a dye sensitizer. A DSSC is comprised of conductive substrate, nanoporous semiconductor TiO2 layer, dye sensitizer, electrolyte with redox couple and a counter electrode with catalyst. In this study, the effect of different light source and different counter electrode are been investigated. However, it is vital to know that further research need to do more on the locally Borneo sourced dyes to evaluate and enhance their performance in Dye Sensitized Solar Cell.

1. Introduction

Dye Sensitized Solar Cell (DSSC) was first published in 1991[1]. This new type of solar cell was reported to have lower production cost with efficiency as high as 12%. Initially, it uses ruthenium dye as light sensitizer for the operation. However, DSSC with ruthenium dyes are facing environment friendly issues due to the toxic chemicals and costly purification in processing ruthenium dye. Regardless of the poor performance in DSSC, natural dyes which are easy to prepare, cheap and environmental friendly still appear to be an alternative as dye sensitizer.

Dye sensitizers are chemicals which can absorb energy from visible light, which then produces excited electrons to be injected to the wide band gap semiconductor [2]. There are few characteristics for an efficient dye sensitizer. Firstly, it must have an intense absorption ability in the visible region under the illumination of the sunlight. Furthermore, the dye molecules should contain hydroxide groups or carboxyl groups which will firmly graft to the Titanium (IV) sites on the TiO_2 surface.

Moreover, it should also have the ability of efficient electron injection into the conduction band of the semiconductor [3].

In 1997, it was found out that the subgroup of flavonoid which is anthocyanin's works well with DSSC system. Anthocyanins are weak acids and occur naturally in all tissues of the plants such as flowers, fruits, leaves, stem and roots. Dabai (Black Olive) is one of the exotic fruit in Sarawak. The major phenolic component found in the peel of Canarium Odontophyllum is anthocyanin [4]. On the other hand, Both Mangosteen and Wild Mangosteen are tropical fruit widespread over Borneo and they are seasonal. Their fruit peel contains anthocyanin [5].

In this work, the photoelectrochemical results obtained by using natural photosensitizers such as Black Olive (Canarium Odontophyllum), Mangosteen (Garcinia mangostana), Wild Mangosteen (Garcinia Hombroniana) are presented. Furthermore, Blueberry (Vaccinium Myrtillus) was prepared as standard natural dye. The objective of this research is to find out the photoelectrochemical properties of the DSSC using anthocyanin as dye sensitizer. Moreover, the effect of different light source, dye concentration and counter electrode on the DSSC efficiency was investigated.

2. Methodology

2.1. Preparation of Photosensitizers

The natural samples shown in Figure 1 were cut into small pieces to the size of about 0.5cm x 0.5cm. The natural dye solutions were prepared by crushing the samples with distilled water and then the solid residues were removed from the solution. Next, the samples were crushed in a mortar and pestle together with distilled water until the solution turns to colour solution. The solid residues were removed from the solution and the remaining solution was prepared as dye sensitizer. N719 dye (0.3mM) was prepared as standard. The anthocyanin sources that used as photosensitizers were shown in Fig 1.

1a) Black Olive	1b) Mangosteen	1c) Wild Mangosteen	1d) Blueberry

Figure 1a-1d. The anthocyanin sources used as photosensitizers in DSSC

2.2. Preparation of Photoanode

The TiO_2 solution is prepared by the adding of 1.0g colloidal P25 Degussa TiO_2 powder and 0.50g of polyethylene glycol 10,000 (PEG 10,000) into the mortar. Then, the paste was blended with 2.0ml nitric acid (HNO_3) solution (0.1M) and 1 ml of nonionic surfactant, Triton X-100. After that, the paste was blended by pestle in the mortar continuously for 15 minutes until the paste becomes thick and free of aggregates. The paste was left to equilibrate in the fume chamber for another 15 minutes. Next, the paste was deposited on the FTO glass using a doctor blade method until the uniform layer was achieved.

The paste was allowed to be dry first before heat at high temperature. When the paste becomes dry, the scotch tape was removed and the FTO glass was heated on a hot plate at the sintering temperature of 450^0C for 30min to achieve nanoporous structure in semiconductor thin film as shown in figure 2. After 30 minutes, the FTO glass was allowed to be cool to room temperature before immersed in the natural dye extract for 24 hours.

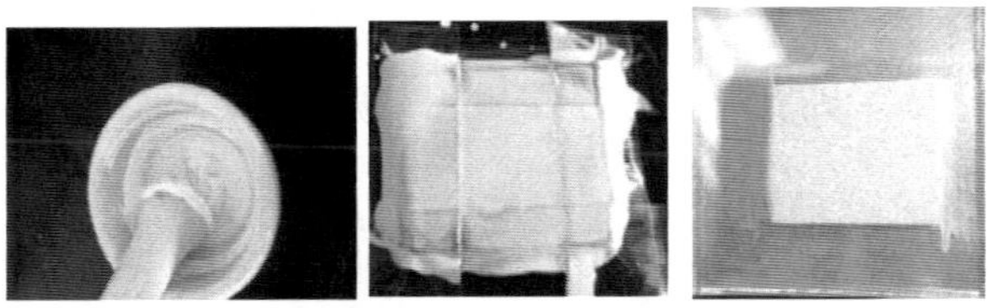

Figure 2. Deposition of the TiO_2 paste on the Fluorine doped Tin Oxide (FTO) glass via doctor blade method.

2.3. Preparation of Electrolyte

The polymer gel electrolyte was prepared by using 3.25g Potassium iodide (KI), 0.5g iodine (I_2) 30ml propylene carbonate (PC) and 20.0 ml of polyethylene glycol 400 (PEG 400). Firstly, KI and I_2 were put in the 250ml conical flask together with 30ml propylene carbonate (PC). The mixture was stirred on the magnetic stirrer for 30 minutes at room temperature then PEG400 was added. The mixture is then reacted at 100°C for 24hr. Finally, the colloidal liquid was cooled to 60°C to form a polymer gel electrolyte after 24 hours. The polymer gel electrolyte was then stored in bottle as shown in figure 3.

Figure 3. Polymer gel electrolyte used in DSSC

2.4. Preparation of Counter Electrode

The FTO glass was coated as carbon soot layer counter electrode by using a candle flame. Then the slide was heated on a hot plate at 150^0C for 15min. For carbon paste counter electrode, the carbon paste was prepared with 0.1g of TiO_2, 1mL of HNO_3 mixed in beaker for 5 minutes. Afterwards, 0.1g of carbon was added and mixed for another 5 minutes. Next, 0.05g PEG10000 was added in the mixture and blended for 5 minutes. Then, 20μL triton-X was added into the mixture. The carbon paste was heated on a hot plate at the sintering temperature of 300^0C for 1hour. Following that, deposition of the carbon paste on the Fluorine doped Tin Oxide (FTO) glass was done by using doctor blade method until the uniform layer was achieved. After the paste becomes dry, the scotch tape was removed from FTO glass. Figure 4 shows the carbon soot and carbon paste counter electrode.

Figure 4. Types of the counter electrode: Carbon soot (left) and Carbon Paste (right)

2.5. Fabrication of Dye Sensitized Solar Cell (DSSC)
The slide was allowed to cool to room temperature before assembly of the cell. The slide that stained with natural dye was left dry and the electrolyte was dropped on the semiconductor slide and heat at 60°C for 30 minutes as shown in figure 5.

Figure 5. Deposition of electrolyte on the semiconductor slide

Subsequently, the two slides that coated carbon and semiconductor respectively were clipped together. Then the alligator clips were used to clip the glass and connected to multimeter as shown in figure 6. The circuit was constructed by using DSSC, potentiometer or variable load resistor and multi-meter. The complete circuit was tested using sunlight and metal halides lamp as light sources.

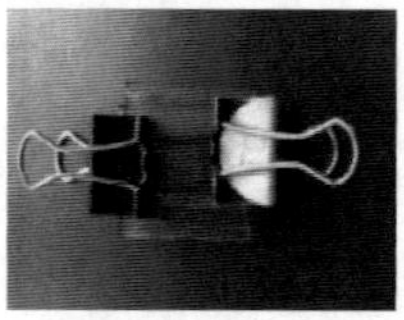

Figure 6. DSSC cell assembly

3. Result and Discussion

3.1. Absorption Spectra of Natural Dye
Figure 7 shows the UV-Vis absorption spectra of Black Olive, Blueberry, Mangosteen and Wild Mangosteen. It has been known that all the extracts absorb a wide range of the visible light. Therefore, they fulfil the characteristics to become potential dye sensitizers which absorption ability in the visible region under the illumination of the light source and also the conversion ability from light energy to electrical energy. The absorption characteristics for each extracts are dissimilar attributable to different types of anthocyanins that contained in the natural extract. It was found out that the absorption of Black Olive is the broadest among other extracts.

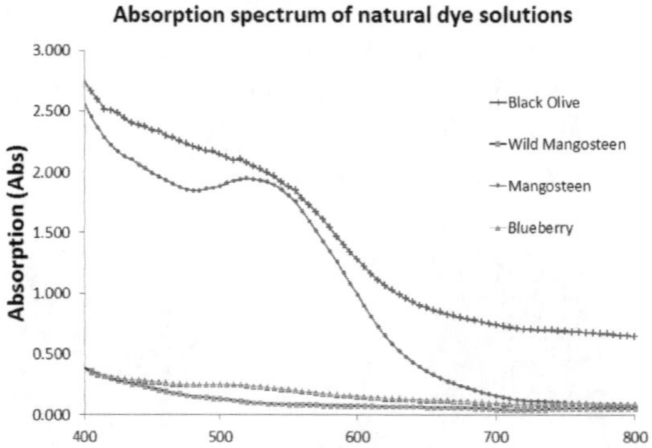

Figure 7. The absorption spectra of the natural anthocyanin sources

3.2. Photocurrent and Photovoltage Measurement

Overall performance of the anthocyanin source as sensitizer in DSSCs was evaluated in term of the open circuit voltage (Voc), short circuit current (Jsc), fill factor (FF) and energy conversion efficiency (ŋ). Dyes obtained from the Black Olive, Mangosteen, and Wild Mangosteen are all capable to promote light harvesting ability and electron injection to the conduction band of TiO_2 (Table 1). Obviously, the efficiency of DSSC sensitized with Black Olive was significantly higher than the standard natural dye, Blueberry. This is due to the higher intensity and broader range of the light absorption ability as referred to Fig 7(a). Moreover, the Black Olive cell has the highest fill factor (in Table 1), thus low equivalent series resistance and promotes efficient electron transfer. Therefore, the Black Olive extract should be an alternative anthocyanin source for DSSC and Black Olive was chosen for further study since it gave a better sensitizing performance compared to other extracts.

Table 1. I-V performance for the metal halides illuminated anthocyanin extract sensitized cell

No.	Samples	Voc(mV)	Jsc (mA)	FF	ŋ, %
1	Ruthenium N719	708	0.54	0.26	0.25
2	Black Olive	543	0.11	0.54	0.08
3	Mangosteen	432	0.12	0.30	0.04
4	Wild Mangosteen	430	0.11	0.24	0.03
5	Blueberry	390	0.11	0.39	0.04

3.3. Effect of different light source

Sunlight has a very broad spectrum of light if compared to the metal halide light which has a narrow spectrum. Therefore, the metal halide light releases lower frequencies of light which means less energy created in the DSSC and resulted in low efficiency. When the photons from sunlight strike the dye molecule, more electrons are released from ground state of dye molecule to excited state. Hence, it produces higher current under illumination of sunlight (in Table 2).

Table 2. I-V performance of the DSSC with the Black Olive extract under illumination of different light source

No.	Light Source	Voc (mV)	Jsc (mA)	FF	ŋ, %
1	Sunlight	569	0.26	0.34	0.13
2	Metal halides lamp	543	0.11	0.54	0.08

3.4. Effect of different counter electrode

The function of the counter electrode is to transfer electrons from the external circuit to triiodide/iodide in the redox electrolyte. The use of carbon paste counter electrode facilitates the electron transfer process to regenerate the electrolyte thus higher efficiency achieved (in Table 3). Consequently, carbon paste counter electrode obtained higher open circuit voltage and short circuit current as well as fill factor over carbon soot layer counter electrode. This may due to the attachment of the carbon paste onto FTO glass substrate are stronger than carbon soot layer. Therefore, the electron transfer process becomes more efficient.

Table 3. I-V performances of the DSSC with the Black Olive extract using different type of counter electrode

No.	Counter Electrode	Voc(mV)	Jsc (mA)	FF	ŋ, %
1	Carbon Paste	543	0.11	0.54	0.08
2	Carbon Soot	507	0.08	0.36	0.04

4. Conclusion

In conclusion, the Black Olive extract has higher photosensitization performance as compared to other extracts. This is due to broader absorption ability and better charge transfer between the Black Olive extract and TiO_2 surface. Black Olive extract obtained higher efficiency under illumination of sunlight due to sunlight has broader spectrum of light which contribute to more energy created in DSSC. The photovoltaic performance of fabricated DSSCs operated with carbon paste coated counter electrode over carbon soot layer coated counter electrode has shown an increase in V_{OC}, Jsc, FF and energy conversion efficiency.

Acknowledgement

The authors would like to express their gratitude to the Faculty of Engineering for their facilities support.

References

[1] O'Regan B, Gratzel, M 1991. *Nature* **353** 737 - 740
[2] Jayaweera P V V, Perera A G U, Tennakone K 2008 *Inorganica Chimica Acta*, **361**(3) 707-711
[3] Smestad G, Bignozzi C, Argazzi R 1994 *Sol Energy Mater. Sol. Cells* **32**, 259-272
[4] Khoo H E, Azlan A, Ismail A, Abas F 2012 *Molecules* **17** 9754-73
[5] Radha T, Mathew L 2007 Fruit Crops Volume 3 of Horticulture science series. (India: New India Publishing)

Failures Analysis of E- Glass Fibre reinforced pipes in Oil and Gas Industry: A Review

Sujith Bobba[a*], Z. Leman[a,b], E.S. Zainuddin[a] and S.M. Sapuan[a,b]

[a]Department of Mechanical and Manufacturing Engineering, Faculty of Engineering Universiti P utra Malaysia, 43400 Serdang, Malaysia.
[b]Laboratory of Biocomposite Technology, Institute of Tropical Forestry and Forest Products (INTROP),Universiti Putra Malaysia, 43400 Serdang, Malaysia.

Sujith.bobba@mail.com

Abstract: A comprehensive review is conducted on the failures in the field of manufacturing and installation of E-glass fiber reinforced pipes (GFRP).Some of the failures which are mainly encountered after the installation of E-Glass fiber reinforced pipes are the for nation of air bubbles in between the polyester resin layer and the surface film, dispersion of moisture in between the tubing outer and inner layers after installation, heat released in between the layers of E-glass fiber reinforced pipes due to exothermic reaction which in turn results in the formation of cracks on the surface of the pipe. The recent findings and challenges performed in conducting research regarding the deterioration caused in glass fiber reinforced pipes are highlighted and each type of failure that was identified was illustrated with an appropriate high resolution photograph .Performing creep resistance and fatigue analysis are new aspects which are still requited to be analyzed which ave not been stated in the literature which are nominated.

Keywords: *E-Glass, composite pipes, Failures, Moisture , Air bubbles, Exothermic, Creep resistance.*

1. INTRODUCTION

Fibre-reinforced polymer (FRP) composites are meeting an increasing demand as construction material due to their excellent properties including light weight, high specific strength, corrosion resistance, and low maintenance cost. Despite the advantages of GFRP pipes, their application is still limited mainly due to inadequate knowledge of failure mechanism and unsatisfactory methods of evaluating the deterioration. Long-term durability of glass-epoxy composite pipeline in service is an industrial important issue mainly because the consequences of failure are severe in oil and gas applications. It is also a challenge because it includes many variables (resins, fibers, design, manufacturing, and environment). The failures which are occurring in the pipes are in the form of short and long term considerations .Some of the those problems that occur during the manufacturing and after installation of FRP are such as formation of air bubbles between the polyester resin layer and the surface film or the mould surface, Moisture formed in between the tubing outer and inner layers after installation, heat released in between the layers of E-Glass FRP Pipes due to exothermic reaction which results in the formation of cracks on the surface of the pipe. The operation temperature is another important variable that determines the degradation resistance of the E-Glass composite pipes. The temperature determines various factors such as moisture, air bubbles, heat produced within the pipes .They are also many other factors for failures in their performance, recent developments in computational techniques have shown the researchers to find more challenging problems. More research in this area has been done with composite materials by using different materials. As per the work done by Roham Rafiee et al, 2016 [1] and his team , constraints in the failure of composite pipes fall into two main categories of short-term and long-term requirements in accordance with international rules and regulations. For the long-term considerations, creep phenomenon, fatigue failure and environmental issues are the most important topics. Figure-1 shows the Layout of GRP Wall construction during the manufacturing process.

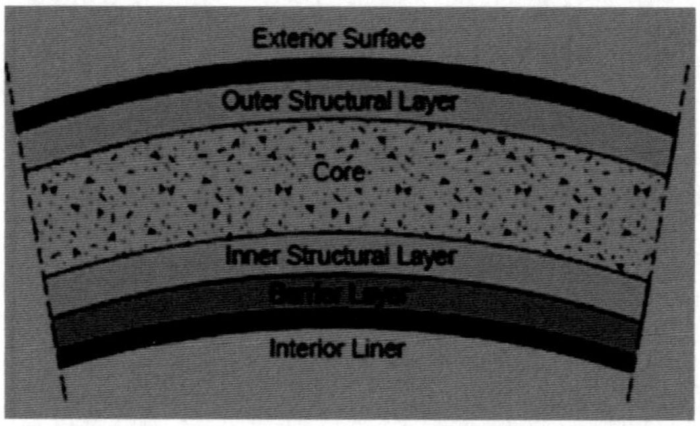

Figure-1: Systematic representation of GRP Wall construction [1]

The methods used for manufacturing GFRP Pipes sometimes may result in the failures in the long time run. The study indicates the failures and possible solution to eradicate the problems caused during the manufacturing and installation process of the pipes.

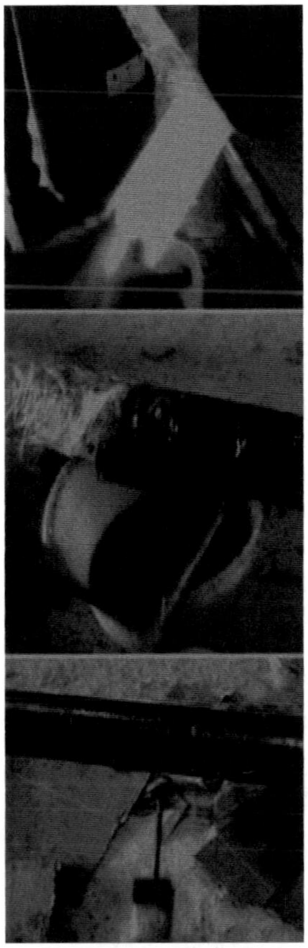

Figure-2:Systematic process of winding and applying resin during the preparation of E-Glass fiber reinforced pipe

2. LITERATURE REVIEW

Failures in GFRP Pipes have been analysed by lot of researchers from various aspects in accordance to growing demand of GFRP pipes in industrial applications. Investigations were done by Shahram Eslami a, Abbas Honarbakhsh-Raouf a , Shiva Eslami b et al. [2] and they found out that composite specimen when subjected to humid conditions underwent mechanical degradation, will cause a decrease in the failure force and maximum displacement in the buckling tests. These reductions became more considerable when the exposure temperature rose from 2°C to 25°C or the ageing environment changed from sea water to water. Yao and Ziegmann[3] have experimentally conducted tests on influence of water absorption to check the durability of GRP pipes.In an experimental Study conducted by Mertiny et al. [4,5] has observed a leakage in a filament winded tube with +_60° at both room and a elevated temperature of 100°,two different resins such as epoxy and bismaleimide were used during the experimental tests with three different loading rates and their leakage stress were recorded and compared and finally it was concluded that bismaleimide have exhibits higher leakage strength that epoxy. In the work done by other researchers buehler and Seferis [6] they found out that diffusion of water may reduce the mechanical properties of epoxy/ glass composites. Hygrothermal degradation can be worsened by absorption and desorption cycles . Arikanat et al [7] has observed crack in the specimen with different crack angle such as 0°, 15°, 30°, 45°, 60°, 75°, 90°, three failure stages of whitening with micro-crack, a pinhole leakage, and burst have been observed. Hull et al. [8] have investigated failures in filament wound GRP pipes designed with helical angle of

54.73°and inside diameter of 50 mm with the wall thickness of 1.5 mm. having both closed-end and unrestrained ends by internal pressure tests and also different modes of failure were observed by the researcher depending on the biaxial or uni axial loadings. For the closed-end situation, non-linear stress/strain response was observed due to resin cracking and seemed that fibre and resin de bonding had occurred at high pressures. For the case of biaxial loading, it was observed that weep age attributed to the transverse cracking of the resin and resin/matrix interface happened at about 20% of the burst pressure. Bakaiyan et al, 2009 [9] have analysed thermo mechanical loading of composite pipes subjected to internal pressure and a temperature gradient using 3-Dthermo-elasticity. Two ply configurations with [±55]2 and [±35/902] were tested. E-Glass Fiber strength reduces sometimes due to tension induced in the fibers during the process of lamination to reduce this tension DaisukeTabuchi, Takaosajima, ToshiroDoi, Hiromichi Onikura, Osamu Ohinisi et al, 2011 [10] have devloped a machine based on the fiber-reinforced plastic(FRP) maunfacturing method and have observed that this machine produces 12-39% more stiffness than conventional method. Xavier Gabrion et al. [11] have observed two types of failures were observed for GFRP material at ambient Temperature, the largest proportion of tested specimens (75%) had a brush-like failure and the second type of failure was cascade-like failure. Jacquemin and Vautrin [12] studied the effect of cyclic hydrothermal degradation in thick composites pipes. They found a strong stress concentration (and therefore fatigue) within a narrow region near to the surfaces where holds the periodic moisture concentration. Davies et al. [13] have studied long term behaviour of epoxy–anhydride/glass composite tubes exposed to hydrothermal ageing. They found that after 7 years of immersion in distilled water at 60°C, the tubes showed a decrease in Tg from124°C to 86°C. They have also found a dramatic fall in the expected lifetime if the temperature of immersion exceeds the 60°C (less than 1 year at 80°C). The lower values of Tg found in the failed tube suggests that the operation temperature could have been higher than the maximum operation temperature (65 °C). The composition of the transported fluid can worsen the operation condition by adding, to the hydrothermal degradation, the plasticizing action of the low molecular weight organic compounds [14]. Ferry et al. [15] has done an experimental device to collect time-to-failure data on filament wound pipes loaded in axial direction. The specimens with winding pattern of [±55°] were subjected to tensile creep tests up to 500 h. Highly scattered data originated from the manufacturing process have motivated them to present the results in the form of a two-parameter log-normal or Weibull statistical distributions. Yao and Ziegmann [16] have done tests to find the equivalence between temperature and moisture in an accelerated test method using short-term creep tests on filament wound GRE mortar pipes. They have conducted short- and long-term creep tests for predicting long-term properties. Relying on this issue that both moisture and temperature have the same effect on the modulus, a moisture aging superposition principle was suggested. The proposed principle was then applied to the results of three-point bending tests to obtain long-term data from short-term properties. The effect of winding angle on fatigue strength was investigated theoretically and experimentally in GRP pipes that were exposed to close-ended internal pressure. They reported that the optimum winding angle for the composite pressure cylinders or vessels under internal fatigue pressure load was obtained as ±45° orientation[17]. Burst strength and impact behaviour of hydrothermally aged glass fibre/epoxy composite pipes were found out practically by using experimental data by A. Hawa a, M.S. Abdul Majid a, M. Afendi a, H.F.A. Marzuki b, N.A.M. Amin a, F.Mat and A.G. Gibsonc[18] and also they found out that Weepage and eruption failures are occured depending on the applied impact energies. A Samanci1, N Tarakc¸iog˘ lu and A Akdemir[19] have worked on pre-crack angle on quasi-static crushing behaviour of filament wound CFRP cylinder was systematically. In the case of failures caused due to moisture content F.Jones et al.[20] have developed a model which shows the diffusion characteristics of two phase polymer material and observed that moisture absorption desisted from Fickian behaviour and approach. I.Mondargon et al. [21] have concluded that Moisture absorption kinetics depends on temperature and observed the weight variation of the specimen until the absorbed moisture has been accepted by all cases of Fick's Law. According to M.Muller et al. [22] penetration of water or moisture within glass filaments or fibers would cause disruption of adhesion which further reduces the strength of the glass fiber reinforced laminate. The work done by researchers Thomas J Barnell,Michael D.Rausher,Rick D.Stienecker,David M.Nickson,Tat H.Tong[23] have found out the during fiber reinforcement with a mixture of uncured resin under vacuum conditions would release exothermic heat of at least 100°c without the use of external heat source which might affect the internal layer of the fiber reinforced material. To find mass diffusion analysis fick's laws have been provided with a formula. The extension of the equations in fick's law makes it possible for non-uniform solubility of the material to occur[24].In order to measure moisture absorption behaviour, the specimen's weight is measured by immersing it upon to saturation period according to ASTM D5229 standards[25].Processing defects such as low matrix content, lack of fiber impregnation, retained air bubbles inside the pipe or on the inner surface are causes of leakage of glass fiber reinforced tubes were noticed by Guillermina Capiel et al. [26] and his team and possible solutions to eradicate the failures were concluded by them. Yang et al. [27] investigated how water absorption and hydrothermal ageing influenced fibre surface on GFRP. Similarly other studies have done on ageing process and the methods to reduce the effects caused due to ageing process. The influence of water ageing on the mechanical properties of flax and glass fibre reinforced composites was done by Assarar et al [28]. The effect of water immersion effects in composites with different types of matrices was studied by Dell'Anno and Lees [29]. studies have be done to eradicate the negative impact of the ageing process on GFRP pipes in the long time run. Studies were done to eradicate these failures are and under research by the researcher. due the presence of water numerous internal defects and reduction in factors such as impact resistance and compression strength were detected by intel nska and Guillaumat[30]. Most recently, Rafiee and Mazhari [31] have developed a general integrated procedure to evaluate long-term creep and predict long-term creep induced failure. Consisting of creep modelling, stress analysis and failure evaluation, the developed modelling is just in need of short-term experimental data of neat resin as the input.

3. FAILURE ANALYSIS

3.1 Air bubble penetration

As per work done, Firstly failures were noticed during the process of Filament winding when a plastic-coated cloth is placed along the longitudinal section of the mandrel, air is vaguely trapped in the reinforcement and resulting in the formation of cells or holes in the wall , So during the forming and curing operations there is a decrease in the strength of the final piping and sometimes making it permeable and so the tubing will not be suitable for all kinds of uses. Also, because when a cloth type reinforcement is applied, it is well oriented into definite layers and is therefore likely to de laminate and also trapping of air in the reinforcement which consequently results in the production of tubing free of passage which promotes leakage and stress concentrations. Water molecules are spread in to fiber-matrix interface and to drifted along the fibers by surface tension reducing fiber-matrix bond. The figure-3 shows the failure caused due the weepage caused by Air bubble formation inside the pipe.

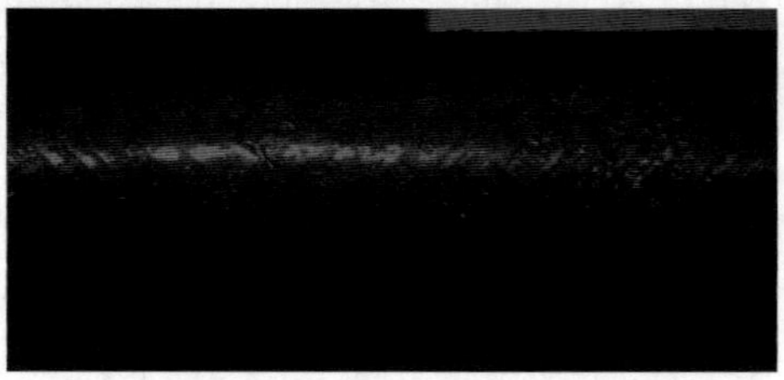

Figure: 3-Failure due the weepage caused by Air bubble formation inside the pipe

According to work done, The Coating resin which is applied must be free from air bubbles and dirt, so that the GFRP Pipe can with stand for long period of time. .

3.2 Moisture Formation

Secondly Moisture that is diffused into polymer matrix composites (PMCs) changed the mechanical and thermo-physical properties of the material and also due to Moisture and temperature increased the ageing process with the reduction of the mechanical performance of fibre, matrix and especially the fibre and matrix interface. Polymeric matrix becomes softened plasticised and swelled when subjected to humidity[33],More over can sometimes moisture can affect the composite material and the bonded joints by two mechanisms, deterioration of the adhesive bulk and deterioration of the interface due to the hydrolysis process. Under some cyclic hygrothermal degradation, thick composites pipes show a heavy stress concentration, a narrow region near to the Surfaces where holds the periodic moisture concentration [34].The amount of moisture absorbed by the glass fiber reinforced composite pipes could be accurately predicted by plotting p.w,g versus t/L where t is the absorption time and L is the thickness of the sample material. The p.w.g is weight gained by the material due the absorbed moisture and is calculated as follows

$$p.w.g = \qquad (1)$$

Where W_O is the initial dry weight of the material and W_t is the weight of the sample at time t,[35] In addition, the diffusion coefficient and percentage of absorbed moisture can also increase with rises in immersed temperature and it improves the ductility of the matrix also. As per the study hydrothermal analysis is the preferred method to predict the moisture content and the movement of moisture before installation so that it can reduce the early degradation and service life can be increased. The temperature of the storage area is important in improving the life of the material. Base life is 6-8 months when stored at -18°C, by controlling the temperature and time moisture inside the GFRP Pipe can be eliminated.

3.3 Heat released by exothermic reaction

Thirdly, when the composite is wrapped with layers heat is released inside the wrapping layer after few months which is termed heat gained by exothermic reaction. The excess heat produced will depend on various factors such as number of layers wrapped across each fiber during filament winding, so due to heat it results in poor fibre-matrix adhesion which then undergoes fibre and matrix de bonding and splitting. In such failure, fibres which are destroyed are located at any place along the specimen. By exothermic reaction in the pipe, thermosetting matrix problems such thick laminates formation, matrix distortion and few other small problems are noticed. In some cases during Manufacturing high exothermic temperature is produced within the layers of the fiber cloth which caused burning and chemical loss in the resin, to control this filler was added. The time to reach highest exothermic heat is also reduced with increasing filler amount, thereby reducing the cure cycle. The graph below shows time versus temperature during curing operation of the GFRP Pipe.

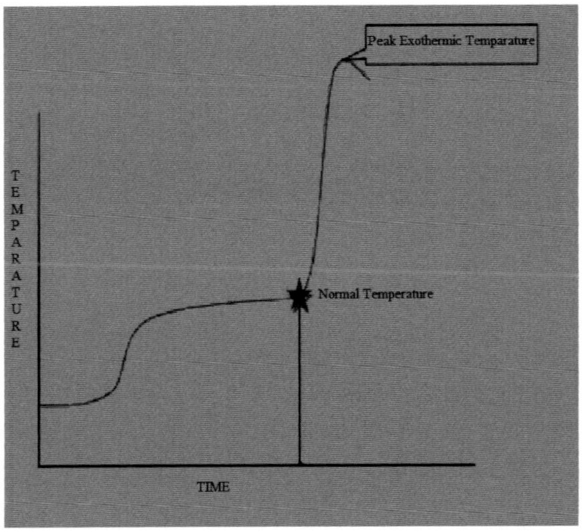

Graph:1-Time vs Temparature during curing operation for optimum temparature during manaufaturing process of Glass fiber reinforced pipes [36]

Sometimes due to high exothermic temperature degradation of resin may occur hence mould cooling is necessary to reduce the heat. The fiber-matrix adhesion reduction which is caused due to heat release will decrease the load transfer from one phase to another degrading the mechanical response of the pipe [37].

4. CONCLUSION

A Comprehensive work was done on the failures of GFRP Pipes and considerable efforts have been made to find the failures of GFRP Pipes. They are suitable precaution methods to control the failures to some extent according the work done:

1. Air bubbles in a surface of the pipe can reduce its strength and may also affect its corrosion resistance properties. Air bubbles in the corrosion liner can be more damaging than those in the structural layer, So try to eliminate the air bubbles cleanly which are residing in the liner. By applying epoxy resin laminate in appropriate ratio can decrease the possibility of snagging air bubbles formation and will make sure that the glass will be pervaded from the bottom.
2. Avoid aggressive mixing of the resin and hardening material which may produce air into resin. However, be sure the catalyst is mixed finely into the resin.
3. The resin mix can be prepared at least one day ahead so that the entrapment of air bubbles escape before the lay-up begins.
4. Mandrels should be applied with resin first, then apply glass, and roll it into the resin. Air bubble problems are unavoidable when resin is applied to dry glass.
5. Roll the laminate from the centre of the mandrel to the edges of it. Roll steadily, but not too hard. Excessive force might break the existing bubbles and make them more difficult to remove.
6. Eliminate all the bubbles from one ply which has been used before starting on the next process.
7. Thoroughly clean the rollers between uses.

When it comes to the point of moisture the possible solution would be , using an outer layer of glass fiber yarn and resin in normal foam insulating material to form a continuous moisture-impervious cover, said outer layer of glass fiber yarn being wound over the ends of the foam insulating material to fully enclose the foam insulating material. The absorbed moisture may lead to the reduction of bending strength due reduction of mechanical properties after moisture absorption. Finally, In the case of failure due to creep induced failure with in the GFRP tubes during long time run so to reduce the failure, analysis should be done by creep modelling, stress modelling and failure evaluation by the designers before implementing it at the work station. In spite of the efforts in the wide range of study done, there are still some issues which are not found out and may be considered for the future studies.

AKNOWLEGDEMENT

I would like to thank Aceon Composite Sdn Bhd Industries, Malaysia for doing my case study over there, especially the site supervisor who helped me throughout my study in the industry.

REFERENCES

[1] Roham Rafiee, on the mechanical performance of glass-fibre-reinforced thermosetting-resin pipes: A review, Composite Structures 143 (2016) 151–164.
[2] Shahram Eslami a, Abbas Honarbakhsh-Raouf a , Shiva Eslami, Effects of moisture absorption on degradation of E-glass fiber reinforced Vinyl Ester composite pipes and modelling of transient moisture diffusion using finite element analysis, Corrosion Science 90 (2015) 168–175.
[3]Yao J, Ziegmann G. Water absorption behaviour and its influence on properties of GRP pipes. J Compos Mater 2007;41(8):993–1008.
[4] Mertiny P, Gold A. Quantification of leakage damage in high-pressure fibre reinforced polymer composite tubular vessels. Polym Test 2007;26(2):172–9
[5] Mertiny P. Leakage failure in fibre-reinforced polymer composite tubular vessels at elevated temperature. Polym Test 2012;31:25–30.
[6] Buehler FU. Scfcrisl, Effect of reinforcement and solvent content on moisture absorption in epoxy composite materials. Compos Part A 2000;30(7):741-S.
[7]Arikan H. Failure analysis of (±55_) filament wound composite pipes with an inclined surface crack under static internal pressure. Compos Struct 2010;92 (1):182–7.
[8] Hull D, Legg MJ, Spencer B. Failure of glass/polyester filament wound pipes. Composites 1978,Voulme 9,Issue-1,17–24.
[9] Bakaiyan H, Hosseini H, Ameri E. Analysis of multi-layered filament-wound composite pipes under combined internal pressure and thermomechanical loading with thermal variations. Compos Struct 2009,Vol-88,pages 532–41.
[10] DaisukeTabuchi,Takaosajima,Toshiro Doi,Hiromichi Onikura,Osamu Ohinisi,Devlopment of a Filament winding machine based on Internal heating by a High Temparature Fluid for composite vessels,sensors and materails,Vol.23,No.6(2011) 347-358.
[11] Xavier Gabrion, Vincent Placet, Fr_ed_erique Trivaudey, Lamine Boubakar, About the thermomechanical behaviour of a carbon fibre reinforced high-temperature thermoplastic composite, Composites Part B 95, 2016. 386-394
[12] Jacquemin F, Vautrin A. The effect of cyclic hygrothermal conditions on the stresses near the surface of a thick composite pipe. Compos Sci Technol 2002;62(4):567–70.
[13] Davies P, Baizeau R, Choqueuse D, Salmon L, Nagot F. Thermoplastic Composite Cylinders for Underwater Applications, Journal of Thermoplastic Composite Materials 2005; 18(5) : 417-443, http://dx.doi.org/10.1177/0892705705054397, 2005 ,SAGE Publications

[14] ASTM D2992-06. Standard test method for obtaining hydrostatic or pressure design basis for reinforced thermosetting resin pipe and fitting. American Society for Testing and Materials ; 2006.
[115] Ferry L, Perreux D, Varchon D, Le Bras J. Tensile failure of filament-wound pipes under long-term creep loading: a probabilistic analysis. Compos Sci Technol 1997;57:1281–8.
[16] Yao J, Ziegman G. Equivalence of moisture and temperature in accelerated test method and its application in prediction of long-term properties of glass fibre reinforced epoxy pipe specimen. Polym Test 2006;25:149–57.
[17] Erkal S, Sayman O and Benli S. Fatigue damage in composite cylinders. Polymer Composites 2010,Voulme - 31, pages-707–713.
[18] A. Hawa a, M.S. Abdul Majid a,M. Afendi a, H.F.A. Marzuki b, N.A.M. Amin a, F.Mat a, A.G. Gibsonc, Burst strength and impact behaviour of hydrothermally aged glass fibre/epoxy composite pipes, Materials and Design 89 (2016) 455–464.
[19] A Samanci1, N Tarakc¸iog˘ lu and A Akdemir , failure analysis of surface-cracked (±45°)3 filament-wound GRP pipes under internal pressure, Journal of Composite Materials,2011,46(9) 1041–1050.
[20] F.Jones, P.Jacobs, Moisture Diffusion into two-Phase Matrix Resins for Fibre Composites,Thrid European Conference on Composite Materails,20-23 March 1989,Bordeaux-France,Pages 627-633.
[21] I.Mondargon, J.Imaz, A.Rubio, A.Valea, Moisture absorption Influence on the mechanical properties of Carbon/Epoxy composites, Thrid European Conference on Composite Materails,20-23 March 1989,Bordeaux-France,Pages 643-648
[22] Manfred Muller,Wuppertal-Barmen,Method For Producing Rotationally Symmetrical Glass Fiber Reinforced Laminated Hollow Structures,United States Patent 1191 July 22, 1971
[23] Thomas J Barnell,Michael D.Rausher,Rick D.Stienecker,David M.Nickson,Tat H.Tong, "Fiber-reinforced epoxy composites and methods of making same without the use of oven or autoclave Original Assignee Comer Stone Research Group", Inc. Patent no- US9370902 B2, June 21, 2016.
[24] ABAQUS/CAE, user's manual, version 6.12,reterived on 23 June 2016.
[25]ASTM D5229, Standard test method for moisture absorption properties and equilibrium conditioning of polymer matrix composite materials.
[26] Guillermina Capiel ,Pablo Fayó,Antonela Orofino,Pablo E. Montemartini, Failure of glass fiber-reinforced epoxy pipes in oil fields, Handbook for materials failure analysis with case studies from oil and gas industry,2016,pg-91-103
[27] B. Yang, J. Zhang, L. Zhou, M. Lu, W. Liang, Z. Wang, Effect of fibre surface modification on water absorption and hydrothermal ageing behaviours of GF/pCBT composites, Compos. Part B 82 (2015) 84–91
[28] M. Assarar, D. Scida, A. El Mahi, C. Poilâne, R. Ayad, Influence of water ageing on mechanical properties and damage events of two reinforced composite materials: Flax–fibres and glass–fibres, Mater. Des. 32 (2011) 788–795
[29] G. Dell'Anno, R. Lees, Effect of water immersion on the inter laminar and flexural performance of low cost liquid resin infused carbon fabric composites, Compos. Part B 43 (2012) 1368–1373
[30] Imichriska K. Guillaumat I, "The effect of water immersion ageing on low-velocity impact behaviour of woven aramid-glass fibre/epoxy composites", Compos Science Technology 2004;64(13-142271-8.
[31] Rafiee R, Mazhari B, "Modelling creep in polymeric composites: developing a general integrated procedure", Int J Mech Sci 2015;99:112–20.
[32] S.A. Grammatikos· B. Zafari,M.C. Evernden, J.T.Mottram,J.M.Mitchels, Moisture uptake characteristics of a pultruded fibre reinforced polymer flat sheet subjected to hot/wet aging, Polymer

Degradatin and Stability, Voulme 121, November 2015, Pages 407-419.
[34] Jacquemin F. Vautrin A. The effect of cyclic hygrothermal conditions on the stresses near the surface of a thick composite pipe. Compos Sci Technol 2002:62(4):567-70.
[35] Sancaktar,David R.Beachtle,The effect of stress Whitening on moisture diffusion in thermosetting polymers,First Canadian International Composites Conference and Exibition(CANCOM 91), September 1991, Pages 81-90. .
[36] http://nptel.ac.in/courses/101106038/mod02lec01.pdf, retrieved on 03 April 2017.
[37] Hodzic A. Kim JK, Stachurski ZH, "Nano-indentation and nano-scratch of polymer/glass interfaces. II: model of inter phases in water aged composite materials", Polymer 2E01;42 :5701-10.

Comparison on characteristic of Mesoparticle Graphene Sand Composite (MGSC) using different types of sugar to remove methylene blue

A W Zularisam, Norul Wahida, Ahmad Alfian

Universiti Malaysia Pahang, Lebuhraya Tun Razak, 26600 Gambang, Pahang, Malaysia

zularisam@ump.edu.my

Abstract. This paper presents the green method to synthesis two types of adsorbent called mesoparticle graphene sand composite (MGSC) by using table sugar ($MGSC_{ts}$) and arenga palm sugar ($MGSC_{aps}$) as different carbon sources to remove methylene blue acted as a dye model. Immobilisations of these materials on sand were introduced by using pyrolysis method without binder usage. Sand was treated by removing deleterious materials before sieved. The solutions of sugar were prepared and heated to 95 °C. The sand and sugar solutions were mixed and constantly stirred before putting them in furnace with nitrogen environment to produce $MGSC_{ts}$ and $MGSC_{aps}$. The composites were activated by using concentrated sulphuric acid to open the pores and maximise the capacity of absorbency. The analyses on the characteristic of both $MGSC_{ts}$ and $MGSC_{aps}$ were conducted through field emission scanning electron microscope (FESEM), elemental dispersive x-ray (EDX) and elemental mapping (EM). FESEM analyses exhibited that the coating process was done uniformly as there were layers of coating sheets formation on the sand particles surfaces. After conducting EDX and EM, there were major elements found in both $MGSC_{ts}$ and $MGSC_{aps}$ which were carbon, oxygen and silica. EM exhibited the distribution of these elements were scattered on the MGSC's surfaces. Removal of methylene blue was successfully carried out by using both $MGSC_{ts}$ and $MGSC_{aps}$, with maximum removal up to 40% at the first hour of contact time.

1. Introduction

The applications of composite materials in contaminated air, water and soil are intensely pursed recently [1], [2] as the released of industrial wastewater to the clean water sources is very difficult to manage and indirectly caused imbalance ecosystem. Thus, it is necessary to find a way to treat [3] and purify waste water [2] to produce cleaner waste before releasing it to the environment. High adsorption of waste in wastewater can be achieved by using a suitable and solid sorbent. The changes in the earth's climate have negative impacts on water resources around the world [4] which lead to the increasing periods of drought, more floods, higher water temperatures, and changes in water chemistry [5] [6]. Hence there is a significant need for the development of efficient and economical wastewater treatment technologies. This paper presents on the synthesisation and characterisation process of green mesoparticle graphene sand composite ($MGSC_{aps}$ and $MGSC_{ts}$) to remove methylene blue from wastewater by using arenga palm sugar and table sugar respectively as carbon sources.

Waste material such as rubber tyres [7] and other adsorbent materials [8], [9], [10] are also been used in adsorbing heavy metals and dyes. The removal is efficient; however the cost is high to synthesis and produces it as a good adsorbent. Thus, the researches on the green composite with facile, lower cost of synthesisation and production are a need for environmental remedy. Furthermore, the additional interesting characteristic of the green and facile adsorbent is the reusability and regeneration of the composite in the next cycle of adsorption process. Recently, there is one composite material that getting paid attention, namely graphene. Hence, the fundamental knowledge of the characteristic, absorbance process and the possible failure of the system are essential before any remediation work is carried out. In this article, the main discussion and review is made on the characteristic and synthesis of graphene, and the potential of MGSC as absorbent to remove methylene blue in wastewater. The objectives of the research are to synthesis MGSC using green technologies and characterise it before being tested as adsorbent. Arenga palm sugar and table sugar were chosen to be acted as the main source of carbon since there were high in sucrose so far. Thus the overall performance of $MGSC_{aps}$ and $MGSC_{ts}$ in removing pollutants needs to be investigated.

2. Methodology

2.1. Materials

The materials used to synthesis $MGSC_{aps}$ and $MGSC_{ts}$ were arenga palm sugar and table sugar as carbon sources, and sand which passed 2 nm sieve. Arenga palm sugar was obtained from arenga pinnata tree which abundantly available regions around the world especially in the tropical such as Malaysia, Indonesia, Thailand, Philippines and Vietnam. In this research, arenga palm sugar was supplied from Pahang, Malaysia whereas table sugar was supplied by Prai Sugar Sdn Bhd. Sand was collected from river located in Pahang, Malaysia.

2.2. Chemicals

Methylene blue and concentrated sulphuric acid were supplied from Permula Chemicals Sdn. Bhd. Distilled water as solvent was used throughout this experimental process.

2.3. Instrumentation for $MGSC_{aps}$ Characterisation

$MGSC_{aps}$ and $MGSC_{ts}$ were subjected to several types of testing including FESEM, EM and XRD.

2.3.1. Field Emission Scanning Electron Microscope. Field emission scanning electron microscopic (FESEM) was analysed by using FESEM JEOL JSM-6500F equipped with MiniFlex2 counter detector. It was also used fixed monochromator of diffracted beam monochromator.

2.3.2. Elemental Mapping. The growth and distribution of carbon on silica oxide surfaces was recorded by using elemental mapping equipment JEOL JSM-6500F.

2.3.3. Elemental Dispersive X-Ray. Elemental Dispersive X-Ray (EDX) was recorded to ascertain elements percentage of graphene sand composite by using 30 kV; 15 mA x-ray with fixed monochromator at 1.0000 sec scan speed.

2.4. Synthesisation Process of $MGSC_{aps}$ and $MGSC_{ts}$

The synthetisation of $MGSC_{aps}$ and $MGSC_{ts}$ followed method mentioned in the literature [11] with a slight modification on the carbon source. For $MGSC_{aps,}$ 150 g arenga palm sugar was used as carbon source and was diluted in 150 g distilled water. The solution was then mixed with 150 g fine river sand before constantly stirred at 95 °C for 3 hours. The mixture then was put in the silica crucibles and burned in the nitrogen atmosphere furnace. The temperature of the furnace was set from the room temperature to 100 °C in 30 minutes to burn deleterious materials. The temperature was changed to 200 °C in another 30 minutes and was held for 1 hour to melt arenga palm sugar to confirm uniform coating on the sand surfaces. The temperature of the furnace then was ramped to 750 °C in 1 hour and held it for another 3 hours to ensure graphenisation process was completed. MGSC $_{aps}$ then allowed cooling to the room temperature gradually. After cooled, MGSC $_{aps}$ activated by using concentrated sulphuric acid for 30 minutes without disturbance to open pore sites which acted as adsorbent sites. MGSC $_{aps}$ is now ready for adsorption process before filtered as shown in Figure 1. The preparation of MGSC$_{ts}$, was followed the same step as MGSC$_{aps}$'s preparation. However the source of carbon was changed from arenga palm sugar to table sugar.

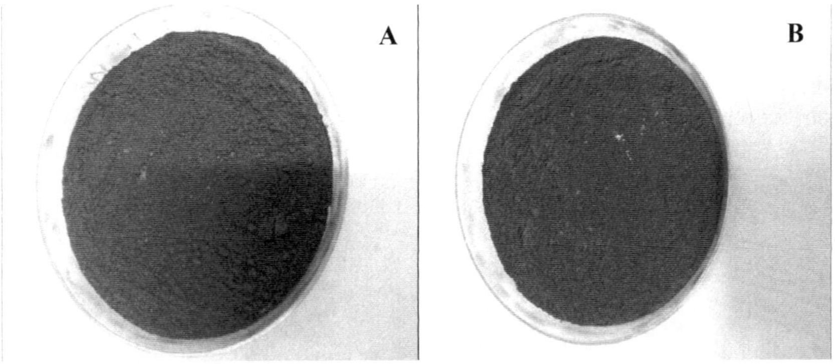

Figure 1. MGSC$_{ts}$ (A) and MGSC$_{aps}$ (B) were successfully synthesised

3. Results and Discussion

3.1. MGSC$_{aps}$ and MGSC$_{ts}$ Characterisation

3.1.1. Field Emission Scanning Electron Microscope. The morphological of MGSC$_{aps}$ and MGSC$_{ts}$ were carried out through FESEM analyses. It was evident that the sand was coated uniformly which confirms the temperature setting for initial burning process was 200 °C successfully melted arenga palm sugar to form a layer of carbon coating. This is in line with the previous researchers [12] which stated that the melting point of sucrose is approximately 186 °C. The graphenic morphology also was detected at the coating sheet's edges where thin wrinkles were formed as shown in Figure 2. The winkled edges of sheets that confirms the production of graphene, even after the formation of the composites, the structure of the composite remained the same where the wrinkles were still present [13].

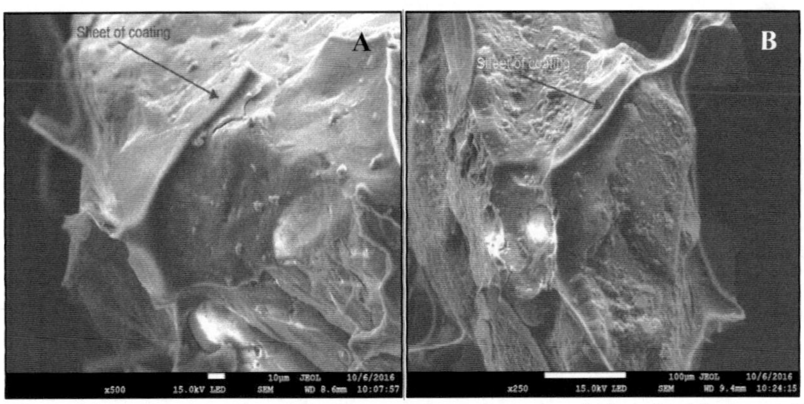

Figure 2. Sheet of coating was formed after arenga palm sugar (A) and table sugar (B) melted

3.1.2. Elemental Mapping. Major elements present were carbon, silicon and oxygen. The presence of large carbon feature was originated from the carbonaceous material which was formed from melted arenga palm sugar and table sugar on sand surface. Previous researchers [14] stated that the elements detected show that the sand was covered by carbon. The other elements; silicon and oxygen were also features in the major elements due to the usage of sand (SiO_2) hence validating the name of this composite in the research, mesoparticle graphene sand composite ($MGSC_{aps}$ and $MGSC_{ts}$).

3.1.3. Elemental Dispersive X-Ray. An EDX analysis was conducted to measure the percentages of each chemicals element in $MGSC_{aps}$ and $MGSC_{ts}$. Figure 3 presents the result of $MGSC_{aps}$ and $MGSC_{ts}$ upon elemental dispersive x-ray test. From the results, there were major chemical elements' peaks in line with EM results which were carbon, silicon, oxygen. Other chemical elements' peaks were ferum and aluminium found on sand surfaces.

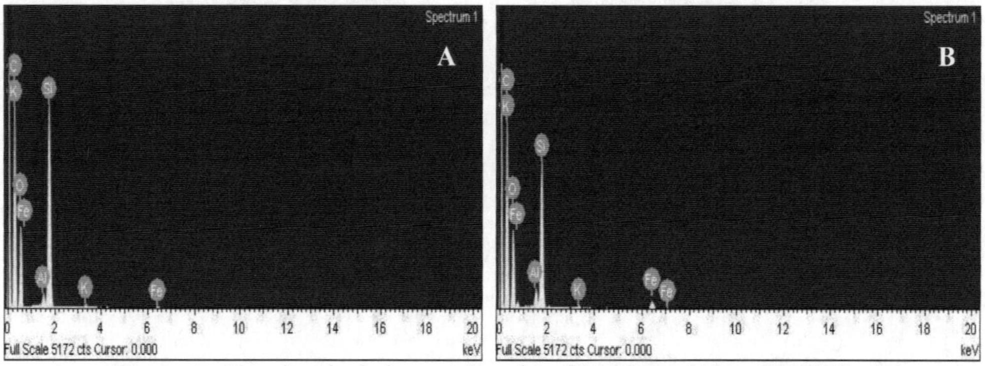

Figure 3. The chemical elements' peaks of $MGSC_{aps}$ (A) and $MGSC_{ts}$ (B)

4. Conclusion

The green adsorbents by using arenga palm sugar and table sugar were successfully synthesised. The chemical elements in the both adsorbent was confirmed the relevance of the composite name which is mesoparticle graphene sand composite (MGSC). Based on the characteristics of the composite, it can be concluded that it has significant character to be acted as a good adsorbent in removing methylene blue.

References

[1] Y. Zhou, O. G. Apul, and T. Karanfil, "Adsorption of Halogenated Aliphatic Contaminants by Graphene Nanomaterials," *Water Res.*, vol. 79, pp. 57–67, 2015.
[2] F. Ahmed and D. F. Rodrigues, "Investigation of acute effects of graphene oxide on wastewater microbial community: A case study," *J. Hazard. Mater.*, vol. 256–257, pp. 33–39, 2013.
[3] M. T. Amin, a a Alazba, and U. Manzoor, "A Review of Removal of Pollutants from Water / Wastewater Using Different Types of Nanomaterials," vol. 2014, 2014.
[4] N. Cai, D. Peak, and P. Larese-Casanova, "Factors influencing natural organic matter sorption onto commercial graphene oxides," *Chem. Eng. J.*, vol. 273, pp. 568–579, 2015.
[5] I. Chowdhury, W.-C. Hou, D. Goodwin, M. Henderson, R. G. Zepp, and D. Bouchard, "Sunlight affects aggregation and deposition of graphene oxide in the aquatic environment," *Water Res.*, vol. 78, pp. 37–46, 2015.
[6] V. K. Gupta and T. a Saleh, "Sorption of pollutants by porous carbon, carbon nanotubes and fullerene- an overview.," *Environ. Sci. Pollut. Res. Int.*, vol. 20, no. 5, pp. 2828–43, 2013.
[7] V. K. Gupta, Suhas, A. Nayak, S. Agarwal, M. Chaudhary, and I. Tyagi, "Removal of Ni (II) ions from water using scrap tire," *J. Mol. Liq.*, vol. 190, pp. 215–222, 2014.
[8] V. K. Gupta, S. K. Srivastava, D. Mohan, and S. Sharma, "Original Contribution Design Parameters for Fixed Bed Reactors of Activated Carbon Developed From Fertilizer Waste for

the Removal of S O M E Heavy Metal Ions," *Pergamon*, vol. 17, no. 8, pp. 517–522, 1998.

[9] A. N. and S. A. Vinod Kumar Gupta, Imran Ali, Tawfik A. Saleh, "Chemical treatment technologies for waste-water recycling—an overview," *RSC Adv.*, no. 16, 2012.

[10] A. Mittal, J. Mittal, A. Malviya, and V. K. Gupta, "Adsorptive removal of hazardous anionic dye 'Congo red' from wastewater using waste materials and recovery by desorption," *J. Colloid Interface Sci.*, vol. 340, no. 1, pp. 16–26, 2009.

[11] S. Sen Gupta, T. S. Sreeprasad, S. M. Maliyekkal, S. K. Das, and T. Pradeep, "Graphene from Sugar and its Application in Water Puri fi cation," *ACS Appl. Mater. Interfaces*, vol. 4, pp. 4156–4163, 2012.

[12] V. P. Parvathi, M. Umadevi, and R. B. Raj, "Improved waste water treatment by bio-synthesized Graphene Sand Composite," *J. Environ. Manage.*, vol. 162, pp. 299–305, 2015.

[13] T. S. Sreeprasad, S. M. Maliyekkal, K. P. Lisha, and T. Pradeep, "Reduced graphene oxide-metal/metal oxide composites: Facile synthesis and application in water purification," *J. Hazard. Mater.*, vol. 186, no. 1, pp. 921–931, 2011.

[14] S. Moradi and S. Azizian, "Preparation of nanostructured carbon covered sand for removal of methyl violet from water," *J. Mol. Liq.*, vol. 219, pp. 909–913, 2016.

IOP Conf. Series: Materials Science and Engineering **217** (2017) 012006 doi:10.1088/1757-899X/217/1/012006

Kinetic and isotherm studies for nickel (II) removal using novel mesoparticle graphene sand composite synthesised from sand and arenga palm sugar

A W Zularisam, Norul Wahida

Universiti Malaysia Pahang, Lebuhraya Tun Razak, 26600 Gambang, Pahang, Malaysia

zularisam@ump.edu.my

Abstract. Nickel (II) is one of the most toxic contaminants recognised as a carcinogenic and mutagenic agent which needs complete removal from wastewater before disposal. In the present study, a novel adsorbent called mesoparticle graphene sand composite ($MGSC_{aps}$) was synthesised from arenga palm sugar and sand by using a green, simple, low cost and efficient methodology. Subsequently, this composite was characterised and identified using field emission scanning electron microscope (FESEM), x-ray diffraction (XRD) and elemental mapping (EM). The adsorption process was investigated and optimised under the experimental parameters such as pH, contact time and bed depth. The results showed that the interaction between nickel (II) and $MGSC_{aps}$ was not ion to ion interaction hence removal of Ni (II) can be applied at any pH. The results were also exhibited the higher contact time and bed depth, the higher removal percentage of nickel (II) occurred. Adsorption kinetic data were modelled using Pseudo-first-order and Pseudo-second-order equation models. The experimental results indicated pseudo-second-order kinetic equation was most suitable to describe the experimental adsorption kinetics data with maximum capacity of 40% nickel (II) removal for the first hour. The equilibrium adsorption data was fitted with Langmuir, and Freundlich isotherms equations. The data suggested that the most fitted equation model is the Freundlich with correlation R^2=0.9974. Based on the obtained results, it can be stated that the adsorption method using $MGSC_{aps}$ is an efficient, facile and reliable method for the removal of nickel (II) from waste water.

1. Introduction

Waste water mainly contains a variety of organic and inorganic compounds of anthropogenics and natural origins, sediments, radioactive materials effluents and heavy metals[1], [2]. These pollutants; especially heavy metal are also contributed to the toxicity, bioaccumulation, non-biodegradable nature and contribute to the larger scale of extraction of metal release of hazardous wastes. Therefore, it is necessary to find a way to treat [3] and purify [4] wastewater contained with toxic metal ions to produce cleaner waste before releasing it to the environment. One of industrial wastewater may contain a large number of heavy metals in it. Nickel (II) is a non-biodegradable and toxic ion which present in wastewater. Increasing accumulation of nickel (II) in industrial wastewater has raised attention and concern. The main source of of nickel (II) pollution derived from processes such as galvanization, smelting, mining, dyeing operations, batteries manufacturing and metal finishing [5].

Most of them were discharged into the water sources without a proper treatment. Trace amounts of nickel (II) are beneficial to human organism as an activator of some enzyme systems. However, if it is beyond the scope of normal levels, different types of diseases such as lung cancer, renal enema, skin dermatitis and gastro intestinal disorder will occur. If nickel adsorbed into the body, it severely irritates gastrointestinal tract leads to circulatory shock and renal damage. Hence, the Water Sanitation and Hygiene under World Health Organization established the toxic limits of permissible concentration of nickel (II) and insoluble compounds of 1.0 mg/m ^{3}, soluble compounds of 0.1 mg/m ^{3}, nickel carbonyl of 0.05-0.12 mg/m ^{3} and nickel sulphide of 1.0 mg/m ^{3} [6]. The conventional method for removing nickel (II) from industrial wastewater includes the chemical oxidation or reduction, chemical precipitation, ion exchange, membrane separation, filtration, electrochemical treatment and adsorption. These methods require high reagent and energy requirements, and the generation of toxic sludge or wastes products which required disposal or treatment. However, the adsorption technique is one of the most popular and preferred methods to remove heavy metals. Absorption by using activated carbon is preferred due to its high specific surface area, highly porous character, multiple adsorption, chemical stability and relatively low-cost. Thus recently, there are many researches on the removal of heavy metal ions from waste water effluents. Numerous materials have been used as adsorbents for the removal of nickel (II) processes including activated carbon, silica, ion exchange resins, rock materials, activated slag, agricultural wastes, microbial and plant derived biomass and chitin. Recently, graphene has become the hottest topic among researchers.

Graphene is a single atomic layer of sp^2-hybridized carbon arranged in a honeycomb structure. Several studies were made in term of heavy metals removal from waste water effluents by using different oxides of composite oxides. There were removal of Cu (II), Co (II), Zn (II), Cr (VI) and Pb (II) previously by using various types of adsorbents. However, it is remain necessary to develop a low cost, easily available, high adsorption capacity material for wastewater treatment to remediate environmental problems. The present research work focuses on exploiting exceptionally high adsorptive potential of graphene and using it for removal of nickel (II) ions from synthetic aqueous solutions. In this research, $MGSC_{aps}$ was produced by using sand and sugar. The performance of $MGSC_{aps}$ in removing nickel (II) from synthetic aqueous solution was investigated throughout the experimental works. $MGSC_{aps}$ was prepared from sugar as it is a highly carbonaceous source [16]. The main objective of the research is to synthesise, characterise and analyse $MGSC_{aps}$ on the effect of pH, contact time and bed depth criteria towards the adsorption capacity in removing nickel (II) from wastewater.

2. Methodology

2.1. Materials

The materials used to synthesis $MGSC_{aps}$ were arenga palm sugar and sand which passed 2 nm sieve. Arenga palm sugar was obtained from arenga pinnata tree which abundantly available regions around the world especially in the tropical such as Malaysia, Indonesia, Thailand, Philippines and Vietnam. In this research, arenga palm sugar was supplied from Pahang, Malaysia. Sand was collected from river located in Pahang, Malaysia.

2.2. Chemicals

For chemicals used in this research, pure metal nickel, concentrated sulphuric acid and concentrated nitric acid were supplied from Permula Chemicals Sdn. Bhd. Distilled water as solvent was used throughout this experimental process.

2.3. Instrumentation for $MGSC_{aps}$ Characterisation

$MGSC_{aps}$ was subjected to several types of testing including FESEM, EM, XRD and Raman Spectroscopy.

2.3.1. Field Emission Scanning Electron Microscope. Field emission scanning electron microscopic (FESEM) was analysed by using FESEM JEOL JSM-6500F equipped with MiniFlex2 counter detector. It was also used fixed monochromator of diffracted beam monochromator.

2.3.2. Elemental Mapping. The growth and distribution of carbon on silica oxide surfaces was recorded by using elemental mapping equipment JEOL JSM-6500F.

2.3.3. X-Ray Diffraction. X-ray diffraction (XRD) was recorded to ascertain crystalline nature of graphene sand composite by using 30 kV; 15 mA x-ray with fixed monochromator at 1.0000 sec scan speed.

2.4. Synthesisation Process of MGSC $_{aps}$

The synthetisation of MGSC $_{aps}$ followed method mentioned in the literature [7] with a slight modification on the carbon source. 150 g arenga palm sugar was used as carbon source and was diluted in 150 g distilled water. The solution was then mixed with 150 g fine river sand before constantly stirred at 95 °C for 3 hours. The mixture then was put in the silica crucibles and burned in the nitrogen atmosphere furnace. The temperature of the furnace was set from the room temperature to 100 °C in 30 minutes to burn deleterious materials. The temperature was changed to 200 °C in another 30 minutes and was held for 1 hour to melt arenga palm sugar to confirm uniform coating on the sand surfaces. The temperature of the furnace then was ramped to 750 °C in 1 hour and held it for another 3 hours to ensure graphenisation process was completed. MGSC $_{aps}$ then allowed cooling to the room temperature gradually. After cooled, MGSC $_{aps}$ activated by using concentrated sulphuric acid for 30 minutes without disturbance to open pore sites which acted as adsorbent sites. MGSC $_{aps}$ is now ready for adsorption process before filtered as shown in Figure 1.

Figure 1. MGSC $_{aps}$ was successfully synthesised using arenga palm sugar as carbon source

2.5. Preparation of Synthetic Nickel (II) Solution

An amount of 5000 mg/L nickel (II) solution was prepared using 5 g pure nickel metal and dissolved in 20 ml of hot concentrated nitric acid. It was cooled and diluted with double distilled water. The stock solution was further diluted with distilled water before used.

3. Results and Discussion

3.1. MGSC$_{aps}$ Characterisation

3.1.1. Field Emission Scanning Electron Microscope. The morphological of MGSC$_{aps}$ was carried out through FESEM analyses. It was evident that the sand was coated uniformly which confirms the temperature setting for initial burning process was 200 °C successfully melted arenga palm sugar to form a layer of carbon coating. This is in line with the previous researchers [8] which stated that the melting point of sucrose is approximately 186 °C. The graphenic morphology also was detected at the coating sheet's edges where thin wrinkles were formed as shown in Figure 2. The winkled edges of sheets that confirms the production of graphene, even after the formation of the composites, the structure of the composite remained the same where the wrinkles were still present [9].

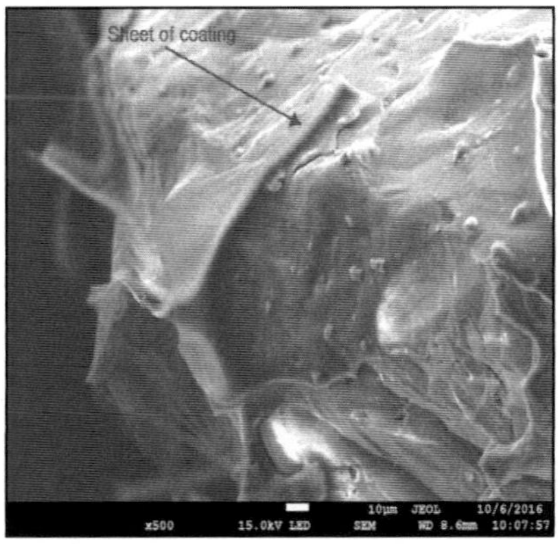

Figure 2. Sheet of coating was formed after arenga palm sugar melted and burned in furnace

3.1.2. Elemental Mapping. Major elements present were carbon, silicon and oxygen as presented in Figure 3. The presence of large carbon feature was originated from the carbonaceous material which was formed from melted arenga palm sugar on sand surface. Previous researchers [10] stated that the elements detected show that the sand was covered by carbon. The other elements; silicon and oxygen were also features in the major elements due to the usage of sand (SiO_2) hence validating the name of this composite in the research, mesoparticle graphene sand composite ($MGSC_{aps}$).

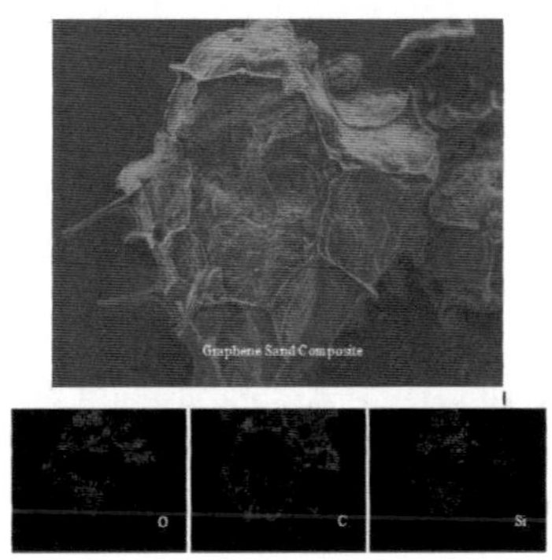

Figure 3. Elemental mapping on $MGSC_{aps}$ surface

3.1.3. X-Ray Diffraction. An XRD analysis was conducted to measure the structure of $MGSC_{aps}$. The space group of $MGSC_{aps}$ is 166: R-3m with hexagonal in shape whereas the figure of merit is 0.269. These two values validated the phase name of the composite where graphite-3R element was found during the test. The strongest diffraction peak located at 26.652 °which indicates the formation of the crystalline of the $MGSC_{aps}$ that gives glowing effect to the composite due to the light reflection. The d spacing of pristine $MGSC_{aps}$ is 0.525 nm. The value in this research of d spacing is higher than previous researcher [8]. This is supported by another researcher [9] which stated that the activation process by using concentrated sulphuric acid on the graphene sand composite resulted in the formation of multiple layers which produced higher d spacing. Due to the reason, the $MGSC_{aps}$ structures were exhibited multilayer of graphene structures simultaneously. The all diffraction peaks and corresponding reflections which at 266.652 ° (0,0,3), 44.36 ° (1,0,1) 54.706 ° (0,0,6) and 77.58 ° (1,1,0) were showing the basic characteristic of graphene similar to the previous research. Figure 4 presents the result of $MGSC_{aps}$ upon x-ray diffraction test.

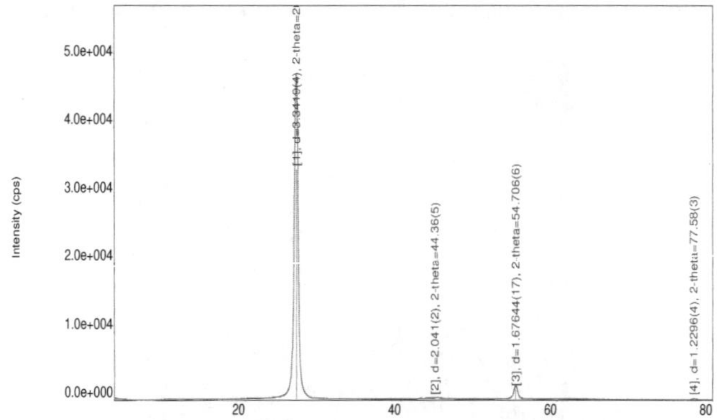

Figure 4. The structure of $MGSC_{aps}$ was determine using x-ray diffraction

3.2. Adsorption Study

The adsorption study is an important stage to determine adsorbency patterns in every different parameter set of $MGSC_{aps}$. Further discussion presented in subchapters as follows:

3.2.1. Adsorption Kinetic. The adsorption data was fitted to two kinetic models namely Pseudo-first-order and Pseudo-second-order. The interaction of nickel (II) and $MGSC_{aps}$ was investigated as a function of time by using column batches experiment to determine its ability to remove nickel (II) at the maximum capacity. The adsorption capacity, q was showed in the Equation 1.

$$q = [(Ci - Cf)V]/W \qquad \text{Equation 1}$$

Where Ci is the initial concentration in mg/L, Cf is the final concentration in mg/L, V is the volume of nickel (II) solution in ml whereas W is the weight of $MGSC_{aps}$ in g, used as adsorbent. The calculations were made to find adsorption capacity as a function of time, q_t. The calculations were repeated to find development curve throughout the adsorption period. In this case, adsorption period was taken at 600 minutes due to the stagnant data starting from 540 minutes which were recorded as adsorption at equilibrium, q_e. The pseudo first order and pseudo second order can be expressed as:

$$\ln(qe - qt) = \ln(qe - k_1.t) \qquad \text{Equation 2}$$

$$\frac{t}{qt} = \frac{1}{k_2 qe^2} + \frac{t}{qe}$$

Equation 3

The values of qe were calculated from the experimental data before plotting $\ln(qe - qt)$ versus t and t/qt versus t to find k_1.and k_2. The value of k_1.and k_2 is calculated from the slope of the function $y = mx + C$. From the curves it was found that k_1 value is -0.003 whereas for k_2 is 1.815. It is also exhibited the increment of adsorption capacity is increasing as the interaction time increase. However, the adsorption is still happening despite of the lower increment on the nickel (II) adsorption at the later time of interaction. It was successfully to adsorb 40% of nickel (II) at the first hour of contact time before the adsorption increment has reached stagnant phase after 960 minutes of interaction, suggesting the whole process of adsorption was achieved the equilibrium phase [11], [12]. The experimental data was compared to the two different models to analyse and determine the most suitable model for the experimental data plotted. The pseudo first order given R^2= 0.9867 whereas pseudo second order given better correlation with R^2= 0.9921. Thus, this experiment showed that the pseudo-second-order equation is more appropriate in indicating better curve fitting to describe experimental data as compared to the other model; pseudo-first order. Longer time was needed for graphene sand composite to completely saturate with the nickel (II) similar with literature review. This is in line with the previous researches[13], where other researchers [14], [15] suggested that saturation of adsorbent and solution influences breakthrough curve. Longer breakthrough is indicated by residual percentages of nickel (II) in waste water where longer point is needed to replace and regenerate graphene sand composite to remove more nickel (II) as shown in Figure 5.

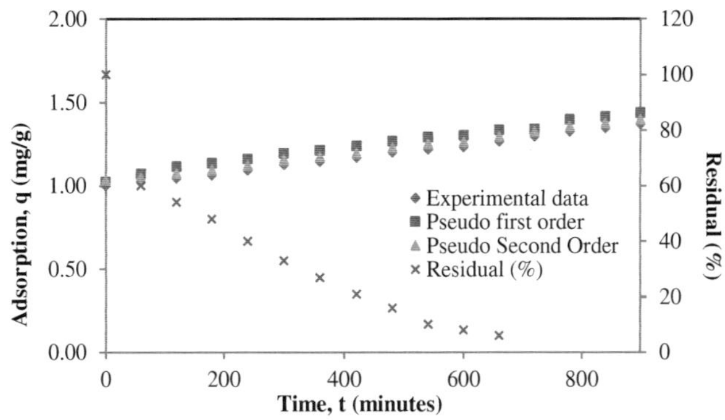

Figure 5. Adsorption of nickel (II) and it's residual in a factor of time

3.2.2. Adsorption Isotherm. For adsorption equilibrium, the data was fitted to two isotherm models namely Langmuir and Freundlich isotherm models. Langmuir model describes the equilibrium of the monolayer of adsorption onto the surface of containing finite identical active sites. This model assumes that there is uniform energy of adsorption onto the surface of adsorbent and no transmigration of the solution in the surface plane. On the other hand, Freundlich model is commonly used to describe heterogeneous surface of the adsorption characteristics. The equations of Langmuir and Freundlich are presented as follow.

Langmuir model, $qe = (qm.K_L.Ce)/(1 + K_L.Ce)$ Equation 4

$$R_L = 1/(1 + K_L Ce)$$ Equation 5

Freundlich model, $qe = K_F.Ce^{1/n}$ Equation 6

Where qe is the amount of nickel (II) adsorbed in mg/g, Ce is the equilibrium concentration of nickel (II) in mg/L, qm is the maximum adsorption capacity in mg/g, K_L is the Langmuir isotherm constant in L/mg, K_F is the Freundlich isotherm constant in mg/g whereas n is the adsorption intensity. The graph of Ce/qe as a function of Ce is plotted to find the parameters of Langmuir isotherm model which are qm, K_L, R_L and R^2. The slope and interception of the curve is equal to $1/qm$, and $1/(qm.K_L)$ respectively. Whereas to find the parameters of Freundlich isotherm model which are K_F, n and R^2, the graph log qe versus log Ce was plotted. The slope of the graph is equal to n value and K_F is calculated based on the fundamental logarithm's principal as follow:

$$log_{10}K_F = 0.8037$$
$$K_F = 10^{0.8037}$$
$$K_F = 6.364$$
Equation 7

Figure 6 exhibits the adsorption equilibrium of nickel (II) adsorbed by $MGSC_{aps}$. From the graph, it is suggested that the nickel (II)'s concentration would reflecting the same towards adsorption equilibrium. The equilibrium data is fitted with Freundlich isotherm model best with $R^2 = 0.9989$ as compared to the other isotherm model, Langmuir isotherm model with $R^2 = 0.9873$ due to the heterogeneous characteristic of $MGSC_{aps}$ consisting different active sites to be adsorbed. This is in complete agreement with the experimental data in kinetic study.

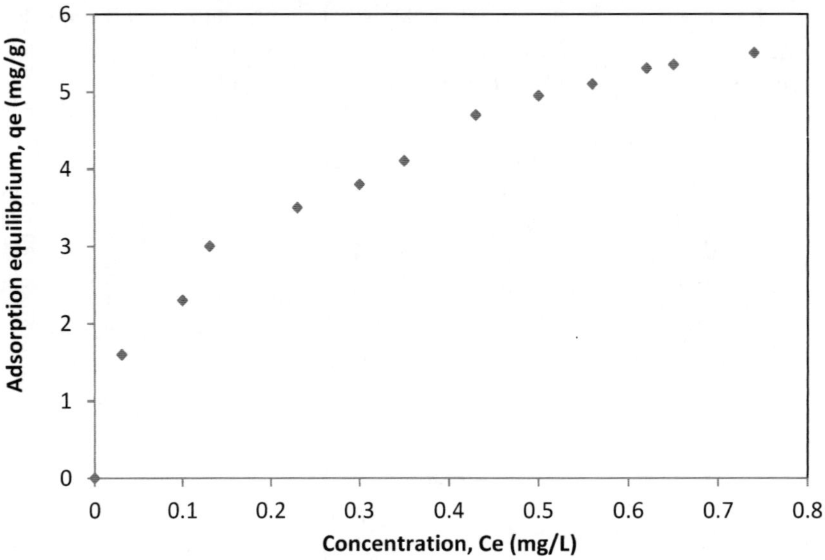

Figure 6. Adsorption equlibrium against concentration of nickel (II) solution

4. Conclusion

The kinetic and isotherm study shows that the surface of the $MGSC_{aps}$ is heterogeneous and favourable to adsorb at the lower temperature due to the exothermic process of it. It is proposed to carry out further research on the effect of reusability and regeneration on the removal nickel (II) or any other heavy metals in waste water.

References

[1] P. S. Goh and a. F. Ismail, "Graphene-based nanomaterial: The state of the art material for cutting edge desalination technology," *Desalination*, vol. 356, pp. 115–128, 2015.
[2] Y. Zhou, O. G. Apul, and T. Karanfil, "Adsorption of Halogenated Aliphatic Contaminants by Graphene Nanomaterials," *Water Res.*, vol. 79, pp. 57–67, 2015.
[3] M. T. Amin, a a Alazba, and U. Manzoor, "A Review of Removal of Pollutants from Water / Wastewater Using Different Types of Nanomaterials," vol. 2014, 2014.
[4] F. Ahmed and D. F. Rodrigues, "Investigation of acute effects of graphene oxide on wastewater microbial community: A case study," *J. Hazard. Mater.*, vol. 256–257, pp. 33–39, 2013.
[5] Z. Al-Qodah, "Biosorption of heavy metal ions from aqueous solutions by activated sludge," *Desalination*, vol. 196, no. 1–3, pp. 164–176, 2006.
[6] A. K. Mishra and S. Ramaprabhu, "Functionalized graphene sheets for arsenic removal and desalination of sea water," *Desalination*, vol. 282, pp. 39–45, 2011.
[7] S. Sen Gupta, T. S. Sreeprasad, S. M. Maliyekkal, S. K. Das, and T. Pradeep, "Graphene from Sugar and its Application in Water Puri fi cation," *ACS Appl. Mater. Interfaces*, vol. 4, pp. 4156–4163, 2012.
[8] V. P. Parvathi, M. Umadevi, and R. B. Raj, "Improved waste water treatment by bio-synthesized Graphene Sand Composite," *J. Environ. Manage.*, vol. 162, pp. 299–305, 2015.

[9] T. S. Sreeprasad, S. M. Maliyekkal, K. P. Lisha, and T. Pradeep, "Reduced graphene oxide-metal/metal oxide composites: Facile synthesis and application in water purification," *J. Hazard. Mater.*, vol. 186, no. 1, pp. 921–931, 2011.

[10] S. Moradi and S. Azizian, "Preparation of nanostructured carbon covered sand for removal of methyl violet from water," *J. Mol. Liq.*, vol. 219, pp. 909–913, 2016.

[11] M. Yari, M. Rajabi, O. Moradi, A. Yari, M. Asif, S. Agarwal, and V. K. Gupta, "Kinetics of the adsorption of Pb(II) ions from aqueous solutions by graphene oxide and thiol functionalized graphene oxide," *J. Mol. Liq.*, vol. 209, pp. 50–57, 2015.

[12] H. Qiu, L. Lv, B. Pan, Q. Q. Zhang, W. Zhang, and Q. Q. Zhang, "Critical review in adsorption kinetic models," *J. Zhejiang Univ. Sci. A*, vol. 10, no. 5, pp. 716–724, 2009.

[13] V. K. Gupta and A. Nayak, "Cadmium removal and recovery from aqueous solutions by novel adsorbents prepared from orange peel and Fe2O3 nanoparticles," *Chem. Eng. J.*, vol. 180, pp. 81–90, 2012.

[14] S. M. M. and T. S. Sreeprasad, Soujit Sen Gupta and T. Pradeep, "Immobilized graphene-based composite from asphalt: Facile synthesis and application in water purification," *Int. Mater. Rev.*, vol. 247, pp. 1–19, 2013.

[15] A. Mittal, J. Mittal, A. Malviya, and V. K. Gupta, "Adsorptive removal of hazardous anionic dye 'Congo red' from wastewater using waste materials and recovery by desorption," *J. Colloid Interface Sci.*, vol. 340, no. 1, pp. 16–26, 2009.

A Review on the Development of Gravitational Water Vortex Power Plant as Alternative Renewable Energy Resources

M. M. Rahman[1,*], J. H. Tan[1], M. T. Fadzlita[1], A. R. Wan Khairul Muzammil[1]

[1]Material and Mineral Research Unit, Faculty of Engineering, Universiti Malaysia Sabah, Kota Kinabalu, Sabah, Malaysia

*Corresponding author: mizanur@ums.edu.my

Abstract. Gravitational water vortex power plant is a green technology that generates electricity from alternative or renewable energy source. In the vortex power plant, water is introduced into a circular basin tangentially that creates a free vortex and energy is extracted from the free vortex by using a turbine. The main advantages of this type of power plant is the generation of electricity from ultra-low hydraulic pressure and it is also environmental friendly. Since the hydraulic head requirement is as low as 1m, this type of power plant can be installed at a river or a stream to generate electricity for few houses. It is a new and not well-developed technology to harvest electricity from low pressure water energy sources. There are limited literatures available on the design, fabrication and physical geometry of the vortex turbine and generator. Past researches focus on the optimization of turbine design, inlets, outlets and basin geometry. However, there are still insufficient literatures available for the technology to proceed beyond prototyping stage. The maximum efficiency obtained by the researchers are approximately 30% while the commercial companies claimed about 50% of efficiency with 500W to 20kW of power generated. Hence, the aim of this paper is to determine the gap in the vortex power plant technology development through past works and a set of research recommendations will be developed as efforts to accelerate the development of GWVPP.

1. Introduction

Electricity plays an important role in human community's daily life especially when most of the household appliances are replaced with mechanized robots or machines. British Petroleum (BP) statistical report 2016 shows that the global primary energy consumption increased by 1.0% from year 2014 to year 2015. Globally, about 32.9% consumed energy was from fossil fuels, mainly from oil. The contribution of renewable energy was still very small which was about 213 terawatt-hours only [22]. Many types of renewable energy sources such as wind, hydro, solar and etc. are used all over the world to generate electricity. The generation cost depends on the energy sources and conversion technology. Among all the sources, most economical abounded source is hydro power but the global hydro power output grew by 1% only, which is below the 10-year average of 3% [22]. The most well-known method of generating electricity from hydro power is by creating an artificial lake and develop it into large dam in river pathway. The environmental impacts of this type hydropower plant are significant and therefore reassessment of construction is necessary [26][27][28]. Micro-hydro and mini-hydro power plant can be used to generate electricity at small scale but this type of power plants requires high potential energy head to run the turbine [29]. An alternate option to harvest energy from low head water sources is essential to meet the growing demand of electricity as well as encourage new investor in this field. One of suitable proposal for electricity generation from hydro source is Gravitational Water Vortex Power Plant (GWVPP) [1].

1.1 Gravitational Water Vortex Power Plant
Gravitational Water Vortex Power Plant is a type of green technology that falls in the category of micro hydro power plant. It is currently being categorized as micro hydropower because the maximum reported power generation had not exceeded 100kW [23]. The main advantage of this power plant is the ultra-low hydraulic head requirement [10] as well as environmental friendly [1]. In this plant, the water passes through a large, straight inlet, which then passes tangentially into a round basin. The water will then form a powerful vortex, which exits the outlet at the centre bottom of the shallow basin. Due to its ultra-low hydraulic head requirement, the plant does not work on the pressure difference but on the dynamic force generated by vortex [1][2]. Hence, the development and power generation costs are very low in the GWVPP compared to other hydro power technologies. This type of power plant has found to be advantageous due to the following properties of water vortex:

- Increases the surface area of water [3]
- Maximizes the velocity of flow on the surface area of water [3]
- Disseminates homogenously contaminants in the water. [3]
- Increases the contact surface of the disseminate contaminants for microorganisms and water plants. [3]
- Aerates the water naturally due to the high velocity of flow on the water surface area and the increased water surface area to support the self-purification of water with microorganisms and water plants. [3]
- Increases the head of evaporation so that water can reduce the temperature itself at rising temperatures in the summer. [3]
- Builds up a peripheral zone of ice in the winter to isolate the centre of the vortex. [3]
- Concentrates dense water in the ring-shaped centre to ensure the survival of microorganisms as long as possible. [3]
- Decelerates the flow of water so that is can be used as an active retention pond [3]
- Improves the dissolved oxygen concentration

2. Literature Review

2.1 Turbine
Turbine is one of the most vital components for GWVPP. It is positioned at the centre, aligned with the central outlet of the basin. The turbine is forced by the water power that came from the water vortex. Most of the researches focused on optimization of turbines to increase the efficiency of the plant [3][4][5][6][7][8][9]. The earliest publication regarding GWVPP is by Marian et. al. in the year 2012 [3]. The main aim of their research is to determine the effects of basin's shape on turbine's performance. Different sizes Francis turbine at different level of depths is a valid option for GWVPP. From the simulation, it is also found that cavern (vortex) formed was proportional with the rotational speed without the presence of turbines in the system but at presence of turbine, the cavern height changed significantly as well as lowering the efficiency [3]. According to Subash Dhakal et. al., as the number of blades of turbine increased from six to twelve, the efficiency of the GWVPP reduced [6]. However, Christine Power et. al. found that the efficiency increased with the increased number of blades from two to four. The opposite findings by both groups of researchers indicate that there might be an optimal number of blades for turbine [9]. The concept of GWVPP is not well established yet therefore different researchers are using different types of vortex turbine plant for their investigation [1][3]. Sagar Dhakal et. al. only simulated different physical parameters of basin of GWVPP therefore validation the results are still questionable. They found that maximum vortex velocity for both cylindrical and conical basin is highest when the position is nearest to the outlet of the basin. However, the argument was made based on CFD modelling results only. The efficiency of conical

basin of GWVPP is also found to be higher than that of cylindrical basin at the same position of turbine [7]. Another researcher, Subash et. al. [6] in the year 2014 also conducted several tests at Manohara River located at Pepsicola, Kathmandu and found that the vortex strength in conical basin is stronger than that of cylindrical basin. In the year 2013, Marian et. al. conducted experiments on vortex turbine power plant by using three different turbines and installed it at different heights in the conical basin [4][30]. According to them, the maximum exergy can be harvested if the turbine is installed at near the outlet of the basin. The experimental results are validated with the theoretical analysis and claimed to be in good agreement by considering free vortex flow where variables are considered from flow, speed and power characteristics [4][30]. Since the blade profile has not been considered during the analysis of the efficiency of the turbine, it is recommended as future research. In the year 2013, Aravind Venukumar generated electricity by using Artificial Vortex power generator which consists of eight inverted cone design blades. The prototype managed to produce more than 150W of electricity [5]. In order to improve the efficiency of GWVPP, study also shows that researchers should focus on turbine's materials to reduce weight, maximize flow rates by optimize canal and notch angle as well as outlet diameter [6]. Sritram P et. al. also managed to prove that GWVPP with aluminium turbine has higher efficiency at different loads and flow rate when compared to steel turbine. Besides that, the higher the electrical loads, the lower the efficiency [8]. In the year 2016, Christine Power et. al. found that parameters including turbines, inlet height and flow rates have significant effects on the efficiency when vertical axis turbine with different blade sizes are used in vortex power plant. A total of nine configurations of turbines were prepared with different sizes and blade numbers. The overall efficiency of the vortex power plant was varied from 15.1% to 25.36% depends on the geometry of the basin, volumetric flow rate, turbine position and blade geometry and number [9].

2.2 Inlet and Outlet Configurations

For GWVPP, the inlet flow rates are the water that is released into a channel connected to the basin. The channel is responsible to direct the water flow into the basin tangentially. It can be horizontal or slanted at desired angle. The channel width between two ends could be different or the same. One of the study that will be mentioned below also shows that the inlet of GWVPP could be in the form of pipe instead of channel [9][19]. The inlet height has two meaning, first one deals with the height of water while another one indicate the height of channel from the bottom of the basin. The inlet and outlet parameters are shown in Figure 1 and Figure 2.

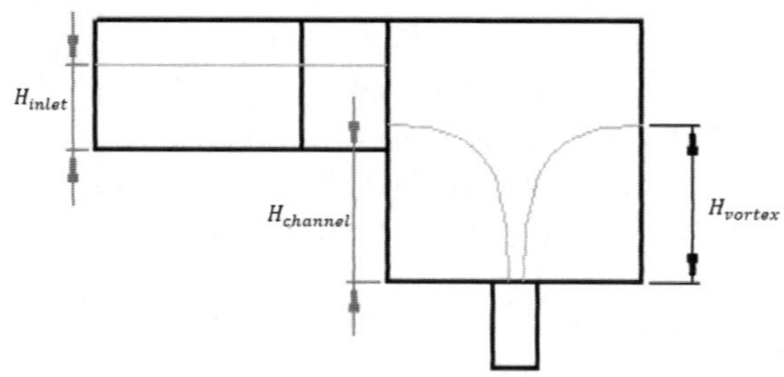

Figure 1. Gravitational Water Vortex Power Plant side view

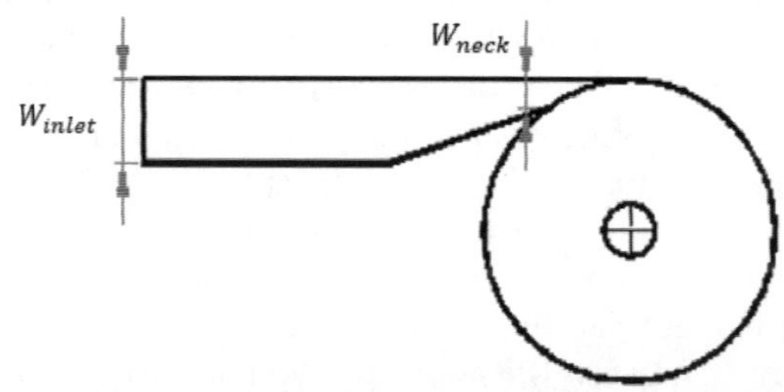

Figure 2. Gravitational Water Vortex Power Plant top view

Figure 2 shows that the outlet was usually at the centre of the basin and its diameter has significant affect on vortex strength as well as efficiency of vortex turbine [10][11][12].
Wanchat et. al. [10] studied the effects of outlet diameter varied from 0.10m to 0.40m on the efficiency of vortex turbine. The inlet velocity was set at 0.1m/s and the inlet channel was converged at the end which was connected to the basin. A five blades vertical axis turbine was used for electricity generation. It is found that the outlet diameter within the range from 0.20m to 0.35m has significant effect on power generation. The overall efficiency was reported as 30%. Shabara et. al. [11][12] conducted simulation and experimental studies. The GWVPP's $H_{channel}$ was zero. The simulation results showed that the outlet speed was inversely proportional to the outlet diameter. For the highest H_{inlet}, the outlet velocity was also maximum, which matched with the experimental results [11][12]. The inlet flow rate has significant effects on efficiency. The efficiency of the vortex turbine is directly proportional to the inlet flow rate and the optimal $H_{channel}$ was one-third of the basin's height [9].

2.3 Basin Configurations
Depending on the design of the basin of GWVPP, the vortex profiles created will be different. Wanchat and Suntivarakorn conducted research through simulations with three proposed designs which were 1) cylindrical basin with central outlet, 2) rectangular basin with pre-rotation and 3) cylindrical basin with inlet guide. It was observed that cylindrical basin with inlet guide was the most suitable basin since it provides better and more uniform velocity compared to others [13]. Dhakal et. al. conducted studies with conical basin instead of conventional cylindrical basin. Vortex velocity was measured for different basin diameter, notch angle, notch inlet width (W_{in}), cone angle and canal height ($H_{channel}$) and it is found that W_{in}, cone angle and $H_{channel}$ have the most significant effects on the vortex velocity. The value of W_{in} was suggested to be as small as possible on the other hand the cone angle and the $H_{channel}$ were suggested to be as high as possible to maximize the performance. Notch length (length of inlet channel) is also recommended to be kept as long as possible to prevent unwanted loses [14].

2.4 Free Surface Vortex
The vortex strength has major effects on the efficiency of the vortex turbine. The strength of the free vortex is weak and creates shallow vortex profile, therefore optimization is required for maximum power generation. The conventional theoretical model can't be used for theoretical analysis since it was developed based on weak air-core vortices above hydraulic intakes [15][16][17].
Sajin and Marian [4] lead the study on free surface vortex by developing theoretical models of the vortex formed in a conical basin. The vortex flow structure can be identified as air cavern (air core of the vortex), forced vortex (area between the surface of the water vortex and free vortex), free vortex (area between forced vortex and boundary layer) and boundary layer (water near the wall of basin). The theoretical model of boundary layer thickness obtained by Sajin and Marian [4] was

$$\delta(R) = \frac{R_0}{\left(\frac{\Gamma}{2\pi \sin(\frac{\alpha}{2})}\right)^{1/2}}\left[4.38\left(1-\frac{R}{R_0}\right)^{1.75} - 9.36\left(1-\frac{R}{R_0}\right)^{1.37} + 5.8\left(1-\frac{R}{R_0}\right)^{0.57} + 1\right] \quad (1)$$

The theoretical model was presented graphically and discussed. Theoretical modelling for forced vortex was described as

$$\frac{p}{\rho} = \frac{\omega^2 r^2}{2} - \frac{\gamma^2}{8} r^2 + \frac{\gamma^2}{2} z^2 - gz - \frac{\sigma}{\rho r} + constant \quad (2)$$

and free surface vortex was modelled as

$$z(r) = \frac{g - [g^2 - 2\gamma^2\left(\frac{\omega^2 r^2}{2} - \frac{\gamma^2}{8}r^2 - \frac{\sigma}{\rho r}\right)]^{1/2}}{\gamma^2} \tag{3}$$

Simulations were carried out to study the water flow in the basin with and without turbine. It was observed that the presence of turbine had huge effect on the vortex flow. Sean Mulligan et. al. [15] found that most of the free-surface vortices were generally in the context of "weak" vortices at hydraulic intakes. Sean et. al. [15] conducted analytical, theoretical and experimental studies on the free-surface vortex. For an ideal vortex model, the flow was classified as irrotational and the tangential velocity profile was described as

$$v_\theta = \frac{\Gamma}{2\pi r} \tag{4}$$

However, for real Newtonian fluid, viscosity must be accounted for. According to Rankines model improved by Vatistas et. al. (1991), Sean et. al. [15] managed to form the expression to describe the radial distribution of the free surface vortex.

$$H_\Delta = \frac{1}{g}\left(\frac{\Gamma}{2\pi}\right)^2 \left(\frac{1}{r_{in}{}^2} - \frac{1}{r_r{}^2}\right) \tag{5}$$

where r_{in} and r_r are radii at the inlet and at an arbitrary radius in the vortex. H_Δ is the height difference between these radii. Sean et. al. [16] [17] continued research on vortex experimentally and develop a surface profile model for open channel flow in the vortex chamber (basin). The theoretical discharge can be calculated from the following equation for a strong free-surface vortex.

$$Q = \frac{\pi}{4}d^2(1-\lambda)\left(2gE - \frac{4\Gamma^2}{d^2\lambda}\right)^{1/2} \tag{6}$$

Sean [16] also designated dimensionless number, β as the ratio of the inlet radius from the shaft center to the inlet flow cross section. Experiments were carried out for different β value and it was found that for larger β value, discharge head rate was lower but the circulation of the vortex increases as the β increases. Circulation of vortex can be determined from the following equation.

$$\Gamma_n = \frac{4.9972}{K_1{}^{1.002}} \cong \frac{5\beta d}{h} \tag{7}$$

Study showed that higher h/d creates stronger tangential velocity as well as stronger vortex. The flow rate and tangential velocities were predicted the greatest reasonable accuracy [16]. Kueh et. al. [18] took a similar approach in the study of vortex formation in vortex turbine. Simulation using x-flow were conducted and compared with the experimental results. Kueh et. al. also found that as the height of the inlet increases, the vortex height increases. The difference between Kueh et. al. [18] and Sean [16] was the basin geometry. Sean et. al. [16] [17] designed their basins for zero $H_{channel}$ whereas Kueh et. al. [18] conduct experiment for basin $H_{channel} = 0.5\ m$..

2.5 Similarity Modelling

Sean [16] carried out intensive reviews on free-surface vortex. In order to validate the theoretical analyses, it is necessary to construct experimental model. Due to the huge size of actual power plant, scaled down model can be designed according to dimensionless similarity such as Froude number (Fr), Reynolds number (Re) and Weber number (We). According to Sean et. al. [16], the most suitable dimensionless number for free surface vortex is Froude number since the gravitation force is a dominating force for this study. Sean [16] noted that both numbers should be kept above the minimum specified values in the model. Study showed that the outlet Weber number should be maintained as 120 (Jain et. al. [24]) and for Reynolds number the criteria for minimum radial Reynolds number of 10^3 in the model can be implemented [25]. Researchers agreed that the design of GWVPP still has many configurations that have yet to be explored. The lacks of technical information such as appropriate methodology for testing, design configuration, performance calculation, etc. also had undesirable impacts on the development of GWVPP. Table below shows the summary of studies made by highlighting their findings.

Table 1. Summary of GWVPP findings by different authors.

Study	Research Methods	Research Parameters	Findings
		Turbine Configurations	
Marian et. al. [3]	Theoretical	Position of multiple turbines	• Vortex∝rotational speed • Vortex height affect the efficiency of GWVPP
Marian and Sajin [4]	Theoretical and Experimental	Position of multiple turbines	• Turbine extracts more energy near the outlet • Experimental results agree well with theories
Aravind Venukumar [5]	Theoretical	Turbine blades design	• Produced more than 150W of power.
Subash et. al. [6]	Experimental	Position of turbine	• Maximum power was extracted near the outlet
		Number of blades of turbine	• Power reduced as number of blades increased • Maximum efficiency = 25.36%
Sagar et. al. [7]	Experimental	Position of turbine	• Optimal position of turbine is 65% to 75% of GWVPP's height
		Basin's Structure	• Conical basin has higher overall power output • Maximum efficiency = 36.84%
Sitram et. al. [8]	Experimental	Turbine materials	• Aluminum is more efficient than steel • Maximum efficiency = 34.79%
Christine et. al. [9]	Experimental	Number of blades of turbine	• Efficiency∝number of blades
		Size of blades of turbine	• Efficiency∝size of turbine • Maximum efficiency = 15.1%
		Inlet and Outlet Configurations	
Wanchat et. al. [10]	Experimental	Outlet diameter varied	• Maximum efficiency = 30% • Outlet diameter between range of

			0.20m and 0.35m
Shabara et. al. [11] [12]	CFD and Experimental	Outlet diameter varied	• Outlet velocity $\frac{1}{\propto}$ outlet diameter • Maximum outlet velocity at highest H_{inlet} • Maximum efficiency = 40%
Christine et. al. [9]	Experimental	Inlet flow rate $H_{channel}$	• Highest efficiency at maximum inlet flow rate • Optimal $H_{channel}$ at one-third of basin's height
		Basin Configurations	
Wanchat and Suntivarakorn [13]	Simulation	Basin's structure	• Cylindrical basin with inlet guide has the best flow field
Sagar et. al. [14]	Simulation	Basin's dominant parameter	• W_{in} should be as small as possible • Cone angle should be as big • $H_{channel}$ should be high • Shouldn't exceed optimum value

3. Research Recommendations

GWVPP is a low hydraulic head hydropower plant. In order to achieve maximum efficiency, a suitable turbine with optimized shape and blade profile should be designed [4][7]. Sitram et. al. [8] proved that in the GWVPP, the efficiency of the Aluminium turbine is higher compared to Steel turbine. Subash et. al. [6] also suggested that lighter and stronger materials can be used to reduce the weight of the turbines to acquire best performance from GWVPP. Cavitation, cracking and corrosion have significant effects on turbine blades but it is not considered in previous study of GWVPP turbine therefore, these parameters should be considered during the design of efficient turbine.

In addition, Christine et. al. [9] found that efficiency of the power plant increased when numbers of turbine blades increased from two to four. Subash et. al. [6] found that efficiency reduced when turbine blades numbers increased, which is the opposite of Christine et. al.'s [9] findings. It is also found that the size of the blade has significant effect on GWVPP's efficiency. Both studies with opposite findings suggest that the optimal number of turbine blades and ratio between blade sizes and basin diameter might exist. Therefore, optimization of the blade numbers and profile of the turbine blade can also be considered for future studies.

The design of channel is also an important parameter for effective vortex generation. The tangential velocity of water in the vortex basin depends on the inlet width (W_{inlet}) and neck width (W_{neck}). The ratio of these parameters can be optimized by doing extensive research on it to enhance the water flow in the vortex basin. The channel height ($H_{channel}$) may has significant effects on flow that may alter the performance of the GWVPP, since its performance depends on volume of water. So, effects of the height of channel can be examined in future studies. Since free surface vortex is the source of power for GWVPP, it has several notable research topics that would form the core of an excellent future hydraulics research study in this field.

4. Conclusion

GWVPP is capable of generating power from low water head. Therefore, this type of power plant is suitable for the areas with rivers. Since the fossil fuels reserves are declining, GWVPP can be an alternative energy source which is environmentally friendly as well as cost effective. In this paper, the past literatures related to GWVPP have been presented to support the present research. The optimal turbine position is 0.65 to 0.75 of the basin's height. The maximum efficiency varied from 30% to 50% depends on the numbers and size of the blade. Study also showed that conical shape basin has better performance than cylindrical shape basin. In addition, the inlet flow rate has significant effects on the efficiency. It increased with increase of flow rate and vice versa. In a nutshell, studies showed that there are significant research gaps in the development of GWVPP. The research recommendations in this paper are efforts to accelerate the development of GWVPP.

5. References

[1] Zotlöeterer F. *Energy Systems* [updated 09 August 2014; cited 12th September 2016]. Available from:http://www.zotloeterer.com/welcome/gravitation-water-vortex-power-plants/zotloeterer-turbine/
[2] Systems P E. *PESWiki* [updated March 2014; cited 13th September 2016]. Available from: http://www.pureenergysystems.com/NEC/conferences/2008/EnvironmentalHallofFame-Chicago/displays/Zotloterer,pdf
[3] Marian G-M, Sajin T, Florescu I, Nedelcu D-I, Ostahie C-N and Birsan C 2012 *World Energy System Conf.* Issue 3 pp 219-226
[4] Marian G-M, Sajin T and Azzouz A 2013 *Applied Mechanics and Materials* **371** 601-605
[5] Aravind V 2013 *IEEE Global Humanitarian Technology Conf.: South Asia Satellite* pp 53-57
[6] Dhakal S, Nakarmi S, Pun P, Thapa A B and Bajracharya T R 2014 *Journal of the Institute of Engineering* **10** 140-148
[7] Dhakal S, Timilsina A B, Dhakal R, Fuyal D, Bajracharya T R, Pandit H P, Amatya N and Nakarmi A M 2015 *Renewable and Sustainable Energy Reviews* **48** 662-669
[8] Sritram P, Treedet W and Suntivarakorn R 2015 *IOP Conf. Series: Materials Science and Engineering* vol 103 012018
[9] Power C, McNabola A and Coughlan P 2016 *Journal of Clean Energy Technologies* **4** 112-119
[10] Wanchat S, Suntivarakorn R, Wanchat S, Tonmit K and Kayanyiem P 2013 *Advanced Materials Research* **805-806** 811-817
[11] Shabara H M, Yaakob O B, Ahmed Y M and Elbatran A H 2015 *Jurnal Teknologi* **74** 77-81
[12] Shabara H M, Yaakob O B, Ahmed Y M, Elbatran A H and Faddir S M 2015 *Jurnal Teknologi* **74** 91-100
[13] Wanchat S and Suntivarakorn R 2012 *Advanced Science Letters* **13** 173-177
[14] Dhakal S, Timilsina A B, Dhakal R, Fuyal D, Bajracharya T R and Pandit H P 2014 *Int. Conf. on Technology and Innovation Management & IOE Graduate Conf.* Kathmandu Nepal pp 380-386
[15] Mulligan S, Casserly J and Sherlock R 2014 *5th IAR Int. Junior Researcher and Engineer Workshop on Hydraulic Structures* Spa Belgium
[16] Mulligan S 2015 *Experimental and Numerical Analysis of Three-Dimensional Free- Surface Turbulent Vortex Flows with Strong Circulation* Institute of Technology Sligo
[17] Mulligan S, Casserly J and Sherlock R 2015 *E-Proc. of the 36th IAHR World Congress* Netherlands
[18] Kueh T C, Beh S L, Rlling D and Ooi Y 2014 *International Journal of Innovation, Management and Technology* **2** 111-115
[19] Kouris Centri Turbine [cited 22nd September 2016]. Available from: http://www.kcthydropower.com/
[20] Genossenschaft Wasserwirbe Schweiz [cited 22nd September 2016]. Available from: gwwk.ch/about/portrait-number-facts/
[21] Gravitational Water Vortex Power Plants [cited 22nd September 2016]. Available from: www.zotloeterer.com/welcome/gravitation-water-vortex-power-plants/
[22] Statistical Review of World Energy 2016 *65th Ed BP Statistical Review of World Energy* 2016 ed 65 p 5
[23] Pimnapat L, Patib T and Bhumkittipichc K 2013 *Energy Procedia* **34** 235-242
[24] Jain A K, Garde R J and Ranga Raju KG 1978 *Journal of the Hydraulics Division* **104** 1429-1445

[25] Anwar H O 1966 *Journal of Hydraulic Research* **4** 1-16
[26] Benchimol M and Peres C A 2015 *PLoS ONE* **10** 1-15
[27] Fearnside P M 2016 *World Development* **77** 48-65
[28] Palmeirim A F, Peres C A and Rosas F C W *Biological Conservation* **174** 30-38
[29] Yah N F, Oumer A N and Idris M S 2017 *Renewable and Sustainable Energy Reviews* **72** 228-239
[30] Marian G-M and Sajin T 2013 *Journal of Marine Technology and Environment* **1** 89-96

Acknowledgments
The authors would like to express their gratitude to the Ministry of Higher Education of Malaysia and Universiti Malaysia Sabah for their financial assistance and facilities support through Fundamental Research Grant (FRG0427-TK-1/2015). In addition, the authors would like to express sincere thanks to the evaluators of ICMTE 2017 for valuable comments and recommendations on this manuscript.

Determination of Temperature Rise and Temperature Differentials of CEMII/B-V Cement for 20MPa Mass Concrete using Adiabatic Temperature Rise Data

GO Chee Siang

30, Jalan 1/27D, Seksyen 6, Wangsa Maju 53300, Kuala Lumpur, West Malaysia.

csgo@loh-loh.com.my

Abstract. Experimental test was carried out to determine the temperature rise characteristics of Portland-Fly-Ash Cement (CEM II/B-V, 42.5N) of Blaine fineness 418.6m^2/kg and 444.6m^2/kg respectively for 20MPa mass concrete under adiabatic condition. The estimation on adiabatic temperature rise by way of CIRIA C660 method (Construction Industry Research & Information Information) was adopted to verify and validate the hot-box test results by simulating the heat generation curve of the concrete under semi-adiabatic condition. Test result found that Portland fly-ash cement has exhibited decrease in the peak value of temperature rise and maximum temperature rise rate. The result showed that the temperature development and distribution profile, which is directly contributed from the heat of hydration of cement with time, is affected by the insulation, initial placing temperature, geometry and size of concrete mass.

The mock up data showing the measured temperature differential is significantly lower than the technical specifications 20°C temperature differential requirement and the 27.7°C limiting temperature differential for granite aggregate concrete as stipulated in BS8110-2: 1985. The concrete strength test result revealed that the 28 days cubes compressive strength was above the stipulated 20MPa characteristic strength at 90 days. The test demonstrated that with proper concrete mix design, the use of Portland flyash cement, combination of chilled water and flake ice, and good insulation is effective in reducing peak temperature rise, temperature differential, and lower adiabatic temperature rise for mass concrete pours. As far as the determined adiabatic temperature rise result was concern, the established result could be inferred for in-situ thermal properties of 20MPa mass concrete application, as the result could be repeatable on account of similar type of constituent materials and concrete mix design adopted for permanent works at project site.

Keywords: Mass concrete, Portland-fly-ash cement (CEM II/B-V), Initial placing temperature, Heat of hydration, Adiabatic temperature rise, Peak temperature, Temperature differential, Compressive strength.

1. INTRODUCTION

In large concrete elements, the interior is considered to be hydrating in an essentially adiabatic process. Because of the low thermal conductivity of concrete, the heat of hydration from its interior is prevented from being released into the environment, thus a negligible amount of heat is lost, compare to exterior of concrete element. The resulting large temperature differential between the interior and exterior of a concrete element may lead to thermal cracking.

The term adiabatic refers to a process occurring without the gain or loss of heat. Adiabatic temperature measurement is used to simulate the condition in the interior of a large concrete element so that the maximum heat development potential in any concrete mix can be measured (Lee MH, Khil BS & Yun HD 2014). This data may then be used as input parameter for subsequent thermal modelling in the pursuit to determine the risk level and thermal cracking potential. Owing to data of this kind is not available locally; therefore adiabatic temperature rise test was carried out to determine the peak concrete core temperature, temperature differential and temperature gradient for this 20MPa mass concrete. This was determined through trial hot-box tests at prescribed initial placing temperature and anticipated concrete section thickness, rather than merely follow project contract specifications requirements (Technical Specifications 2000; 2007; 2009, 2010 & 2013), as stipulated below:

(1) The fresh concrete placing temperature shall not exceeding 20°C;
(2) Maximum core-surface temperature differential shall not exceeding 20°C

To avoid surface cracking caused by heat generated in the concrete, European Standard ENV 206: 1992 suggests that the limit on the temperature differential between core and surface is 20°C and the average maximum temperature limit of concrete at core is 60°C. Based on UK experience (BS 8110-2: 1985), by limiting temperature differential to 20°C in *gravel* aggregate concrete, cracking can be avoided. This represents an equivalent restraint factor R of 0.36. However, MS523-2: 2011, Clause A9.3 has specified peak placing temperature of 30°C for mass concrete structure construction. As for the case of HTP project, the proposed quarried rocks for mass concrete production was *granite* aggregate whereby the limiting temperature differential adopted should be capped at 27.7°C instead of 20°C; BS 8110-2: 1985 (pp. 32), Table 3.2- Estimated limiting temperature changes to avoid cracking has specified that limiting temperature differential when R = 0.36 is 27.7°C when granite aggregate concrete was adopted for mass concrete production and construction. However, BS 8110-2: 1985 does not specify the fresh concrete placing temperature and peak temperature requirements in tropical environment.

Concrete temperature rise test under adiabatic condition is normally an approach in laboratory to evaluate the exothermic characteristics of the hydration of the concrete. In this case, 20MPa plain concrete trial hot-box with an array of thermocouples and strain gauges are used to assess the characteristics of the temperature rise, maximum temperature difference and temperature distribution profile in mass concrete were conducted with due consideration given on the selection of concrete block size and geometry, insulation, thermal couples and strain gauges locations, and the duration of temperature monitoring. Due to the reduced rate of hydration, the use of CEMII/B-V cement to replace OPC (CEM 1) is one of the recommended practices to considerably reduce the temperature rise in mass concrete. Since hydration occurs at the surface of cement particles, therefore finely ground cement will have a higher rate of hydration. As finely ground cement has a higher specific surface area, which means there is more area in contact with water for more complete hydration. The finer particles will also be more fully hydrated then coarser particles. However, the total heat of hydration at very late ages is not significantly affected.

This research is intended to overcome the above-mentioned 20MPa mass concrete specification requirements for the construction of Tembat Hydropower dam by adopting more prudent temperature control requirements through subsequent thermal modelling rather than arbitrary determined basing on designer's temperate country experience and practices, which is overly conservative, costly method, and non-sustainable due to carbon emission by running huge capacity of chillers and flake-ice plant to bring down the temperature of water for making concrete and aggregates substantially.

The experimental program was carried out in accordance with the procedure as shown in below Table 1. This experiment is to determine the temperature rise characteristics of Portland-fly-ash cement concrete under adiabatic conditions as well as the compliance to 20MPa characteristic strength in 90 days. The established adiabatic temperature rise test results (Adiabatic Temperature Rise Test Results 2011 & 2012) allow better understanding of the temperature rise characteristics and its time-base temperature distribution profiles in the mass concrete, whereby the establish result could be adopted as input parameter for subsequent thermal simulation purposes.

Table 1. Procedure for Adiabatic Temperature Rise Modelling Test

Step:	Procedure
1	Analysis of raw material test results, i.e. cementitious, aggregates, admixtures, water for making concrete (similar to site material) and etc, complying with HTP project technical specifications and/or equivalent international codes and standards prior to Hot-Block casting and testing.
2	Key material selection and factors affecting the temperature rise as per "ACI305R-99 on Hot Weather Concreting" and temperature distribution profile in mass concrete.
3	Establishment of concrete mix design and concrete production for 20MPa Class A mass concrete.
4	Casting of Hot-Block using raw materials established from step 1 above
5	Temperature probing, slump measurements and concrete sampling using cube and cylinders as per testing plan carried out by established and accredidated testing laboratory.
6	Adiabatic temperature rise monitoring tests: (i) March 2011: 12 days; (ii) June 2012 : 15 days.
7	Compilation of testing results and data analysis.
8	Temperature analysis and interpretation of results.
9	Conclusions/Recommendations

2. EXPERIMENTAL SET-UP

For Stage 1 testing, two (2) hot boxes were prepared as part of the trials. The first hot box (HB1) was prepared with full insulation (100mm thick MRB-32 Insulation Board) with block size of dimension 1.5m x 1.5m x 1.5m and two sets of thermocouples (thermocouples wire type K) installed, one set (3 thermocouple each) at the center and the other set (3 thermocouple each) at the corner of the hot-box and two strain gauges installed at the centre of the block. A second hot block (HB2) of dimension of 1.3m x 1.3m x 1.3m was prepared with two sets of thermocouples installed, one set (3 thermocouple each) at the center and the other set (3 thermocouple each) at the corner of the block. In this case, the base and sides were insulated but the top surface of the box was exposed to the environment. Based on the completed trials conducted on 11/2/2011 and 12/2/2011, an identical concrete mix design was selected for the hot box casting as given in Table 5 (see Page 3 of 11). For stage 2 testing, one hot box was prepared at Hulu Terengganu project site with similar block size and configuration with stage 1 HB, comply with full insulation (100mm thick MRB-32 insulation Board) and two sets each of thermocouples and strain gauges installed using identical concrete mix design as indicated in Table 3.3, whereby the trial was conducted on 21/6/2012.

2.1 Test Specimen

For the evaluation of the temperature rise in mass concrete trial block, a 1.5m x 1.5m x 1.5m concrete block was cast. Blended cement (CEM II/ B-V) containing 25% fly ash was used for the Grade 20 mass concrete hot block trial. The cement content and water/cement ratio was 200 kg/m3 and 0.70 (Stage 1 & 2), respectively. The schematic drawing of the block is described in Figures 1-3, with the locations of the thermocouples, 100mm insulation and timber formwork indicated. 100mm thick MRB-32 insulation board was adopted as the insulation to the concrete mass around the six sides. The top of the concrete was only covered with the polyfoam. Tc is the core temperature at center of the block while Ts and Tn represent the temperature at surface and corner location as indicated in the schematic drawing. The peak core temperatures obtained from stage 1-HB1, stage 1-HB2 and stage 2-HB1 were also used to determine the repeatability of the experiments.

2.2 Equipment

(i) Thermocouple wires Type K manufactured by Tokyo Sokki Kenkyujo Co. Ltd of Japan, and vinyl insulated 0.32 x 1p K-G, with 20mm diameter protective PVC pipe for thermocouples wires.
(ii) Vibrating Wire Embedment Strain Gauge [VWSG] (Model: 5265099)
(iii) Data-logger model CR10X from Campbell Scientific Inc
(iv) AM 16/32multiplexer for multiple data input from Campbell Scientific Inc
(v) VWSG interface module, AVWI and Laptops.

2.3 Casting of Hot-Box

Hot-box is casted with 20Mpa temperature control concrete (average 25°C) with MSA 75mm whereby compaction was done by using 50mm vibrator pokers.

2.4 Data Acquisition

Campbell scientific CR10X data-logger was program automatically to record concrete temperature (°C) and VWSG's strain (με) data at every hourly interval simultaneously for a period of 7 days and the data are downloaded to laptop. The maximum temperature is defines as the readings when no further appreciably rise in temperature over a period of 3 – 4 successive hours).

3. CONCRETE MIX DESIGN AND STRENGTH ANALYSIS

The physical properties and chemical compositions as well as the fineness of the Portland-fly-ash cement (PFAC) used in the study are shown in Tables 2-4 respectively. Table 5 presents the concrete mix design for grade 20 mass concrete applications for all three concrete hot-box tests. The measured initial placing temperature of the mixes was 23°C to 25°C respectively.

Table 2. Typical physical properties of PFA Cement (CIMA, 2011 & 2012)

	PFAC (CEM II/B-V)
Standard	EN197-1 – CEM II/B – V 42.5N – LH
Typical Properties	
Blaine fineness (m^2/kg)	420
Specific gravity	2.9
Initial Setting time (min.)	157
2d strength (MPa)	22
7d strength (MPa)	37
28d strength (MPa)	50

Table 3. Typical chemical analysis of PFA cement (CIMA, 2011& 2012)

Chemical Properties	Chemical Composition (%)
SiO_2	26.57
Al_2O_3	8.84
Fe_2O_3	5.40
CaO	51.89
MgO	0.63
SO_3	2.31
Na2O	0.18
K2O	0.72
Na2O + 0.658 x K2O	0.65
Cl-	0.009

Table 4. Cement test result on properties of specific surface area and heat of hydration (CIMA, 2011 & 2012)

Issue Date	Blaine (cm^2/g)	Heat of Hydration @7D, 270 Max (J/g)
Jan 2011	4186	258
June 2012	4446	274

Table 5. Class A mass concrete C20 mix design (Mass Concrete Class A, 2011)

Description	Value
Cement (kg/m^3)	200
Water (kg/m^3)	140
Sand (kg/m^3)	756
Coarse Aggregate (20mm) (kg/m^3)	1287
Realset 233 (l/m^3)	1.6
Realflow 610 (l/m^3)	0.8
Slump (mm)	20-50
Characteristic Strength (MPa)	20
S/A Ratio (Sand/ Aggregates)	0.37
W/C Ratio(Water/Cement)	0.70

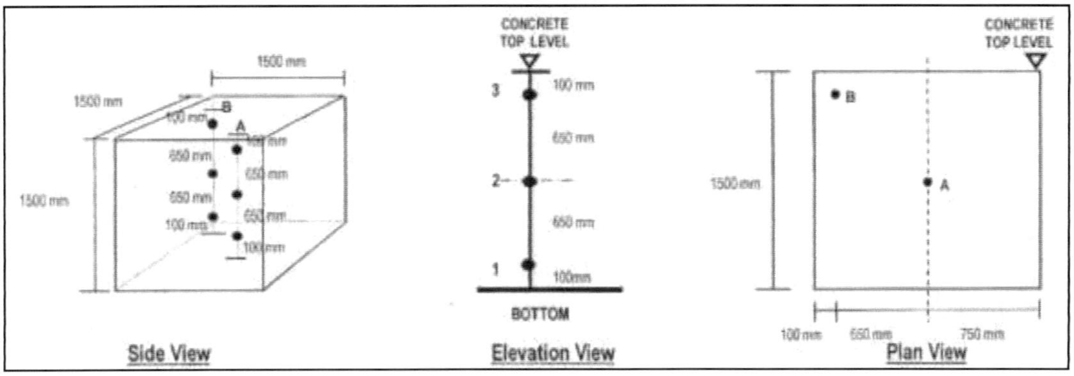

Figure 1. Sketch showing locations of thermocouples in 1.5m x 1.5m x 1.5m hot-box for Stage 1-HB1 and Stage 2-HB1 m^3

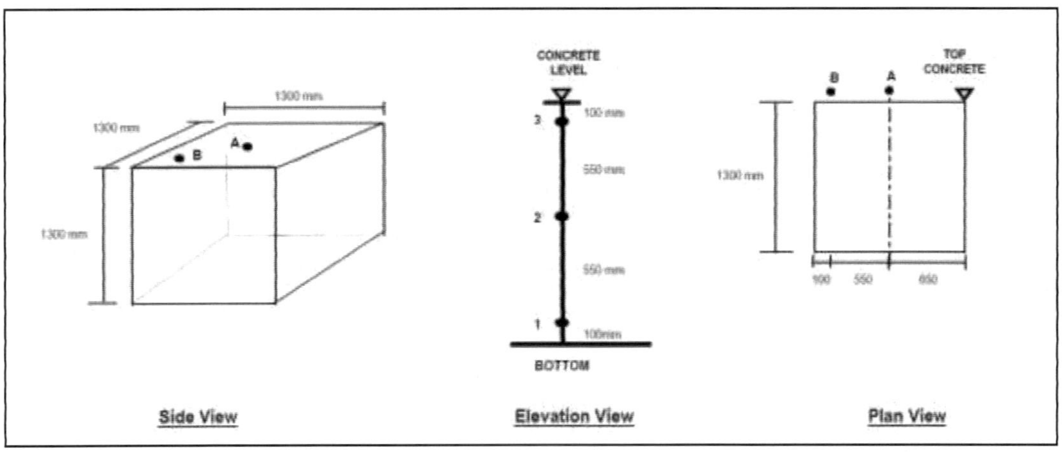

Figure 2. Sketch indicating locations of thermocouples in 1.3m x 1.3m x 1.3m hot-box for Stage 1-HB2

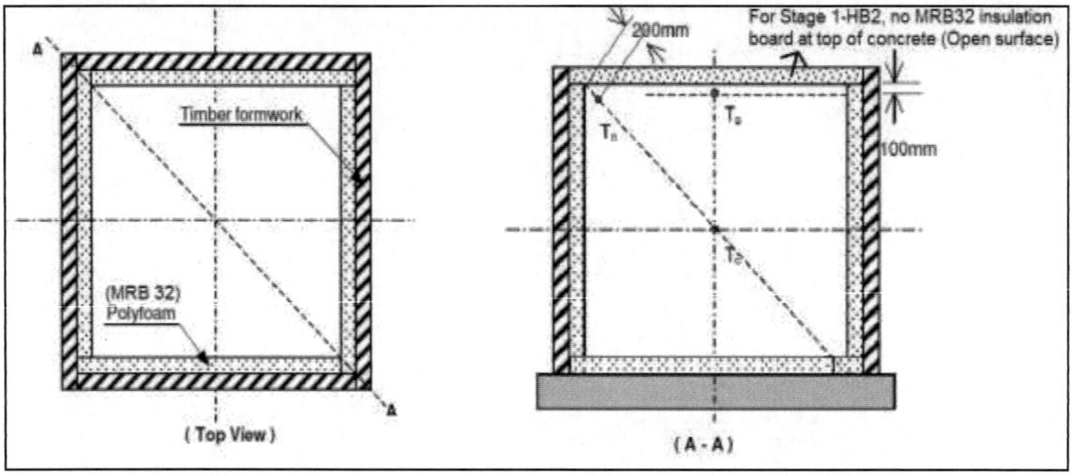

Figure 3. Sketch indicating locations of thermocouples in 1.5m x 1.5m x 1.5m hot boxes for Stage 1-HB1 & Stage 2-HB1 and 1.3m x 1.3m x 1.3m hot-box for Stage 1-HB2 (except top surface without insulation)

Table 6. Summary of concrete compressive strength, tensile strength and elastic modulus test results (Mass Concrete Class A, 2011)

	7Days			28Days			56Days		
Description	Stage 1-HB1 (2011)	Stage 2-HB1 (2012)	Δ	Stage 1-HB1 (2011)	Stage 2-HB1 (2012)	Δ	Stage 1-HB1 (2011)	Stage 2-HB1 (2012)	Δ
Compressive Strength (MPa)	25.25	23.43	1.82	33.05	31.77	1.28	38.15	39.77	1.62
Tensile Strength (MPa)	1.70	1.61	0.09	2.43	2.19	0.24	2.81	2.59	0.21
Modulus of Elasticity (MPa)	19.25	20.33	1.08	25.60	25.03	0.57	28.90	28.17	0.73

Based on the strength results shown in above Table 5, all concrete cubes tested at 7 days had exceeded the contract documents stipulated 90 days characteristic strength of 20MPa confirming that the required design strength of 20MPa is achievable within 28 day. Therefore, further optimization on content of cementitious material should be considered for lower heat of hydration.

Table 7. Summary of in-situ core test results (Mass Concrete Class A, 2011)

Description		FF1 50-200	FF1 500-650	FF1 50-200	FF1 500-650
	Unit	Aged 34 days		Aged 56 days	
Maximum load of failure	kN	667	643	741	687
Core compressive strength	MPa	37.7	36.4	41.9	38.9
In-situ compressive strength (as per BSEN 13791: 2007)	MPa	38.8	37.5	43.2	40.1
Cube compressive strength (see Table 6)	MPa	32.41		38.96	
Ratio of estimated in-situ cube strength to measured		1.20	1.16	1.10	1.03

In addition, internal cores were taken to study in-situ strength as shown in Figure 4. As expected with concrete containing flyash, the heat cycling had a beneficial effect on strength and the core strength is about 6.5 per cent higher comparing to the cube strength (see Table 7) measured at 56 days. The compressive strength test results confirm that the concrete is well above the specified requirements in terms of strength.

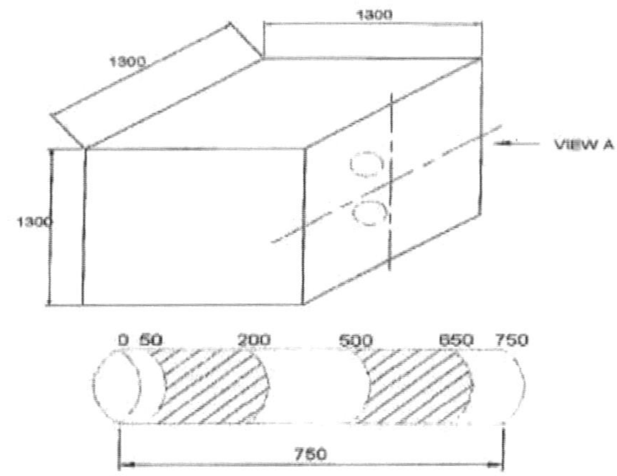

Figure 4. Sketch showing the in-situ strength with cores taken from 1.3m x 1.3m x 1.3m hot-box for Stage 1-HB2 (Mass Concrete Class A, 2011)

It was also observed that the differential for Stage-1 & Stage-2 strength tests at effective age of 7 days for the compressive strength, tensile strength and elastic modulus are 1.82 N/mm^2, 0.08 N/mm^2, and 1.08 N/mm^2 respectively.

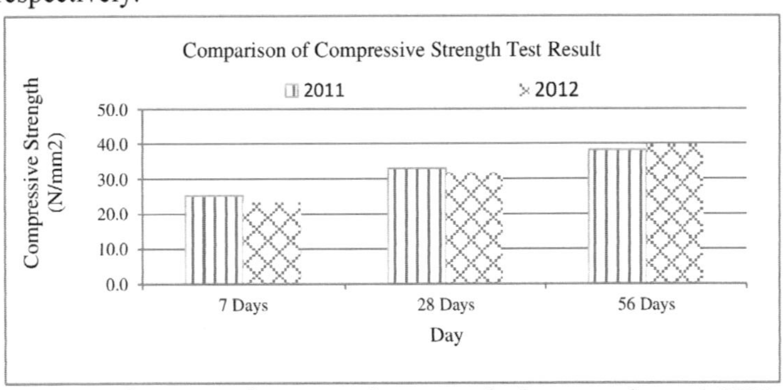

Figure 5. Comparison of compressive strength test results between Stage 1-HB1 and Stage 2-HB1 (Mass Concrete Class A, 2011)

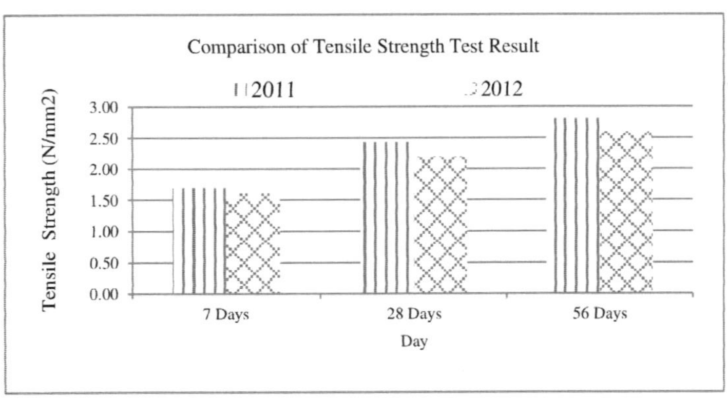

Figure 6. Comparison of tensile strength test results between Stage 1-HB1 and Stage 2-HB1 (Mass Concrete Class A, 2011)

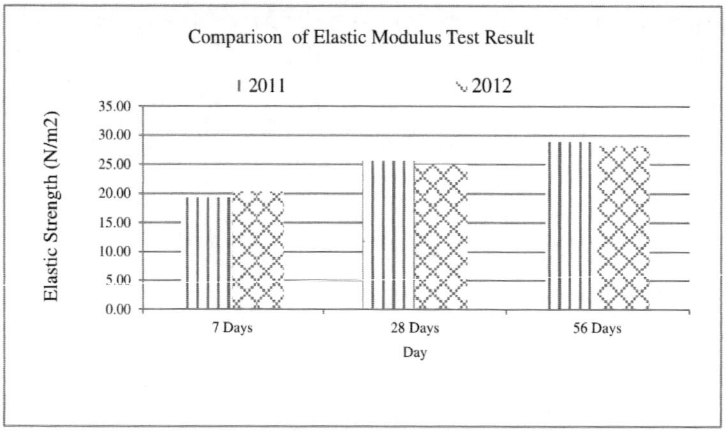

Figure 7. Comparison of elastic modulus test results between Stage 1-HB1 and Stage 2-HB1 (Mass Concrete Class A, 2011)

4. MODELLING OF ADIABATIC TEMPERATURE RISE

4.1 Modelled Heat Generation Curve using CIRIA C660 Formulas (Bamforth 2007)

CIRIA C660 formulas using British cementitious material is applied for modelling of heat generation curve basing on extensive study carried out in University of Dundee, United Kingdom. A two part two component equation for heat generation has been developed on this basis of the form:

Component 1 equation = the exponential curve, starting at time zero;

Component 2 equation = being delayed to start at time t2

The total heat generation for above two components, Q = Q1 + Q2, expressed in Equation 1, are presented below:

$$Q = Q_1 + Q_2 = \left[\left(\frac{Q_{ult}}{2}\right) \times \left(1 - e^{[(-B)\times(t^C)]}\right)\right] + \left[\left(\frac{Q_{ult}}{2}\right) \times \left(\frac{t-t_2}{t-t_2+D}\right)\right] \quad \text{(Eq.1)}$$

In order to establish values for the coefficients, analysis has been carried out on adiabatic test data from tests carried out at University of Dundee (Dhir et al. 2006) for cement with a start temperature of 20°C.

(a) Ultimate Heat Generation, Q41

Relationships between the CEM I and proportion of addition with flyash for the heat generated after 41 hours could be established basing on following expressions.

$Q41_{ult}$ (CEM I) = 351 kj/kg; Q41 (CEM I) = 330.11 kj/kg

Q41 (flyash) = Q41 (CEM I) - 2.989 * (%flyash) = 294.79 kj/kg

(b) Ultimate Heat Generation Qult

For flyash, Qult = Q41 (flyash) / (0.00002*(% flyash) ^2 - 0.0034 (% flyash) + 0.925)

(c) The Coefficient B

B = 0.011724, regardless of the cement type

(d) The Coefficient C

For flyash, C = 1.6-0.001*(% flyash)

(e) The coefficient D

For flyash, D = 6.2 + 0.2131 (% flyash)

(f) Activation time t2 for component 2

For fly ash, t2 = 3.5 + 0.0236(% flyash)

The constants and coefficients of heat generation for above item a), b), c), d), e) and f) are indicated in below Table 8.

Table 8. Constant & coefficient of heat generation of CEM II/B-V by method of CIRIA C660 (Bamforth 2007)

Constant	Unit	Value
% of Fly Ash	%	20
Q 41	kJ/kg	294.79
Q ult	kJ/kg	330.11
B	-	0.01172
C	-	1.59
D	-	8.331
t2	hrs	3.736

The modelled heat output (kj/kg) in the semi-adiabatic test and a Q_{ad} exponential curve measured at Q_{41} is 294.79 and 296.77 (See Figure 8b for exponential curve, Q_{ad}) respectively, which showing reasonably good correlation/ consistency has been obtained between prediction of temperature rise and the measured semi-adiabatic temperature rise derived from hot-box test, as graphically represented in below Figure 8a and 8b respectively.

4.2 Estimated Adiabatic Temperature Rise, Qad

The thermal model as provided in this paper uses the semi-adiabatic temperature rise data from published sources based on an extensive study carried out in University of Dundee, United Kingdom as the input. The hot-box tested as per the present trials was semi-adiabatic in nature as they are not perfectly insulated. This involves the establishment of different values of heat loss coefficient and adding incremental heat losses to the semi-adiabatic curve. A comparison between the estimated adiabatic curves is shown in below Figure 8a & 8b.

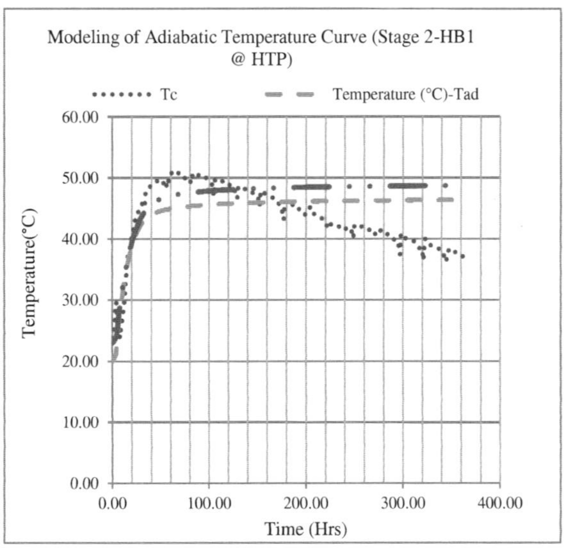

Figure 8a. Modelling of adiabatic temperature rise curve for CEMII-B-V by method of CIRIA (Bamforth 2007)

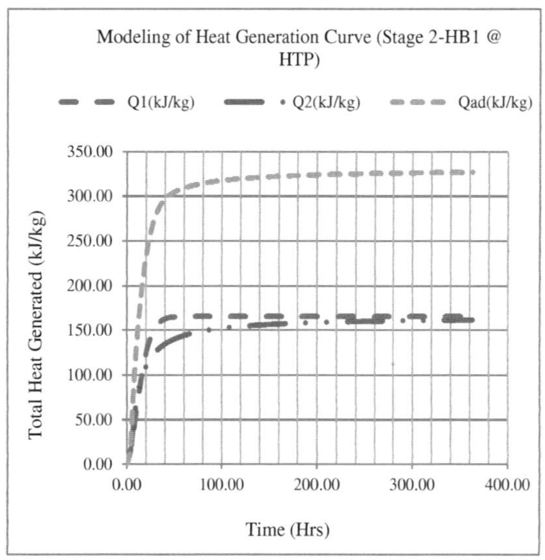

Figure 8b. Modelling of heat generation curve for CEMII-B-V by method of CIRIA 660 (Bamforth 2007)

The estimated maximum adiabatic temperature rise (MATR) measured at 72 hours is 23.72°C using data by method of CIRIA C660 could be used as input parameter for subsequent thermal modelling/ simulation i.e. heat of hydration analysis, from an initial placing temperature of 25°C. The aforesaid MATR could be further validated by using another empirical adiabatic hydration model (Tanabe et

IOP Conf. Series: Materials Science and Engineering **217** (2017) 012008 doi:10.1088/1757-899X/217/1/012008

al. 1985), as presented in below Equation 2, whereby the estimated MATR is 23.18°C, as detailed in Table 9.

$$Q(t) = Q_{\infty} \times (1 - e^{-\gamma t}) \quad \text{(Eq.2)}$$

Q_{∞} = Ultimate adiabatic temperature rise,
γ = Constant on rate of temperature rise,
t = Age, in days
$Q(t)$ = Adiabatic temperature rise at an age of t, in °C

The constant and coefficient of adiabatic heat generation using above Equation 2, is demonstrated in below Table 9.

Table 9. Constant & coefficients of heat generation of PFAC (Tanabe et al. 1985)

C	=	200	kg/m^3
t	=	3	days
Tp	=	20	°C

Table C15.3.1 Standard Value of Q

Type of cement	Placing Temperature	Q(C) = AC+B		Q_{∞}	$(1-e^{-\gamma t})$
	Tp (°C)	A	B		
OPC	10	0.11	9.00	31.00	0.60
	20	0.10	10.00	30.00	0.89
	30	0.07	16.00	30.00	0.95
MHPC	10	0.10	4.00	24.00	
	20	0.08	11.00	27.00	0.78
	30	0.11	5.00	27.00	0.86
PFAC	20	0.13	-0.40	25.60	0.91

Table C15.3.2 Standard Value of γ

Type of cement	Placing Temperature	γ(C)= aC+b		γ
	T_p (°C)	a (x 10^-3)	b	
OPC	10	0.0016	-0.017	0.303
	20	0.0016	0.409	0.729
	30	0.0037	0.273	1.013
MHPC	10	-	-	-
	20	2.40	0.032	0.512
	30	2.50	0.160	0.660
PFAC	20	0.20	0.746	0.786

Therefore, for PFAC:

$Q_{(t)}$	=	Q_{∞} (1- e -γt)	
Q_{∞}	=	25.60	
$Q_{(t)}$	=	23.18	°C

Legend: (1) MHPC denotes Moderate Heat Portland Cement, (2) PFAC denotes Portland Fly Ash Cement (The blend ratio of flyash is 20%)

The standard values of Q and γ development calculations used are explained in detail in Standard Specification for Design and Construction of Concrete Structures, Part 2 Construction, Chapter 15: Mass Concrete (Page 141-142). Judging from the established values of 23.72°C and 23.18°C, it could be concluded that both prediction model has demonstrated good correlation in terms of adiabatic temperature rise of Portland flyash cement.

5. ADIABATIC TEMPERATURE RISE ANALYSIS AND INTERPRETATION OF RESULTS

5.1 Graphical Plots for Adiabatic Temperature Rise Monitoring Results (Adiabatic Temperature Rise Test Results 2011 & 2012)

From the adiabatic temperature rise monitoring test results (see Figure 9, 10 and 11), it could infer from VWSG's thermistor that the peak core temperature recorded for both phases of the trials i.e. Stage 1-HB1 @ point A2 (48.30°C) and Stage 2-HB1 (49.53°C) only showing minor differential of 1.53°C, whereby the established average peak core temperature is 48.915°C. This is well within the stipulated peak core temperature 50°C requirement.

a) Stage 1-HB1

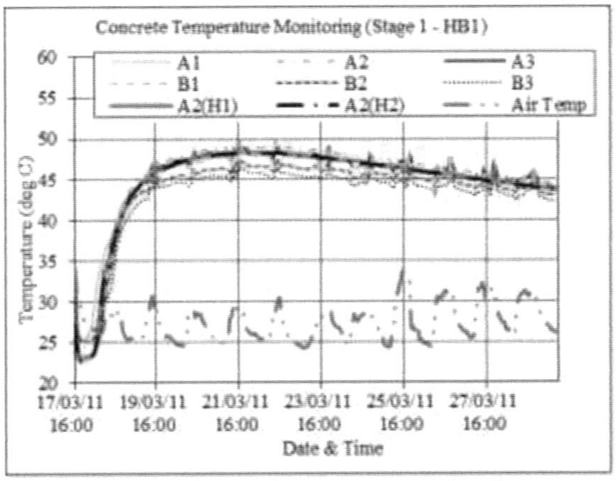

Figure 9. Experimental temperature rise data data for Stage 1-HB1

b) Stage 1- HB2

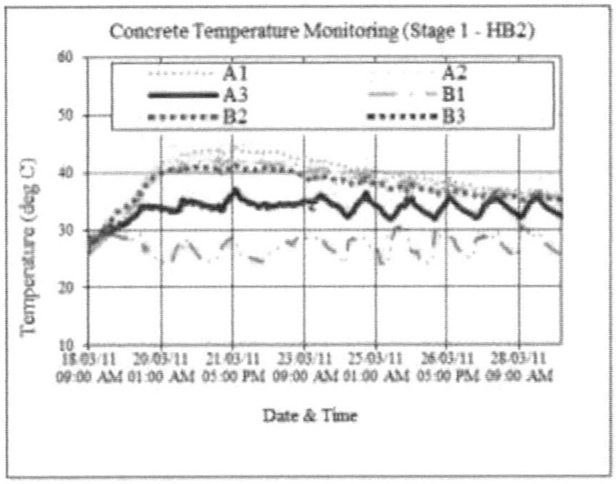

Figure 10. Experimental temperature rise for data for Stage 1-HB2

c) Stage 2-HB1

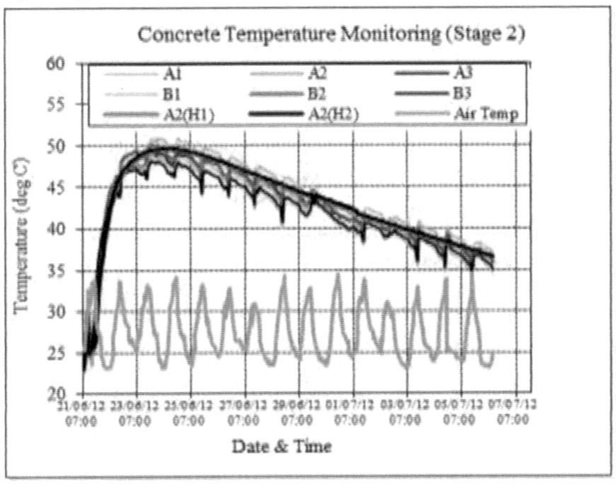

Figure 11. Experimental temperature rise data for Stage 2-HB1

The respective maximum temperature of T_c , T_n and its corresponding time of occurrence for Stage 1-HB1, Stage 1-HB2 and Stage 2-HB1 are summarised in below Table10.

Table 10: Comparison between peak temperature & time of occurrence

Stage	Peak Temperature at line A (°C) by VWSG			Peak Temperature at line B (°C) by thermocouples		
	Temp (°C)	Location	Time (Hrs)	Temp (°C)	Location	Time (Hrs)
1-HB1	48.30	A2 (Core)	1-HB1	48.30	A2 (Core)	1-HB1
1-HB2	44.43	A1 (Bottom)	1-HB2	44.43	A1 (Bottom)	1-HB2
2-HB1	49.53	A2 (Core)	2-HB1	49.53	A2 (Core)	2-HB1

Legend:
Line A located in center; Line B located at corner

T_c = Temperature at core, at Point A2 (center core)
T_s =Temperature at surface, at Point A1 (near bottom surface)
T_{cn} =Temperature at center, at Point B2 (at center)
T_n = Temperature at corner, at Point B1 (near top corner)

It has been demonstrated during mock-up test (Test Report for Mass Concrete, 2013); the core reaches its maximum temperature in about 72 hours after concrete placement.

5.2 Effects of Size of Concrete Block and Boundary Condition
The maximum temperature at center heavily depends on the insulation and the size of the hot-block. The results indicate that the center of a 1.5m x 1.5m x 1.5m mass concrete hot-box with proper insulation could produce the maximum core temperature T_c very close to that under full adiabatic condition. This implies that the maximum temperature at the center of a concrete block is little affected by the size of the concrete if the dimension is big enough.

The adiabatic temperature rise test simulation by method of CIRIA C660 (using identical mix design and input parameter as indicated in below *Table 11*) showing that hot-box size of 3m x 3m x 3m and greater having proper 50mm thick polyfoam insulation will yielded Tc at full adiabatic temperature. Furthermore, for a concrete hot-box of dimension 4m x 4m x 4m, the temperature recorded has exhibited full-adiabatic behaviour even without any insulation as the block size is big enough. However, this is non-practical for any laboratory mock-up test to be considered in such huge block dimension unless it is cast in-situ for thick raft foundation.

Table 11. Input parameters for thermal simulation

Item	Description	Value	Unit
I	Input Parameters - Concrete		
	Type of cement	CEM II:B-V	
	Binder content	200	kg/m^3
	Aggregate type	Granite	
	Density	2380	kg/m^3
	Pour thickness	1.5	m
	Insulation	0.05	m
	Insulation removal	168	hours
II	Input Parameters - Environment		
	Concrete placing temperature	25	°C
	Ambient temperature (mean)	27	°C
III	Input Parameters - Heat Exchange		
	Surface conductance	1.30	W/m^2.K
	Concrete thermal conductivity	3.028	W/m.K
	Concrete specific heat	1.037	kJ/kg.°C
	Thermal diffusivity	0.0051	m^2/h

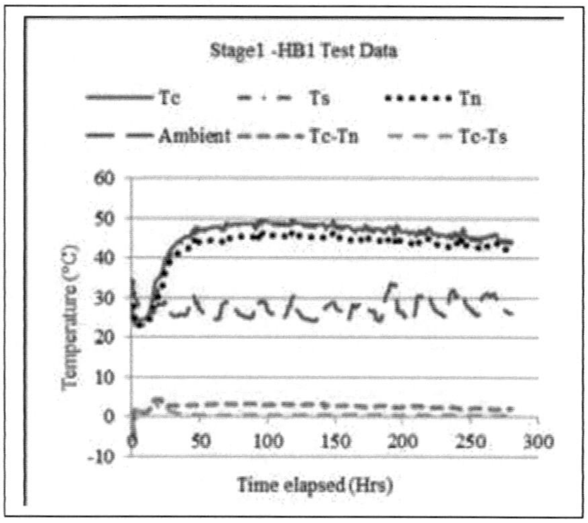

Figure 12. Temperature rise curve and differential in 1.5m x 1.5m x 1.5m hot box for Stage 1-HB1

Figure 12 presents the maximum temperature difference of ($T_c - T_n$) and ($T_c - T_s$) in a 1.5m x 1.5 x 1.5m concrete hot-box. The results reveal that the effective insulation using polyfoam of low conductance material plays a dominant role in controlling the maximum temperature difference across the block, even though the core temperature varies marginally.

Referring to below Table 12, for 1.5m x 1.5m x 1.5m hot box tests (Stage 1-HB1 and Stage 2-HB1), the adiabatic temperature rise differential of 1.23°C (49.53°C minus 48.30°C) at center is little affected by the environment due to good insulation on concrete blocks; while the temperature differentials ($T_c - T_n$) and ($T_c - T_s$) are affected to some degree between 1.79°C to 4.53°C. Whereby the recorded maximum temperature differentials between ($T_c - T_n$) and ($T_c - T_s$) are 4.66°C and 4.53°C respectively. As for the case of 1.3m x 1.3m x 1.3m hot box (Stage 1-HB2), the direct exposure of concrete surface to air is measured whereby the recorded maximum temperature differential between ($T_c - T_n$) and ($T_c - T_s$) is 9.22°C and 8.88°C respectively, which was about two-fold than the fully insulated hot block as in the case of 1.5m x 1.5m x 1.5m hot box. Therefore, direct exposure of concrete surface to air or the use of steel formwork would expect even higher temperature differentials for ($T_c - T_n$) and ($T_c - T_s$).

Table 12. Maximum T_c, $T_c - T_n$ and $T_c - T_s$

Stage	Stage 1-HB1		Stage 1-HB2		Stage 2-HB1	
	Temperature (°C)	Time (Hrs)	Temperature (°C)	Time (Hrs)	Temperature (°C)	Time (Hrs)
Peak T_c (°C)	48.30	96	42.81	72	49.53	66
Peak (T_c - T_n) (°C)	4.46	19	9.22	47	3.14	56
Peak (T_c - T_s) (°C)	4.53	17	8.88	47	1.79	76

Legend:
T_c =Temperature at core, at Point A2 (center core)
T_s = Temperature at surface, at Point A3 (near top surface)
T_n = Temperature at corner, at Point B3 (near top corner)

5.3 Temperature Rise and Temperature Distribution for Stage 1 & Stage 2 Hot-Box Tests

The two hot-box trials have exhibited slightly different rates of temperature rise at some period during the experiments, whereby their peak core temperature rise values recorded by thermistors (strain gauges) only differed by around 1.225°C as shown in below Table 13. Likewise the difference in peak core temperature recorded by thermocouples is 1.25°C as shown in Table 14.

The differences in aforesaid peak core temperatures could be attributed to lower initial concrete placing temperature for Stage 1: HB1 hot box (average 23°C) as compare to Stage 2: HB1 hot box (average 25°C). This slight difference in temperature rise rate was also attributed to the rate of heat gain during phase 2 (Stage 2- HB1) is faster, resulting from different fineness and age of cement used during phase-1 and phase-2 test resulting in different rate of hydration.

Table 13. Summary of Test Results for Peak Core Temperature @ A2 recorded by Strain Gauges

Hot-Block	Stage 1-HB1 (°C)	Stage 2-HB1(°C)	Difference (°C)
Strain Gauge No.1 (°C)	48.3	49.34	-
Strain Gauge No.2 (°C)	48.24	49.65	-
Average (°C)	48.27	49.495	1.225

Table 14. Summary of Test Results for Peak Core Temperature @ A2 & B2 recorded by Thermocouples

Hot-Block	Stage 1-HB1 (°C)	Stage 2-HB1 (°C)	Difference (°C)
Thermocouple @ A2 (°C)	49.69	50.94	1.25
Thermocouple @ B2 (°C)	49.26	50.22	-
Average (°C)	49.475	50.58	1.105

The various temperature rise curves, temperature differentials, and temperature distribution profiles are shown in Figures 12-16. Due to the greater heat loss rate at corner and side of the block, the temperatures at corner and side reach the peak before the maximum core temperature due to heat exchange to the environment. The center temperature begins to decrease when the heat generated in block is not enough to compensate the heat loss to the environment (as for the case of adiabatic condition). After the decrease of the core temperature, the temperature difference ($T_c - T_n$) and ($T_c - T_s$) gradually climb to the peak. This is normally regarded as the critical period for the surface

IOP Conf. Series: Materials Science and Engineering **217** (2017) 012008 doi:10.1088/1757-899X/217/1/012008

concrete. Situations would become even worse if the formwork is removed during this stage. Also, the poor insulation could significantly reduce the time to reach the maximum temperature differences of ($T_c - T_n$) and ($T_c - T_s$). This infers that poorly insulated mass concrete could develop thermal cracks shortly after the placement.

The temperature at center is little affected by the environment variation due to the good insulation and big enough size of the hot box i.e. 1.5m x 1.5m x 1.5m, while the temperature difference ($T_c - T_n$) and ($T_c - T_s$) is affected in some degree. Essentially, during the course of insulation selection, it is necessary to consider the environment heat loss variation to minimize the effect. It is thus inferred that temperature gradient of actual concrete structures could be improved by additional insulation at edges and corners. The temperature gradient along the regions close to the side is much steeper than that in the central region. This suggests that the locations of the thermocouples in the concrete mock-up block should be carefully defined so that the comparable results can be obtained.

Figures 12-14 showing how the temperature gradients formed after the concrete were cast into hot box. Temperature differential formed from the very beginning of the temperature rise in the concrete mass, whereby the temperature is also affected by the ambient temperature fluctuation, which is formed by the temperature change from daytime to night-time. The peak value of ($T_c - T_n$) and ($T_c - T_s$) occurred after T_c.

This implies that the maximum temperature difference was encountered in the cooling down period of the concrete mass. The temperature differences of between centre core versus surface ($T_c - T_s$) and between core versus corner ($T_c - T_n$) were also demonstrated in below Figure 15 and Figure 16, which showing how the temperature at the corner of the block had the lowest value while the temperature at center gave the highest. ($T_c - T_n$) is always greater than ($T_c - T_s$). This may be explained as that the heat loss in the corner of the block is multi-dimensional. The peak value of ($T_c - T_n$) and ($T_c - T_s$) occurred after T_c.

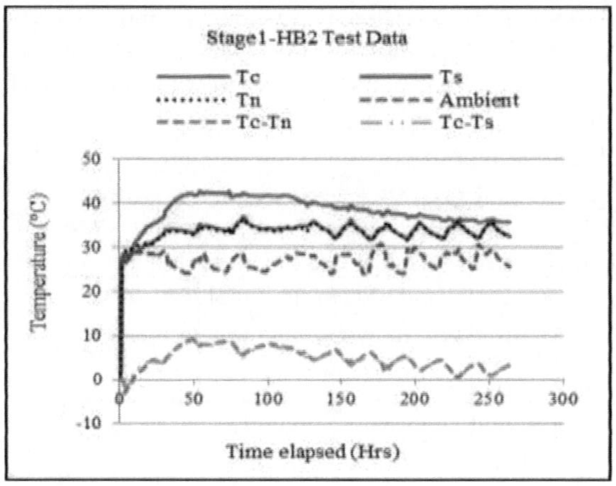

Figure 13. Temperature rise curve and temperature differential in 1.3 x1.3x1.3m hot-box for Stage 1-HB2

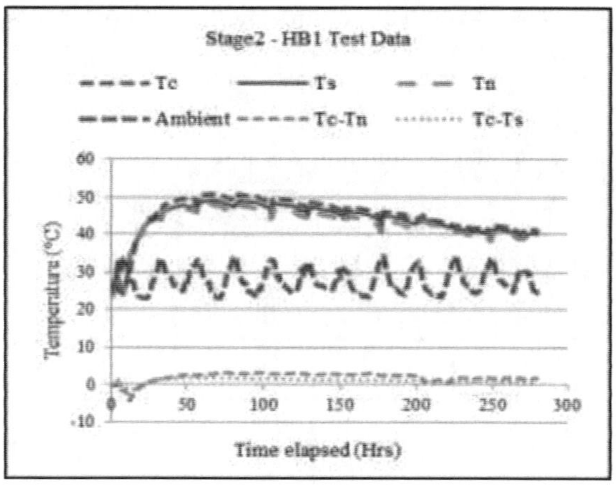

Figure14. Temperature rise curve and temperature differential in 1.5x1.5 x 1.5m hot-box for Stage 2-HB1

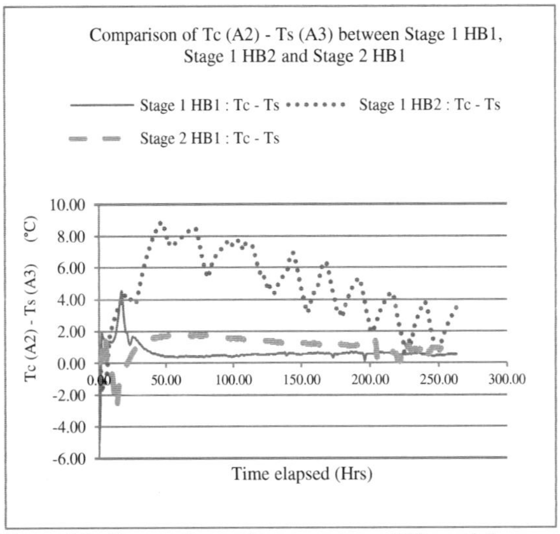

Figure 15. Comparison of temperature differential between Tc-Ts for Stage 1-HB1, Stage 1-HB2, and Stage 2-HB1

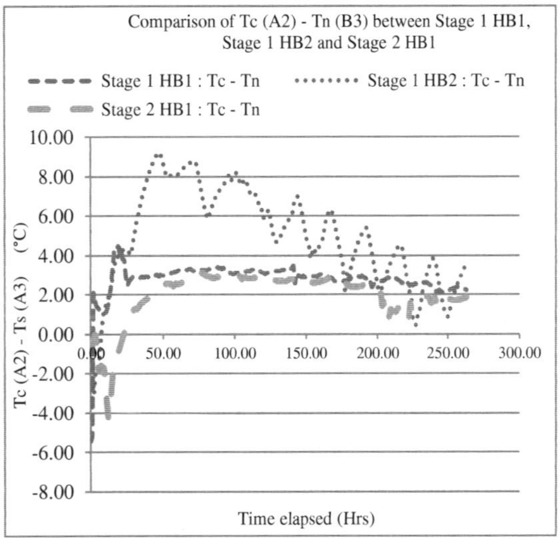

Figure 16. Comparison of temperature differential between Tc-Tn for Stage 1-HB1 and Stage 2-HB1

5.4 Temperature differential

The maximum differential temperature (Temperature recorded at point A1 minus temperature recorded at point A3) for stage 1-HB2 at the open top surface was found to be ~ 10.13°C, measured on 21st March 2011 at 09:00 am, about 72 hrs after the concrete placement. The top thermocouple was located 100mm below the surface. However an adjustment is required to determine the true differential between the centre and the surface. This adjustment is made by assuming the temperature profile follows a parabolic curve to the cooling surface as shown in Figure 17. With a peak temperature of 44.43°C as recorded, the difference between the surface and the point of location of the thermocouple at 100mm depth is about 2.3°C. This must therefore be added to the measured differential. Hence the true differential is 10.13 + 2.3 = 12.43°C ≈ 12.5°C. This is significantly less than the Contract Documents specified 20°C and the proposed differential of 27.7°C for granite aggregate concrete as in accordance with BS8110-2: 1985, Table 3.2- Estimated limiting temperature changes to avoid cracking.

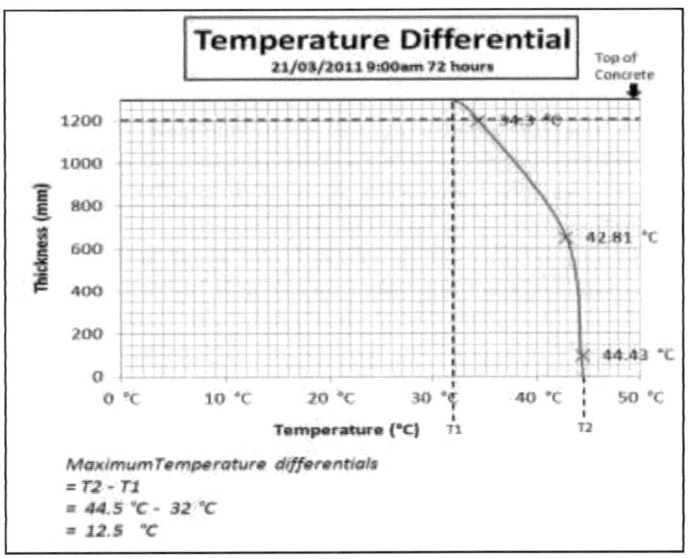

Figure 17. Adjustment for temperature differential at top surface of 1.3m x 1.3m x 1.3m hot-box, Stage 1- HB2

6. DISCUSSIONS

1. The two hot-box trials have exhibited slightly different rates of temperature rise at some period during the experiments, whereby their peak core temperature rise values differed by merely 1.225°C only. The difference in peak core temperature could be attributed to lower initial concrete placing temperature for Stage 1: HB1 hot-box (average 23°C) as compare to Stage 2: HB1 hot-box (average 25°C). This slight difference in temperature rise rates may be attributed to the rate of heat gain during Phase 2 (Stage 2- HB1) is faster, which could be further attributed to different fineness and age of cement used during Phase-1 and Phase-2 test resulting in different rate of hydration.
2. From the two experiments performed, the difference in peak core temperature rise for both trials was merely 1.225°C and 1.25°C, as recorded by strain gauges (built-in thermistor) and thermocouples respectively. This implies that the concrete temperature during the entire duration of the tests varying only by a small amount.
3. As expected with concrete containing flyash, cores extracted from Stage 1-HB2 demonstrated that the in-situ heat cycles had a beneficial effect on fly-ash concrete, with the average estimated in-situ strength being about 6.5 per cent higher than the test cube strength at 56 days. This compressive strength results confirm that the proposed flyash concrete is well above the specified requirements in terms of strength and hence compliance with the technical specification requirements.
4. The computer thermal modelling/ simulation revealed that for a concrete hot-box of dimension 2.5m x 2.5m x 2.5m coupled with good insulation (50mm thick polyfoam), the maximum core temperature recorded by thermal probe is almost near full-adiabatic state whereby no loss of heat to the environment. Likewise for a concrete hot-box of dimension 4m x 4m x 4m, the temperature recorded has exhibited full-adiabatic behaviour without any insulation as the block size is big enough. However, this is non-practical both technically and commercially for any laboratory mock-up test to be considered in such huge block dimension unless it is cast in-situ for thick raft foundation.

7. CONCLUSIONS

1. From the adiabatic temperature rise monitoring test, it could infer from strain gauges that the peak temperature values recorded for both phases of the trials i.e. Stage 1-HB1 (48.30°C) and Stage 2-HB2 (49.53°C) is merely difference by 1.23°C, whereby the established average peak core temperature and maximum core-surface differential was 49.53°C and 12.5°C (worst case scenario for Stage 1-HB2) respectively are well within stipulated 50°C and 20°C requirements as specified in various project technical specifications, from a maximum initial placing temperature of 25°C. As demonstrated during mock-up test that the core reaches its maximum temperature in about 72 hours after concrete placement.
2. Significant reduction of temperature rise and maximum temperature rise rate of PFAC concrete under adiabatic condition was successfully demonstrated in this experimental investigation. This indicates that the application of CEM II/B-V cement is effective and beneficial in mass concrete pours in minimizing the mass overheating inducing thermal cracking. Thus the use of flyash as partial replacement for Portland cement resulted in lower adiabatic temperature rise.
3. The outcome of the hot-box test suggests that good insulation is critical in minimising the temperature difference in mass concrete. In mass concrete pouring, additional insulation at edges and corners can help to improve temperature gradient in the critical regions. Moreover, hasty removal of the formwork during the concrete cooling down period may greatly increase the temperature gradient, and this could be detrimental to the surface of concrete mass.

4. All concrete cubes compressive strength results tested at 7 days had exceeded the contract documents stipulated 90 days characteristic strength of 20MPa confirming the compliance of proposed 20MPa concrete mix design.
5. Both empirical adiabatic hydration models (CIRIA C660, 2007 and Tanabe et al. 1985) have demonstrated good agreement on the maximum adiabatic temperature rise values of 23.72°C and 23.18°C respectively. Thus, the heat of hydration model by method of CIRIA C660 can be successfully used in modelling the characteristics of semi-adiabatic temperature rise data of 20MPa mass concrete adopted in this study.
6. As far as the determined adiabatic temperature rise monitoring results is concern, the established results could be inferred for in-situ thermal properties of 20MPa mass concrete application, as the results could be repeatable on account of similar types of constituent materials and concrete mix design adopted for permanent works at project site.

References

[1] ACI Committee 207.2R-07, 2007. Report on Thermal and Volume Change Effects on Cracking of Mass Concrete. Reference.
[2] ACI Committee 305.1-06, 2007. Specification for Hot Weather Concreting.
[3] ACI Committee 207.1R-96, 1996. Specification for Mass Concrete.
[4] Adiabatic Temperature Rise Test Result (from 17/3/2011 to 29/3/2011), Stage 1, for 265MW TNB Hulu Terengganu Hydroelectric Project, Malaysia.
[5] Adiabatic Temperature Rise Test Result (from 21/6/2012 to 6/7/2012), Stage 2 for 265MW TNB Hulu Terengganu Hydroelectric Project, Malaysia.
[6] Bamforth P. B. 2007. CIRIA C660 Early-Age Thermal Crack Control in Concrete, CIRIA, London.
[7] BS 8110-2: 1985, Structural Use of Concrete – Part 2: Code of Practice for Special Circumstances. BSI, London.
[8] BS EN 13791: 2007, Assessment of In-situ Compressive Strength in Structures and Precast Concrete Components. BSI, London.
[9] BS EN 196-9: 2010, Methods of Testing Cement. Heat of Hydration. Semi-Adiabatic Method, BSI, London.
[10] CIMA Cement Test Certificates on NS Ecocrete-LH, CEM II/ B-V 42.5N-LH (January 2011 and June 2012), CIMA, Malaysia.
[11] DD ENV 206:1992, Concrete: Performance, Production, Placing and Compliance Criteria, BSI, London.
[12] Guo D S et al. 2001. Experimental Modeling of Temperature Rise of Mass Concrete by FDM Method, 26th Conference on Our World in Concrete & Structures, 27-28 August 2001. Singapore.
[13] Japan Society of Civil Engineers (JSCE), Standard Specification For Design And Construction Of Concrete Structures (First Edition 1986), Part 2 (Construction), Chapter 15: mass Concrete, 141-142, Tokyo, Japan.
[14] Lee MH, Khil BS, and Yun HD 2014. Influence of Cement Type on Heat of Hydration and Temperature Rise of Mass Concrete (Accepted 9 June 2014).
[15] Mass concrete Class A/Grade III Concrete Mix Trials and Hot Block Monitoring, Overview of Results, Report 1202/11/7322, (2011) for 265MW TNB Hulu Terengganu Hydroelectric Project, Malaysia.
[16] MS 523-2: 2011, Concrete – Part 2: Method of Specifying and Guidance For The Specifier (Second Revision), Department of Standards Malaysia, Malaysia.
[17] Neville, A.M., Brooks, J.J., Advanced Concrete Technology (Second Edition 2010), Chapter 9: Temperature Problems in Concreting, 161-168, Publisher, London.
[18] Tanabe T, Kawasumi M and Yamashita M. Thermal Stress Analysis of Massive Concrete: Finite Element Analysis of the Reinforced Concrete Structures, (ASCE), 1985, 407-421.

[19] Technical Specification (Jan 2000) [Gutteridge Haskins & Davey (GHD)] for Construction of Sg. Kinta Dam and Associated Works, Section 10, Clause 10.2.9.

[20] Technical Specification (2007) [Sepakat Setia Perunding (SSP)] for Construction and Completion of Sg. Batu Hampar Dam and Associated Works, Section 10: Clause 10.2.9.

[21] Technical Specification (2009) [SMEC International] for Lot CW2 – Main Civil and Associated Works, Volume 3: Part 1 of 2, Section 9: RCC, Section 10: Clause 10.10.1 & Section 11: Clause 11.2 – Concrete Production and Construction.

[22] Technical Specification (2010) [SMEC International] for Lot CW2 – Main Civil and Associated Works, Volume 3: Part 1 of 2, Section 10: Clause 10.10.1 & Section 11: Clause 11.2 – Concrete Production and Construction.

[23] Technical Specification (2013) [SNC–Lavalin Power] for Chenderoh Unit 5 Hydroelectric Project, Section 7.3.3.: Clause 4 & 5.

[24] Test Report for Determination of Heat of Hydration on Portland Cement at & Days using BS EN 196-8: 2003, Ref.: CT-16925/JAS dated 25/07/2011.

[25] Test Report for Mass Concrete Temperature Monitoring for Mock-up of 1.5m x 1.5m x 1.5m Hot Box (2013), Report No.: TEM 003/12/R0855 for 265MW TNB Hulu Terengganu Hydroelectric Project, Malaysia

Determination of Coefficient of Thermal Expansion (CTE) of 20MPa Mass Concrete Using Granite Aggregate

GO Chee Siang

30, Jalan 1/27D, Section 6, Wangsa Maju 53300, Kuala Lumpur, West Malaysia.

csgo@loh-loh.com.my

Abstract. Experimental test was carried out to determine the coefficient of thermal expansion (CTE) value of 20MPa mass concrete using granite aggregate. The CTE value was established using procedure proposed by Kada et al. 2002 in determining the magnitude of early-ages CTE through laboratory test which is a rather accurate way by eliminating any possible superimposed effect of others early-age thermal deformation shrinkages such as autogenous, carbonation, plastic and drying shrinkage. This was done by submitting granite concrete block samples instrumented with ST4 vibrating wire extensometers to thermal shocks. The response of the concrete samples to this shock results in a nearly instantaneous deformation, which are measured by the sensor. These deformations, as well as the temperature signal, are used to calculate the CTE. By repeating heat cycles, the variation in the early-ages of concrete CTE over time was monitored and assessed for a period of upto 7 days.

The developed CTE value facilitating the verification and validation of actual maximum permissible critical temperature differential limit (rather than arbitrarily follow published value) of cracking potential. For thick sections, internal restraint is dominant and this is governed by differentials mainly. Of the required physical properties for thermal modelling, CTE is of paramount importance that with given appropriate internal restraint factor the condition of cracking due to internal restraint is governs by equation, $\Delta T_{max} = 3.663 \varepsilon_{ctu} / \alpha_c$. Thus, it can be appreciated that an increase in CTE will lower the maximum allowable differential for cracking avoidance in mass concrete while an increase of tensile strain capacity will increase the maximum allowable temperature differential.

Keywords: CTE, granite aggregate, early-age thermal cracking, thermal shock, temperature differential, tensile strain capacity, internal restraint

1. INTRODUCTION

Experimental test was carried out to validate the actual value of coefficient of thermal expansion (CTE) of 20MPa mass concrete using granite aggregate quarried from borrow hill located within project site for the construction of 265MW TNB Hulu Terengganu Hydroelectric Dam.

"The coefficient of thermal expansion (CTE) of a material is defined as the unit change in length per degree of temperature change. The CTE measures expansion or contraction of a material with temperature. For a given temperature change, the free thermal expansion of the concrete is determined by the CTE, α_c. Thus, concrete, the most important composite construction material, is not an exception. At a simplified level, concrete has paste and aggregate phases. These two main constituents of concrete is dependent on material factors such as type of cement paste, aggregate type, moisture conditions, age, and environmental factors such as temperature fluctuations as well as relative humidity; whereby aggregates constitute approximately 75% of the volume of concrete which therefore is a major factor influencing the CTE of concrete. Therefore, thermal properties of the coarse aggregate can have a significant effect on mass concrete. Concrete containing low-thermal-expansion aggregates such as granite and limestones generally permit higher maximum allowable temperature differences than concretes made using high-thermal-expansion aggregates such as gravel, as indicated in below Table 1. This means that selecting an aggregate with a low thermal expansion will reduce the potential for thermal cracking (ACI 207.2R-07, 2007)".

"The CTE in concrete is important in numerous engineering applications including design of Dams & Hydropower, Highway Infrastructures such as bridges and pavements. It plays a key role in design, performance, and service life of concrete structures because it affects critical stresses and joint and crack openings. The CTE in aggregates influence the performance of concrete especially mass concrete structure (MCS) mainly because of its effect on dimensional change attributable to the change in temperature condition (ACI 207.2R-07, 2007)".

Researchers (Naik et al. 2011; Jahangirmejad et al. 2009; Mallela et al. 2005; Won M 2005; Ndon & Bergeson 1995) have investigated that "the CTE of concrete with various aggregates and reported a wide variation in the values of CTE among different aggregates depending on type, origin, and geographic location. To minimize the effect of thermal expansion or contraction of concrete on the performance of concrete infrastructures in particularly MCS, the measured value of the CTE should be incorporated in design equations. A higher CTE of concrete may adversely affect the durability of concrete because of thermal incompatibility of its components (Naik et al. 2011)."

A concrete with a low CTE can significantly reduce the risk of early-age thermal cracking as well as the percentage of reinforcement required for the control of thermally-induced cracks. Therefore the selection of aggregate of low αc (coefficient of thermal expansion of concrete) can have a significant impact in preventing, minimizing or controlling early-age thermal cracking. Generally, the definitions of early-age and long-term thermal cracking are defined as i) Early-age: Typically up to seven days and ii) Long-term: Beyond the period of the early-age temperature cycle, typically beyond 28 days (Bamforth 2007).

Table 1. Coefficients of thermal expansion (CTE) for different kinds of aggregates (Bamforth 2007)

Coarse aggregate/ rock group	Thermal expansion coefficient (microstrain/ °C)		
	Rock	Saturated concrete	Design value
Chert or flint	7.4–13.0	11.4–12.2	12
Quartzite	7.0–13.2	11.7–14.6	14
Sandstone	4.3–12.1	9.2–13.3	12.5
Marble	2.2–16.0	4.4–7.4	7
Siliceous limestone	3.6–9.7	8.1–11.0	10.5
Granite	1.8–11.9	8.1–10.3	10
Dolerite	4.5–8.5	Average 9.2	9.5
Basalt	4.0–9.7	7.9–10.4	10
Limestone	1.8–11.7	4.3–10.3	9
Glacial gravel	-	9.0–13.7	13
Lytag (coarse and fine)	-	5.6	7
Lytag coarse and natural aggregate fines	-	8.5–9.5	9

BS EN1992-1-1:2004 also recommends that unless more reliable information is available, the coefficient of thermal expansion should be assumed to be 10με/°C. While this is a representative value for a wide range of aggregates there are some commonly used materials in the UK, in particular flint gravels that lead to higher values in the order of 12 με/°C, as shown in Table 1. The higher value should be assumed if no data are available. In the absence of project data, the value of CTE can be estimated from knowledge of the CTE of the aggregates using aforesaid Table 1 where proposed design values are given. While there is insufficient data to provide values with a defined level of statistical significance, the proposed values are at the high end of the observed range and represent safe values could be used in design. Precisely, CTE itself is a physical property of the concrete depending on a certain concrete mix designer's fixed proportions, exclusively on the selected/ predetermined age of concrete which corresponds to the peak temperature results derived from the established semi-adiabatic temperature rise test. Therefore, it is customary whenever possible to rely on the physical properties of the concrete such as CTE based on actual tests carried out on the proposed concrete mix rather than pick published values that however still would provide sound reference as to which limits (maximum/ minimum) a certain physical property under investigation should lie within.

In other words, once the mix design is assigned and the age of the CTE testing is determined; then the value obtained by any representative testing becomes a 'constant' property of the concrete regardless whether it is measured on site or in the laboratory as there is not any known factor to convert the value of CTE as determined in the laboratory to a specific notional size on-site once the conditions for the testing have been fixed. It has been demonstrated and experience by author that in the determination of an accurate value of CTE for the thermal modelling and design purposes by means of large scale mock-up has some limitations in the sense that it require to determine very accurate instruments in a very "gross" environment where any little modification of the test set up and boundary conditions for the strain gauges can amplify errors beyond control. In addition, there is possibility that strain induced by early-age thermal deformations such as autogenous, plastic, carbonation, and drying shrinkage would also sum-up the value of total strain measured by the gauges on-site and therefore a large scale pour used for determining CTE would not be recommended. This is because a certain degree of aforesaid measured thermal induced strains would be captured by the strain gauge as well which should be avoided in order to have an accurate and genuine CTE value.

Thus, there is less volume and time consuming processes thought from which CTE of a certain mix design can be determined in controlled environment by choosing an appropriate value of CTE as determined by testing at the ages where peak/ cooling cycles, based on simple method established by (Kada et al. 2002) and correlating with the semi-adiabatic temperature rise test results derived from hot box test. This method is relatively inexpensive but rather accurate way by eliminating any possible superimposed effect of others early-age thermal deformation shrinkages as mentioned in above paragraph. Furthermore, choosing an appropriate range of ages for the determination of CTE on the designated concrete mix will allow (in conjunction with thermal data obtained from large scale volume mock-up tests) the selection of those values of CTE corresponding to the time where concrete is rigid enough to crack and accurate input value for the modelling determined experimentally under controlled conditions on the proposed mix will be eventually available.

Of the required physical properties for thermal modelling, CTE is of paramount importance as it can be proven (Bamforth 2007 and TxDOT 2011) that with given appropriate internal restraint factors (the main type of restraint acting on mass concrete is actually and relaxation factors for the creep of concrete at early ages) the condition of cracking due to internal restraint is governs by equation, $\Delta T_{max} = 3.663\ \varepsilon_{ctu} / \alpha_c$. From the aforesaid equation it can be appreciated that an increase in the CTE will lower the maximum allowable differential for cracking avoidance in unreinforced concrete while an increase of the tensile strain capacity will increase the maximum allowable temperature differential.
BS 8110: Part 2: 1985 Clause 3.8.4 has stipulated that by limiting temperature differential to 20°C has been successful to avoid early-age thermal cracking, but this apply to gravel type of aggregate at the corresponding restraint factor, R = 0.36. However, when referring to Table 2 and Table 3 of BS 8110: Part 2: 1985, the granite type of aggregate is typically used in local Malaysia construction industry. This granite aggregate has increased limiting temperature differential of 27.7°C at the same corresponding restraint value, R= 0.36. This can be conservative value as studies have found that massive pour cast into blinding concrete has typical restraint factor of R= 0.1 to 0.2 (See Table 3 for values of external restraint recorded in various structures).

Table 2. Values of limiting temperature changes to avid cracking from BS 8110-2: 1985

Table 3.2 — Estimated limiting temperature changes to avoid cracking

Aggregate type	Thermal expansion coefficient	Tensile strain capacity (10^{-6})	Limiting temperature drop for varying restraint factor (R)				Limiting temperature differential when R = 0.36)
			1.00	0.75	0.50	0.25	
	(10^{-6}/ °C)		°C	°C	°C	°C	°C
Gravel	12.0	70	7.3	9.7	14.6	29.2	20.0
Granite	10.0	80	10.0	13.3	20.0	40.0	27.7
Limestone	8.0	90	14.1	18.8	28.2	56.3	39.0
Sintered p.f.a.	7.0	110	19.6	26.2	39.2	78.4	54.6

Table 3. Values of external restraint recorded in various structures from BS 8110-2: 1985

Table 3.3 - Values of external restraint recorded in various structures	
Pour configuration	Restraint factor (R)
Thin wall cast on to massive concrete base	0.6 to 0.8 at base 0.1 to 0.2 at top
Massive pour cast on to existing mass concrete	0.3 to 0.4 at base 0.1 to 0.2 at top
Suspended slabs	0.2 to 0.4
Infill bays, i.e. rigid restraint	0.8 to 1.0

Owing to the acceptable field data of this kind is not available locally i.e. peak core temperature, limiting differentials and temperature gradient were thus be established for conditions at the job site. This was validated basing on trialled hot-box tests at prescribed initial placing temperature as well as the anticipated concrete section thickness, rather than merely follow seemingly arbitrarily values of 50°C maximum peak core temperature and 20°C maximum allowable core-surface temperature requirements as stipulated in various local project contract documents as experienced by author.

Whereby, the maximum temperature differential is a function of concrete mechanical properties such as CTE, tensile strength, elastic modulus, as well as the size and restraints of the element. Therefore, the established maximum allowable core-surface temperature differential can significantly reduce the amount of time that protective measures, such as surface insulation, must be kept in place. Thermal cracking will occur when contraction due to cooling at the surface causes tensile stresses that exceed the tensile strength of the concrete.

It has to be pointed out, these temperature criteria is merely a general guideline based on experience with unreinforced mass concrete placed in Europe more than 50 years ago. In many situations, limiting temperature difference to 20°C is overly restrictive and thermal cracking may not occur even at higher temperature differences as in the case of for the completed Sg. Kinta roller compacted concrete dam under above-mentioned similar temperature restrictions.

Local experience has also shown that for the case of thin wall cast on to massive concrete base, temperature differential of around 12°C to 14°C may lead to early thermal cracking in lower 1m of wall or higher due high restraint factor, R (0.6 to 0.8). For pilecaps and raft foundation slabs, restraint factor, R is generally low (0.1 to 0.4) and with granite aggregate, temperature differential up to around 35°C to 40°C have been recorded without early thermal cracks.

Thermal modelling requires amongst many input parameters and regardless whether it would be made by means of a British oriented approach such as CIRIA C660 (Bamforth 2007 or ACI 207.2R-07, 2007) or through the application of finite element software such as LUSAS, FEMMASSE, MIDAS, ASTEA-MAX and et cetera whereby the foreseeable temperature rise within the concrete as a function mainly of: (i) member shape, (ii) size and thickness, (iii) environmental conditions, (iv) concrete temperature, (v) properties of cementitious materials type and maturity, as well as dosage,

(vi) formwork type and insulation, (vii) degree of restraint, and et cetera. The scope of the thermal modelling will be mainly to predict the temperature and stress development from overlapping/ adjacent mass pours along with the prediction of the spacing and width of cracking induced by the thermal (plus hydraulic and autogenous) shrinkage strains (or actions) to which the tensile strain capacity of the concrete will oppose as a reaction (whereby tensile strain capacity could be approximated by the ratio of concrete flexural strength and its elastic modulus).
The CTE value established as per procedure highlighted in below Table 4 and presented in this report could be adopted as important input parameter for thermal modelling analysis whereby the pour thickness, placing interval, casting sequence (time-based temperature profiles) and et cetera of mass concrete structure construction could be assessed and established.

Table 4. Procedure for the determination of CTE

Step:	Procedure
1	Selection of appropriate CTE testing method
2	Procedure to experimentally determine the CTE
3	Actual project concrete design mix & raw materials used for test specimens casting
4	Laboratory testing: Mould preparation, instrumentation & samples casting in controlled condition
5	Data acquisition & observation thermal heat cycles (heating/cooling)
6	Derivation of equation, $\Delta T_{max} = 3.663\ \varepsilon ctu / \alpha c$
7	Analysis & evaluation of test results corresponds to strength & temperature results derived from semi-adiabatic temperature rise test
8	Interpretation of results
9	Conclusions/ recommendations

2. CONCRETE MIX DESIGN

The physical properties and chemical compositions of the Portland-fly-ash (PFA) cement and concrete mix design used in the study are listed in below Tables 5-7 respectively.

Table 5. Typical physical properties of PFA Cement (CIMA, 2011 & 2012)

CEM II/B-V	
Standard	EN197-1 – CEM II/B - V 42.5N - LH
Typical Properties	
Blaine fineness (m^2/kg)	420
Specific gravity	2.9
Initial Setting time (min.)	157
2d strength (MPa)	22
7d strength (MPa)	37
28d strength (MPa)	50

Table 6. Typical chemical analysis of PFA Cement (CIMA, 2011 & 2012)

Chemical Properties	Chemical Composition (%)
SiO_2	26.57
Al_2O_3	8.84

Fe_2O_3	5.40
CaO	51.89
MgO	0.63
SO_3	2.31
Na2O	0.18
K2O	0.72
Na2O + 0.658 x K2O	0.65

Table 7. Class A Mass Concrete C20 concrete mix design (Mass Concrete Class A, 2011)

Description	Value
Cement (kg/m3)	200
Water (kg/m3)	140
Sand (kg/m3)	756
Coarse Aggregate (20mm) (kg/m3)	1287
Realset 233 (l/m3)	1.6
Realflow 610 (l/m3)	0.8
Slump (mm)	20-50
Characteristic strength (MPa)	20
S/A Ratio	0.37
W/C Ratio	0.70

3. ESTIMATING THE RISK OF CRACKING AND CRACK-INDUCING STRAIN

3.1. Internal restraint

Internal restraint is a result of differential temperature changes within an element. It may lead to both surface cracking and internal cracking that may not be observed from the surface. The development of cracking due to internal restraint is shown in below Figure 1. As heating occurs, the surface is subject to tensile stresses as the centre of the pour gets hotter and expands to a greater extent. At elevated temperature and early-age, creep is high and the stresses generated are relieved, at least in part. As cooling occurs there is a stress reversal and the surface cracks generally reduce in width. At the same time tension is generated at the centre of the pour as it cools more than the surface and internal cracking may occur (Bamforth 2007).

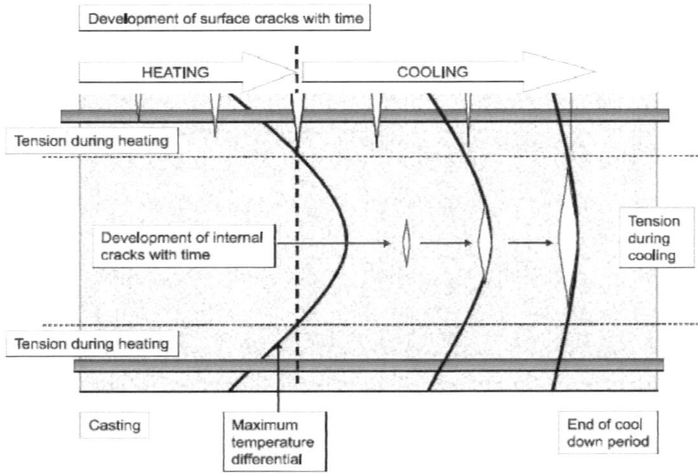

Figure 1. Schematic representation of the development of crack in a massive element due to temperature differentials assuming no external restraint (such cracking may occur in both the vertical and horizontal orientation)(Bamforth 2007).

Where property data are available for a particular mix, the limiting temperature differential, ΔTmax, i.e. the differential at which cracking may occur, can be calculated using equation 7 (see page 6 of 11) in which with the restraint factor R = 0.42 and assuming K1 = 0.65 (for creep) and K2 = 0.8 (for strength reduction due to sustained loading). In a previous investigation the value of R was determined to be 0.36 (Bamforth 1982). This value was derived to match the recommended limiting temperature values with those observed in practice and based on property data available at the time. The adjustment in R has been made to maintain these limits with the application of property data from EN1992-1-1:2004 (Bamforth2007).

3.2. Internal restraint

As demonstrated in literature (Bamforth 2007), the CTE can be determined by putting the concrete through relatively sudden temperature shocks and measuring, by mean of embedded strain gauges, the induced strains. Of course, especially at early ages, thermal deformations will not be the only ones measured. Thus, particular care must be put in excluding all the other strain components. The total deformation that can occur at early ages can be expressed as follow:

$$\varepsilon_{tot} = \varepsilon_{plastic} + \varepsilon_{drying} + \varepsilon_{carnonation} + \varepsilon_{thermal} + \varepsilon_{autogemeous}$$

Where:

$\varepsilon_{plastic}$: is the shrinkage due to the evaporation of water from the fresh concrete or due to the adsorption by the formwork when it is still at the plastic state. This deformation can be neglected since each sample has been firmly wrapped by mean of polystrene sheets.

εdrying: is the shrinkage caused by the evaporation of water from capillary pores in the hardened concrete. This deformation is also considered to be negligible in this study since evaporation was prevented.

$\varepsilon^{carbonation}$: this component of the strain is caused by the reaction of the hydrated cement paste with carbon dioxide present in the atmosphere when humidity is present. It can also be neglected considering the PE wrapping and the fact that it is a typical long term phenomenon.

εthermal: this is the component of the strain object of this study. It must be pointed out that this deformation also includes the thermal strain of the strain gauge, which will have to be subtracted to the measured strain.

$\varepsilon_{autogenous}$: is the strain due to the consumption of water during the hydration process. As already mentioned, autogenous shrinkage is not expected to occur for high water-binder ratio concrete. In this research only Grade 20 mass concrete is analysed hence this deformation is neglected (as autogenous shrinkage is an issue solely deals with high strength, very low water/ binder concrete and can be neglected at any stage when mass concrete for dams is considered). Therefore, under experimental conditions of this research, the total deformation measured by the strain gauge corresponding to the thermal shock, during which a temperature variation of ΔT occurs, can be presented in below Equation 1:

$\varepsilon_{total} = \varepsilon_{thermal} = \alpha_c \Delta T + \alpha_e \Delta T$ (1)

Where:

α_c = CTE of concrete, to be determined, με/°C

α_s = CTE of the steel strain gauge, which equals 12.2 με/°C [Bamforth 2007].

ΔT = amplitude of the thermal cycle °C.

Whereby αc and αe are the thermal expansion coefficients of the concrete and steel, respectively, thus:

$$\alpha_c = \frac{\alpha_{total} - \alpha_e \Delta T}{\Delta T} \quad (2)$$

The following paragraphs depicting the condition of cracking potential as well as the approach employed in deriving the internal restraint induced strain for cracking avoidance i.e. . $\Delta T max = 3.7\ \varepsilon_{ctu} / \alpha_c$. In particularly, the maximum allowable core-surface temperature differential will be determined and the parameters that affect it would be analysed. When a concrete member is subjected to thermally induced volume changes under restrained (internal restraint or external restraint) condition, a certain amount of internal strain is built-up and cracking may occur when the tensile strain capacity of the concrete is exceeded. In mathematical terms, external restraint induced thermal strain can be formulated as follow (Bamforth 2007):

$$\varepsilon_r = K_1\{[\alpha_c T_1 + \varepsilon_{ca}]R_1 + \alpha_c T_2 R_2 + \varepsilon_{cd} R_3\} \quad (3)$$

Where:

T_1 - Difference between the peak temperature and the mean ambient temperature

T_2 - Long-term fall in temperature which takes into account the time of year at which concrete was cast

α_c - Coefficient of thermal expansion of concrete

ε_{ca} - Autogenous shrinkage

ε_{cd} - Drying shrinkage

R_1 - Restraint factor applies for early-age thermal cycle

R_2, R_3 - Restraint factors applying to long-term thermal movement and drying shrinkage

K_1 - Coefficient for the effect of stress relaxation due to creep under sustained loading (= 0.65)

As far as massive concrete structures are concerned, as in the case for Tembat Hydropower Dam construction, internal restraint becomes the bottleneck condition for thermal cracking. In this case, it is the core-surface temperature differential that can cause concrete to crack. Indeed, the core is prevented from exchanging heat with the surroundings, thus remaining hot for long time, while the surface tends to reach equilibrium with the environment and thus cooling. At this point, the following assumptions must be pointed out:

- Considering the concrete mix design, autogenous shrinkage can be neglected, as evidences in literature (Kada et al. 2002) that this phenomenon occurs only for low water/binder ratio {typically below 0.4} concrete {i.e. for high to very high strength concrete};
- Drying shrinkage plays a major role in the medium and long-term deformations and can be neglected for early stage determination of total strain;
- Restraint factors for external restraints are negligible for thick sections where – as mentioned above – internal restraint is dominant;

- Ignoring the long-term effects of temperature, T2 in consideration of early-age thermal cracking as in this case.

Taking into above consideration, the internal restraint induced strain is given by:

$$\varepsilon_r = k1\Delta T_{\alpha c}R \qquad (4)$$

Therefore, the risk of cracking and crack width should be assessed at both early-age and in the long-term. For the early-age calculation of εr using above equation 4, include only the terms for early-age thermal contraction at three days, i.e. assume that T2, εca, and εcd are zero. Where ΔT now becomes the core-surface temperature differential of the concrete element, the condition for cracking is:

$$\varepsilon_r > \varepsilon_{ctu} \qquad (5)$$

Where, ε_{ctu} is the tensile strain capacity of the concrete under sustained loading. Thus it is assumed that to avoid cracking, the restrained-strain εr should be less than the tensile strain capacity εctu of the concrete. The tensile strain capacity of plain concrete represents its ability to elongate before rupture and could be approximated by the ratio between the tensile strength of concrete and its elastic modulus (Bamforth 2007). Rearranging equation 4 and equation 5 for the condition of cracking, $\varepsilon_r = \varepsilon_{ctu}$, then equation 4 becomes:

$$T_{1,max} = \frac{\varepsilon_{ctu}}{k_1 \alpha_c R} \qquad (6)$$

For the condition of internal restraint it has been estimated that R = 0.42 and K1 = 0.65, it can be defined that the quick check for internal restraint induced strain for avoidance of cracking is expressed by (Bamforth 2007):

$$\Delta T_{max} = \frac{3.7\, \varepsilon_{ctu}}{\alpha_c} \qquad (7)$$

In summary, restrained volume changes can be caused by the superimposed effect of:

- Restrained (internal, external) thermal expansion/contraction cycles;
- Shrinkage (early, late).

Therefore any thermal modelling phase should attribute appropriate input parameters in order to closely model thermal behaviour and maximum strain developed by a certain structure.

3.3. Temperature Rise and Temperature Differential ($T_{1,max}$, ΔT_{max})

For the control of early-age thermal cracking, there are two important temperature considerations:

- The temperature rise above that of the adjoining concrete or substrate. It depends on the location of the point being considered and is applicable to conditions of external restraint {i.e. casting a wall onto a rigid mature base};
- The maximum temperature differential and the thermal gradient within the section. This is the condition of internal restraint.

In the calculation of restrained early-age temperature movements in concrete, it is necessary to define the basis upon which the temperature drop from the peak temperature is calculated. This requires a definition of the placing temperature of the concrete, the maximum temperature rise and the temperature to which the element will return after cool down, see below Figure 2.

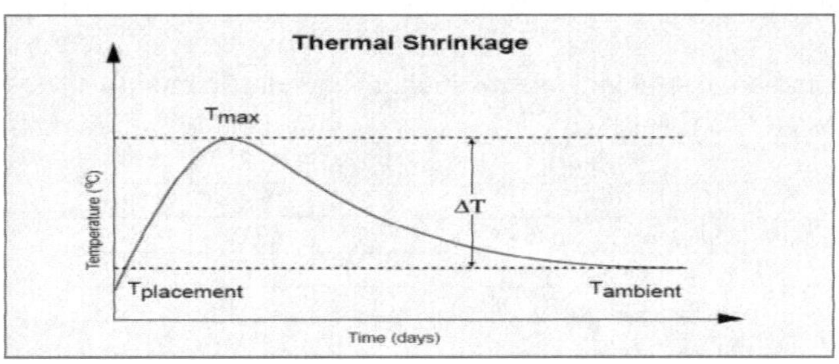

Figure 2. Typical heat development's plot of a concrete member

The placing temperature is defined as the temperature of the concrete when placed into the forms. This may differ from the temperature at the mixing plant. In this analysis, the maximum value for the project concrete at the delivery point is assumed to be 23°C {based on data gathered during mock ups} (Mass Concrete Class A 2011). The temperature rise is assumed to be the difference between the placing temperature and the maximum temperature achieved during the early-age thermal cycle.

The temperature rise depends upon the rate at which heat is generated, the total heat output and the rate of cooling. The heat generating characteristics are influenced by both the binder chemistry and its fineness. In thin sections, the rate of heat evolution is dominant in determining the temperature rise. In massive sections, where the heat loss is relatively slow compared with the heat being generated, the temperature rise is more dependent on the total heat evolved. The main factors governing the temperature rise and fall are:

- Binder content & types and sources of cementitious material;
- Concrete placing temperature & section thickness;
- Formwork and insulation and its time of removal;
- Ambient conditions.

The temperature drop, T_1, which determines the thermal contraction during cool down, is defined as the difference between the peak temperature and the mean ambient temperature and for the local conditions of average environmental temperature of 27°C of Hulu Terengganu can be also taken to equivalent to the temperature rise with the only difference that during temperature rise heat is built-up relatively quick while it would be desirable to limit to a maximum rate at the cooling phase. The T_1 values are based on modelling using semi-adiabatic data (BS EN196-9) derived from earlier test (Mass Concrete Class A 2011). As only the binder produces heat of hydration, the higher the binder content, the greater the heat evolved per unit volume and the greater the temperature rise. Along with the cement, combination and addition types the binder content has a major effect on the temperature rise.

3.4. Determination of Maximum Allowable Core–Surface Temperature Differentials

The mechanical properties of 20MPa mass concrete are shown in below Table 8.

As demonstrated during mock-up tests (Mass Concrete Class A 2011), the core reaches its maximum temperature at about 72 hours. It is therefore reasonable to assume that the maximum core-surface temperature differential occurs in the same period within the actual lifts. Hence, it is necessary to estimate the maximum allowable tensile strain εctu at early ages {in particular, at 3 days}. The tensile strain capacity εctu is the maximum strain that the concrete can withstand without a continuous crack forming. The aforesaid εctu can be measured directly or derived from measurements of the tensile strength and the elastic modulus of the concrete at 3 days. Therefore, these two variables must be estimated as well for correlation. Concrete tensile strength: at 3 days calculated as 68% of the tensile strength at 7 days available in (Bamforth 2007) as shown in below Tables 9-10:

Table 8. Concrete compressive strength, tensile strength and elastic modulus test results for 20MPa mass concrete (Mass Concrete Class A 2007)

Description	*Duration*		
Stage 1-Hox Box 1	7 day	28 day	56 day
Compressive Strength (MPa)	25.25	33.05	38.15
Tensile Strength (MPa)	1.70	2.43	2.81
Modulus of Elasticity (MPa)	19.25	25.60	28.90

This assumption is validated by the percentage increase of concrete tensile strength shown in below Table 9.

Table 9: Values of early-age and long-term tensile strength estimated in accordance with EN 1992-1-1: 2004

Strength class	C20/25	C25/30	C30/37	C35/45	C40/50	C45/55	C50/60	C55/67	C60/75
Early-age tensile strength $f_{ctm}(3)$ (MPa)	1.32	1.53	1.73	1.92	2.12	2.27	2.44	2.52	2.61
Long-term tensile strength, f_{ctm} (MPa)	2.21	2.56	2.90	3.21	3.51	3.80	4.07	4.21	4.35

Table 10. In-situ concrete tensile strength at 3 and 28 days – EN 1992-1-1: 2004

Concrete Tensile Strength At Early Ages in accordance with EN 1992-1-1: 2004						
Item	*Early-age (3 days)*			*Long-term (28 days)*		
	In-situ values		*EN1 992-1-1*	In-situ values		*EN1 992-1-1*
Strength class	95%	99%		95%	99%	
C20/25	1.16	1.30	1.33	1.92	2.17	2.28
C25/30	1.30	1.46	1.51	2.17	2.49	2.58
C30/37	1.52	1.70	1.73	2.48	2.81	2.90
C35/45	1.72	1.94	1.98	2.84	3.19	3.30
C40/50	1.83	2.05	2.12	3.06	3.41	3.54
C45/55	1.95	2.17	2.26	3.25	3.73	3.77
C50/60	2.06	2.34	2.39	3.43	3.90	4.00
C55/67	2.15	2.44	2.51	3.63	4.05	4.20
C60/75	2.27	2.52	2.62	3.76	4.23	4.38

Therefore, it can be seen that the assumed value of 1.16 MPa (0.68x1.7Mpa) for tensile strength at 3 days is reasonable.

Concrete Young modulus: calculated as 90% of the Young modulus at 7 days available in [BS8110-2: 1985]. This assumption is validated by the increase of the concrete Young modulus as shown in below Figure 3.

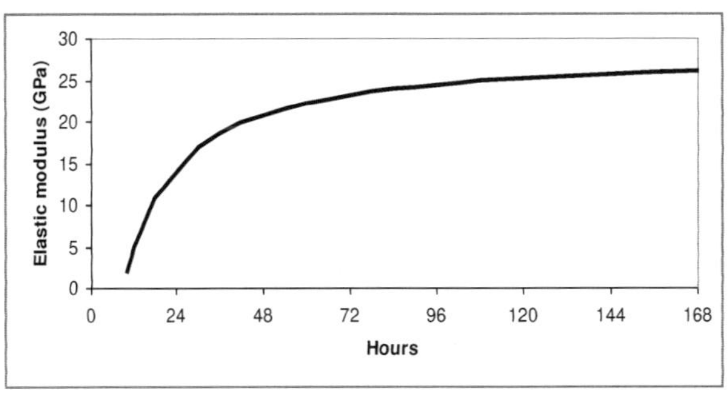

Figure 3. Increase of the Young modulus with time

In this case, a conservative value of 17.33GPa for the Young modulus at 3 days has been assumed. From the aforesaid determined concrete Young modulus and tensile strength, by applying equation 8, it is thus possible to calculate the maximum allowable tensile strain at 3 days as:

$$\varepsilon_{ctu} = \frac{f_{ctm}}{E_{cm}} \times 10^6\ \mu\varepsilon \qquad (8)$$

ε_{ctu} *(3 days) =* f_{ctm} *(3 days) /* E_{cm} *(3 days)* = 66.96 µε

Considering the value reported in below Table 11 for granite aggregate (ε_{ctu} @ 3 days = 75 µε, as extracted from [Bamforth 2007], the strain capacity for concrete aged at 3 days could then be estimated. This value is in line with the requirements stipulated in the literature (Bamforth 2007).

$$\varepsilon_{ctu}(f_{ck,cube} = 25) = \varepsilon_{ctu}(f_{ck,cube} = 37) \times \left[0.63 + \left(\frac{25}{100}\right)\right] \quad (9)$$

By using above equation 9, we obtain ε_{ctu} = 66.19 με. At this stage, by adopting above ε_{ctu}, ΔT_{max} could be calculated by means of $\Delta T_{max} = 3.7\ \varepsilon_{ctu}/\ \alpha_c$.
Hence, it is possible to calculate the maximum allowable core-surface temperature differential for different values of the coefficient of thermal expansion (CTE). Referring to Table 14 {(8.648+8.834)/2 = 8.741, Says 9.0 με/°C} and by applying above *Equation*-8, for a CTE of 9.0 microstrain/C° (Aged 3 days) as experimentally determined, the maximum differential is 27.53°C, as described in below Figure 4.

Aggregate type	Estimated ε_{ctu} under sustained loading for strength class C30/37	
	Early-age 3 days	Long-term 28 days
Basalt	63	90
Flint gravel	65	93
Quartzite	76	108
Granite, gabbro,	75	108
Limestone, dolomite	85	122
Sandstone	108	155
Lightweight aggregate (sintered fly ash) with natural sand	115	165

Table 11. Estimated values of ε_{ctu} for strength class C30/37 under sustained short-term loading using different aggregate types in accordance with EN 1992-1-1: 2004

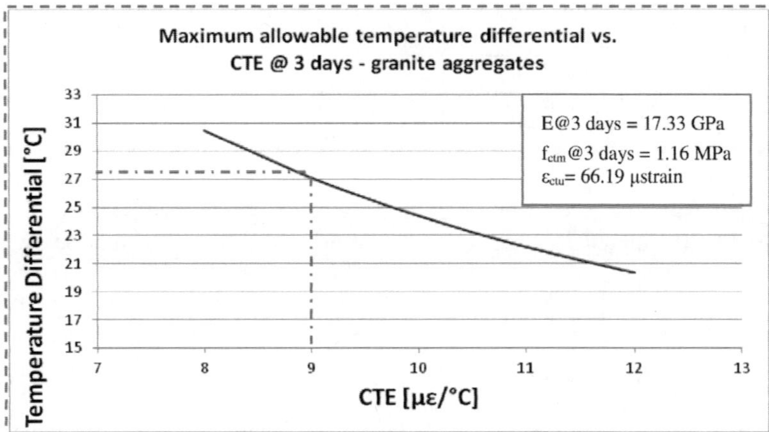

Figure 4. Maximum allowable core-surface temperature differentials as a function of the CTE; For a CTE of 9.0 as experimentally determined, the maximum differential is 27.53°C

The above calculated value of maximum temperature differential is perfectly in line with the published value i.e. 28°C as stipulated in BS 8110: Part 2 - Table 3.2 {See Table 3} and EN 1992-1-1: 2004 which provides estimates of temperature differentials applicable for a variety of conditions of external restraint and for internal restraint due to temperature gradients as shown in below Table 12.

Table 12. Limiting temperature drop T1, max and temperature differentials ΔTmax, to avoid early-age cracking, based on assumed typical values of αc and εctu as affected by aggregate type (for K_1 = 0.65 and ε_{ca} = 15με) [Bamforth 2007]

Aggregate type	Gravel	Granite	Limestone	LWAC with natural sand
Thermal expansion coefficient (με/°C)	13	10	9	9
Tensile strain capacity (με) under sustained loading	65	75	85	115
Limiting temperature change in °C for different external restraint factors:				
1.00	6	9	12	20
0.80	8	12	16	25
0.70	9	14	18	29
0.60	11	17	22	34
0.50	14	21	27	42
0.40	18	27	34	53
0.30	24	36	46	71
Limiting temperature differential (°C) for internal restraint R = 0.42	20	28	35	53

4. EXPERIMENTALLY DETERMINATION OF COEFFICIENT OF THERMAL EXPANSION (CTE)

Even though in literature (Bamforth 2007) it is possible to find data regarding the concrete CTE for different kinds of aggregates {see Table 1}, these are more deem-to-satisfy values provided when an experimental determination of the CTE cannot be performed**.** The design values in Table 1 may overestimate the actual value of the CTE, causing a large and unnecessary decrease in maximum allowable core-surface temperature differential. In the followings, an experimental procedure to accurately calculate the real CTE value will be proposed based on (Kada et al. 2002).

4.1. Experimental Procedure and Set-up

An identical concrete mix design previously adopted for the hot box trials as shown in Table 7. Class A Mass Concrete C20 Concrete Mix Design was applied on the casting of CTE block samples. The cement content and water/cement ratio were 200 kg/m3 and 0.70 respectively. The CTE was carried out under controlled laboratory conditions in accordance with the established procedure [Kada et al. 2002]. For the establishment of CTE value for grade 20MPa granite concrete, six (6) specimens of 0.2m x 0.2m x 0.4m concrete block {See below Figure 6} was cast whereby two results for each age were be made available so that a mean value will be calculated. The procedure of obtaining an accurate value of the CTE for the mix design under consideration is summarized below:

- Preparation of moulds;
- Samples casting and ageing;
- Subjection of samples to a thermal cycle (cooling/heating/cooling);
- Data acquisition and interpretation;
- Calculation of the CTE value.

Age of testing will be chosen based on the following criteria:

i. Before 24 hours concrete dimensions are dominated by stress relief by creep, hence no cracking would occur.
ii. After 7 days the value of CTE will tend to be constant whereby between 1 day and 7 days CTE is expected to be time dependent (decreasing in value with time)
iii. Based on the thermal data obtained during the hot box test the most critical age of for thermal cracking will be evaluated and at that age a suitable value of CTE will be chosen as input parameter for calculating of maximum differentials acceptable for the concrete under condition of internal restraint which is the dominating restraint for thick sections.

Therefore, the determination of the coefficient of thermal expansion (CTE) will be based on 1-Day, 3-Days, and 7-Days respectively.

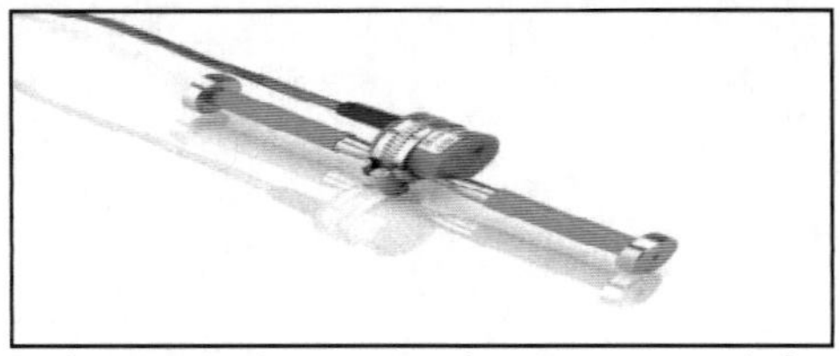

Figure 5. The ST4 Vibrating Wire Embedment Strain Gauge used for measuring strain in mass concrete (ST4 VWESG from itmsoil, UK).

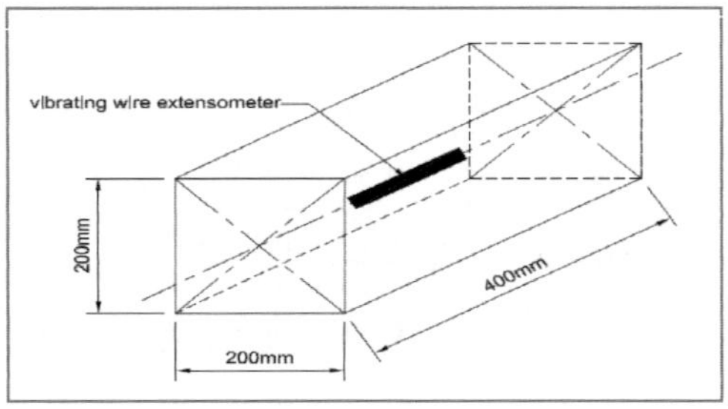

Figure 6. Schematic representation of the test sample

4.1.1. Methodology

1) Test was conducted on all six specimens using methodology proposed by Kada et al. 2002, for which the specimens were further divided into two specimens each to facilitate three different ages of testing and monitoring.
2) The samples were cast in steel moulds of 200 x 200 x 400mm and a vibrating wire extensometer was installed at the center of the sample along its longitudinal axis before concrete placement in the mould (For this case the height will be about 3 times the maximum aggregate size, hence 3 x 65mm MSA = 195mm, Says = 200mm).
3) Each strain gauge has then been tied to the steel frame by mean of plastic cable ties and connected to a remote data acquiring system. Finally, a thermocouple has been attached to the strain gauge in order to measure the core temperature of the samples.
4) The extensometer is connected to a data acquisition system that monitored the strain and temperature variations within the specimen.
5) Subsequently, concrete has been cast inside the moulds.
6) The specimen is demoulded just as the concrete has been set and reached the right age (right consistency for demoulding).
7) The samples obtained were then wrapped in plastic bags to prevent from hygrometric exchange with environment (prevent water evaporation). Thus allowing excluding different components of the total strain from the measured strain as described in preceding paragraph.
8) Each sample was then immersed in a heat-controlled water bath tank which then subject to alternate heat cycle at temperatures of 50°C±3°C and 10°C±3°C, respectively.
9) Each one of the samples is submitted to a thermal shock in the order of 40°C when stable temperature is reached within samples.
10) The samples from the 50°C±3°C is then switched to the 10°C±3°C water bath and vice versa until the intended duration of test is achieved for all 6 specimens.
11) In order to determine the CTE values at 1 Day, 3 days and 7 days, 2 samples for each aging time have been tested. Identification codes for such samples are shown in Table 13.

IOP Conf. Series: Materials Science and Engineering **217** (2017) 012009 doi:10.1088/1757-899X/217/1/012009

12) All the specimens are subjected to hours of alternating heat cycles (cooling/heating/cooling) which separating each thermal shock as illustrated in Table 14, 15 and 16: Duration of Heating and Cooling Cycles for 24 hours (1 Day), 72 hours (2 Days) and 168 hours (7 Days).

Table 13. Samples identification code and aging time

Description		Sample Aging		
		24 Hou	3-Days	7-Days
Samples Identification	SG1	√		
	SG2	√		
	SG3		√	
	SG4		√	
	SG5			√
	SG6			√

Note:* In order to measure very accurately the deformation induced by thermal cycles, it has been decided to construct steel frames to be firmly screwed to each of the casting mould as a rigid support for the strain gauge (see below Figures 7a-7b).

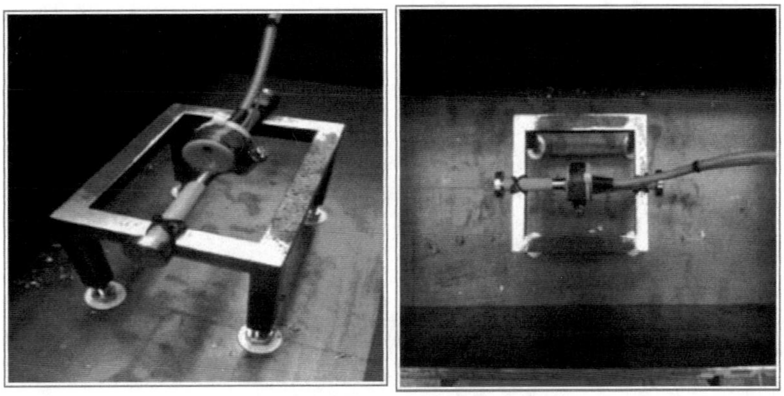

7a 7b

Figure 7. Front and top view of the system composed by steel frame and strain gauge inside the mould [7a and 7b showing the installation of ST4 vibrating wire embedment strain gauge. Size of specimen is 200mm x 200mm x 400mm]

Subsequently, the samples obtained were then wrapped in plastic bags to prevent from hygrometric exchange with environment (prevent water evaporation) as illustrated in below Figures 8a-8b.

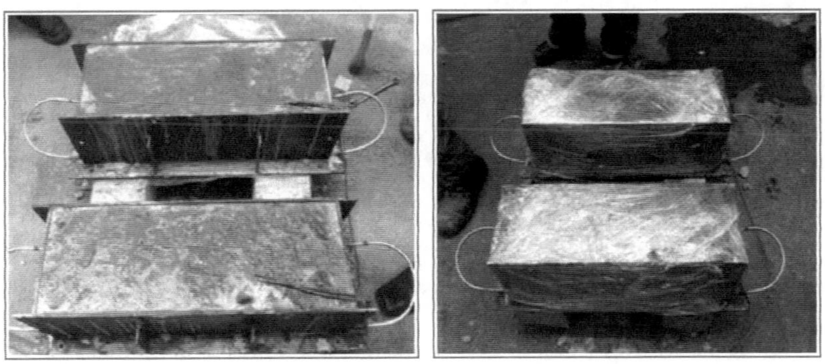

8a 8b

Figure 8. CTE specimens using Grade 20MPa [8a and 8b indicating the casting of CTE specimens using Grade 20MPa granite concrete and wrapped up with multilayer plastic sheet prior to CTE testing.(Samples as casted and after wrapping)]

5. THERMAL CYCLE

At the end of each thermal cycle step, the CTE can be determined as follow:

$$\alpha_c = \frac{\varepsilon_{concrete}}{\Delta T_{cycle}}$$

From above expression, it can be noted that it is possible to determine one CTE for the cooling phase and one CTE for the heating phase. These two values should coincide, at least theoretically. Experimentally, the closer these two values are, the more accurately the test has been performed. The below Figure 9 showing sample that have been put through a thermal cycle.

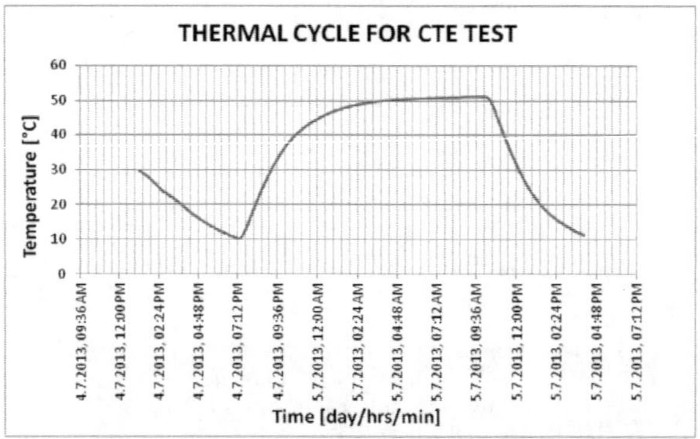

Figure 9. Typical Thermal cycle for the CTE determination

Thermal cycle shown in above Figure 9 refers to sample SG3. Anyway, all the other samples have been put through an identical thermal cycle. To reach the temperatures of 10°C and 50°C a chilled bath (chilled with ice cubes) and a hot bath (with a thermostat able to keep temperature constant) have been used, respectively (see below Figures 10a-10b).

10a

10b

Figure 10. Chilled and hot thermal baths

An alarm (see below Figure 11) has been set to ring when the core temperature of the sample (measured by the thermocouple fixed to the strain gauge) reaches 10±3°C in the cooling phase or 50±3°C in the heating phase). When this happens, the sample is immediately removed from one bath and put in the other one, depending on the thermal cycle period according to above Figure 9. Finally, to prevent water to flow inside the sample, samples themselves have been further wrapped with thick plastic bags (black colour) as shown in below Figures 11a-11b.

IOP Conf. Series: Materials Science and Engineering **217** (2017) 012009 doi:10.1088/1757-899X/217/1/012009

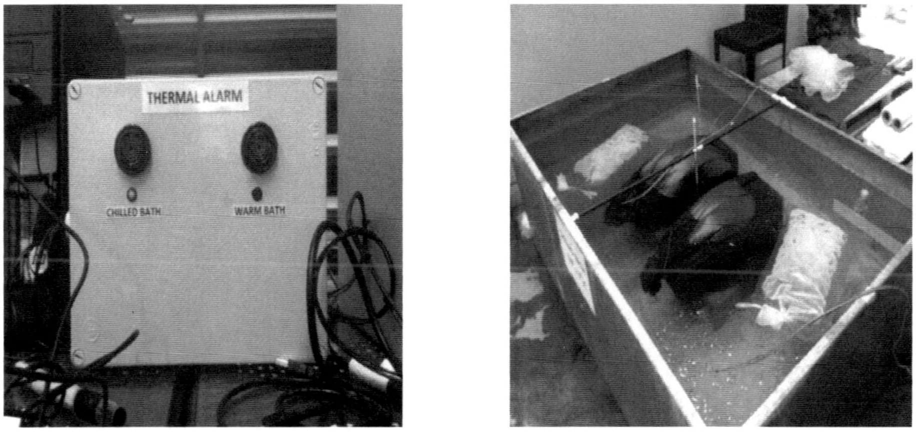

11a 11b

Figure 11. Thermal alarm and wrapped samples in the chilled bath

6. INTEPRETATION OF TEST RESULTS & ANALYSIS

The results for test performed on samples aged 24 hrs, 3 days and 7 days (Raw Data for CTE Test 2013), are shown in below Table 14. It must be pointed out that, because the test duration is about 24 hours, the age of samples at the time of CTE measurement will be of 2 days, 4 days and 8 days, respectively.

Table 14: The values of CTE recorded for heating and cooling cycles for each specimen during the effective age of concrete at 2 days, 4 days and 8 days

SG	Heating (με/°C)	Cooling (με/°C)	CTE average (με/°C)	Effective Age of Concrete	CTE average (με/°C)
SG1	8.882	8.764	8.823	2-Days	8.858
SG2	8.825	8.958	8.892		
SG3	8.764	8.531	8.648	4-Days	8.741
SG4	9.116	8.552	8.834		
SG5	8.954	8.349	8.652	8-Days	8.924
SG6	9.570	8.822	9.196		

From the above analysis, it is possible to deduce that the CTE remains virtually constant after 48 hours from samples casting. For the design purpose, the value of the CTE will take as 9.0με/°C [(8.648+8.834) = 8.741, Says = 9.0με/°C] between the period of 2-days and 8 days. Hence the CTE value at 3 days to be used for calculation.

6.1. Interpretation of r and r2 for Concrete at Effective Age at 4-Days

As per analysis shown in above Table 15, it could be concluded that the r & r2 for heating and cooling cycles for respective SG1, SG2, SG3, SG4, SG5 and SG6 for concrete aged at 3-days is of about 99.48% versus 99.87% and 98.96% versus 99.73% respectively, in agreement is achieved between the recorded temperature and strain values which implied strong linear relationship.

Table 15. Coefficients, r & r^2 at Age of 4-Days

Item	Cooling r	Heating r	Cooling r^2	Heating r^2
SG1	0.9876	0.9959	0.9754	0.9919
SG2	0.9966	0.9980	0.9932	0.9960
SG3	0.9985	0.9989	0.9971	0.9979
SG4	0.9973	0.9995	0.9947	0.9989
SG5	0.9934	0.9998	0.9869	0.9996
SG6	0.9950	0.9998	0.9901	0.9996
Mean	0.9948	0.9987	0.9896	0.9973
%	99.48	99.87	98.96	99.73

7. DISCUSSIONS

1. It has to be pointed out that variation of coefficient of thermal expansion occurred on all six specimens during the initial hours of strain measurements while concrete is having effective age of 1-day and whereby the strain is significantly dependent on the age of testing following concrete casting. The CTE values will gradually decreasing until the end of concrete setting and remain constant thereafter which implied that appropriate crack control and protection measure need to be taken in order to avoid early-age thermal cracking.
2. From the CTE analysis, CTE values remains virtually constant after 48 hrs from samples casting. The average CTE value for samples SG3 and SG4 cured for 72 hours and tested during a 72 hours cycle (hence for a resulting concrete effective age of 4 days) can be taken as 8.741 με/°C. For design purpose, the CTE value of granite aggregate could then be assumed as 9.0 με/°C between 2 and 8 days.
3. Besides thermal deformation, this proven test method is rather accurate way in eliminating any possible superimposed effect of others early-age thermal deformations that can occur during early-age strain measurements such as autogenous, plastic, drying and carbonation shrinkages.
4. Therefore, experimental determined value of CTE of a concrete mixture is of paramount importance in avoiding variation in CTE owing to a change in aggregate types. In contrast, any investigation aiming at evaluating early-age strain development in high strength concrete should take into consideration of the sensitivity of other thermal deformations such as autogenous, drying and plastic shrinkages to the total thermal expansion coefficient variations during concrete hardening.
5. From the established statistical analysis of r and r^2, both coefficients showing very strong positive linear correlation as r and r^2 is very close to unity i.e. 1. This could be observed that CTE value is virtually identical for both samples (SG1 and SG2) tested and regardless of the heating/cooling phase, meaning that the test is repeatable and test conditions have been controlled accurately.

8. CONCLUSIONS

6. Based on the actual CTE value of granite aggregate of 9.0 με/°C, it could be inferred that a simple CTE test method [Kada et al. 2002] is proven effective in determining early-age value of CTE, as compare to CTE value of concrete made with granite aggregate which varied from 8.1-10.3με/°C [CIRIA C660 and BSEN 1991-1-1:2004]. Therefore, the CTE value derived from this test is perfectly coherent (consistent/harmonious) with data available in scientific literature for concrete containing large volumes of granite aggregates [BSEN 1991-1-1:2004 and TxDOT2011].

7. Maximum allowable core-surface differential for the present concrete mix design with design CTE of 9.0 με/°C → 27.53°C, which is perfectly in accordance with the published values as stipulated in Bamforth 2007 and BS 8110-2: 1985.
8. Recommendations on optimum fresh concrete placing temperature, core-surface temperature, and maximum peak concrete core temperature, could be further verified and validated during further thermal modeling studies using suitable finite element software such as LUSAS and MIDAS FEA for heat of hydration and thermal stress analyses, with the experimental determined input parameters established in this paper along with other key parameters such as maximum adiabatic temperature rise data, convection heat transfer coefficients, environmental temperature and et cetera.

References

[1] ACI Committee 207.2R-07, 2007, 'Report on Thermal and Volume Change Effects on Cracking of Mass Concrete.'

[2] BS 8110-2: 1985, Structural Use of Concrete – Part 2: Code of Practice for Special Circumstances. BSI, London

[3] BS EN 1992-1-1:2004 Eurocode 2: Design of concrete structures - Part 1-1: General Rules and Rules for Buildings. BSI, London

[4] BS EN 196-9: 2010, Methods of Testing Cement – Heat of Hydration - Semi Adiabatic Method. BSI, London

[5] Bamforth P. B, 2007, 'Early-Age Thermal Crack Control in Concrete', CIRIA C660.

[6] CIMA Cement Test Certificates on NS Ecocrete-LH, CEM II/ B-V 42.5N-LH (January 2011 and June 2012) for 265MW TNB Hulu Terengganu Hydroelectric Project, Malaysia.

[7] Jahangirmejad, S Buch, and Kravchenko A, 2009, 'Evaluation of Coefficient of Thermal Expansion Test Protocol and Its Impact on Jointed Concrete Performance', ACI Material Journal, Vol 106 (1), 64-71.

[8] Kada et al 2002, 'Determination of the Coefficient of Thermal Expansion of High Performance Concrete from Initial Setting', Materials and Structures, Vol.35: 35-41.

[9] Mallela, J Abbas, Harman T, Rao, C Liu and Darter, M I, 2005, Measurement and Significance of Coefficient of Thermal Expansion of Concrete in Rigid Pavement Design', Transportation Research Record 1919, Transportation Research Board, Washington DC, 38-46.

[10] Mass concrete Class A/Grade III Concrete Mix Trials and Hot Block Monitoring, Overview of Results, Report 1202/11/7322 (2011) for 265MW TNB Hulu Terengganu Hydroelectric Project, Malaysia.

[11] Naik. T.R. et al 2011, 'Influence of Types of Coarse Aggregates on the Coefficient of Thermal Expansion of Concrete', Journal of Materials in Civil Engineering, Vol.23: 467 & 472.

[12] Ndon U J, Bergeson K L 1995, Thermal Expansion of Concretes: Case Study in Iowa, Journal of Materials in Civil Engineering, Vol 7(4), 246-251

[13] Raw data for Coefficient of Thermal Expansion Test Result (from 2/7/2013 to 9/7/2013) for 265MW TNB Hulu Terengganu Hydroelectric Project, Malaysia.

[14] ST4 Vibrating Wire Embedment Strain Gauge from itmsoil, United Kingdom for 265MW TNB Hulu Terengganu Hydroelectric Project, Malaysia.

[15] TxDOT Designation, Tex-428-A (June 2011). Test Procedure for Determining the Coefficient of Thermal Expansion of Concrete.

[16] Won M 2005 Improvements of Testing Procedures for Concrete Coefficient of Thermal Expansion, Transportation Research Record 1919, Transportation Research Board, Washington DC, 23-28.

High Energy Vibration for Gas Piping

Gary Y H Lee[1], K B Chan[2], Aylwin Y S Lee[2], ShengXiang Jia[3]

[1]Shell Global Solution (M) Sdn. Bhd., Menara Shell, No. 211 Jalan Tun Sambanthan, 50470 Kuala Lumpur, Malaysia,
[2] Sarawak Shell Berhad, Locked Bag No. 1, Lutong, 98100 Miri, Sarawak, Malaysia.
[3]Wood Group - Vibration Dynamics & Noise (formerly SVT), 432 Murray Street, Perth, Western Australia 6000.

E-mail: Gary.Y.Lee@shell.com, Kheng-Boo.Chan@shell.com, Aylwin.Lee@shell.com, ShengXiang.Jia@WoodGroup.com

Abstract. In September 2016, a gas compressor in offshore Sarawak has its rotor changed out. Prior to this change-out, pipe vibration study was carried-out by the project team to evaluate any potential high energy pipe vibration problems at the compressor's existing relief valve downstream pipes due to process condition changes after rotor change out.

This paper covers high frequency acoustic excitation (HFAE) vibration also known as acoustic induced vibration (AIV) study and discusses detailed methodologies as a companion to the_Energy Institute Guidelines for the avoidance of vibration induced fatigue failure, which is a common industry practice to assess and mitigate for AIV induced fatigue failure. Such detailed theoretical studies can help to minimize or totally avoid physical pipe modification, leading to reduce offshore plant shutdown days to plant shutdowns only being required to accommodate gas compressor upgrades, reducing cost without compromising process safety.

Keywords: Acoustic-induced vibration; AIV; high-frequency acoustic excitation; HFAE; high-energy vibration

1. Introduction

Vibration induced fatigue failure of pipework is an area of concern in the oil & gas industry due to negative impact towards process safety, production down time, cost and environmental impact. Acoustic induced vibration (AIV) is a type of vibration induced fatigue failure caused by high pressure drop devices in a gas system such as pressure relief valve, blowdown valve, orifice restriction plates and control valves which can generate high sound power levels in the pipework.

The time to failure from AIV is short (typically a few minutes to hours) due to high frequency response and high tonal noise levels external to the pipe, forming a high frequency vibration which are commonly between 500 to 2000Hz [1].

The AIV phenomenon was first reported by Carucci and Mueller in 1982. Curucci and Mueller showed the relationship between pressure drop and pipe diameter based on actual 36 cases where high vibrational levels had resulted in either pipe failure, non-failure or no abnormal experience [2]. In 1997, Eisinger further explored relationship between pipe diameter and pipe wall thickness (D/t) and the expected sound power level at which failure could occur [3]. The Energy Institute (EI) published a document known as "Guidelines for Vibration Induced Failure in Process Pipework" in 2008. The EI guidelines introduced empirical formula to calculate the Likelihood of Failure (LOF) at each pipe discontinuity (e.g. branch fittings, tee joints). If the LOF score = 1, the design changes / physical modification is mandatory which include one or more of the following mitigation measures to improve the LOF score:

a. Eliminating stress concentrations (e.g. removing pipe discontinuity)
b. Increasing pipe wall thickness, to reduce stress concentration at the pipe wall discontinuity.
c. Use of contoured fittings such as forged tees or sweepolets instead of weldolet fittings to reduce stress concentration at the pipe wall discontinuity.
d. Reduce mass flow rates
e. Use of low noise valve trims (applicable only for control valves)
f. Full wrap around reinforcement to reduce stress concentration at the pipe wall discontinuity. Case-by case basis evaluation, which requires finite element analysis to quantify the effectiveness of the reinforcement design.

However, in today's low-oil price environment, such physical modification is best kept minimal especially for operating plants due to production revenue loss (viz., plant shutdown) to accommodate for the safe site execution.

This paper presents a detailed theoretical methodology to AIV study – via dynamic stress analysis and finite element analysis (FEA) to determine the pipe fatigue life (time to failure), should the EI guidelines LOF calculation method demand a major re-design of the existing pipework.

2. Methodology & Result

2.1 AIV Screening

There are 3 units of relief valves (RV-102, RV-103, RV-104) at the compressor discharge line. Similar to EI Guidelines, process data for each relief valve was obtained (Table 1) to calculate the sound power level (PWL) generated by each relief device for AIV screening as shown in the equation [1] below:

$$PWL = 10\log_{10}\left[\left(\frac{P_1 - P_2}{P_1}\right)^{3.6} W^2 \left(\frac{T}{Mw}\right)^{1.2}\right] + 116.2 + SFF \quad (1)$$

Where:
P_1 = is upstream pressure (bara)
P_2 = is downstream pressure (bara)
W = is flowrate (kg/s)
T = is upstream temperature (K)
Mw = is molecular weight (grams / mol)

SFF = is a correction factor to account for sonic flow. If sonic conditions exist then SFF = 6; otherwise SFF = 0

If the calculated PWL exceeds 155dB, then the screening fails.

Table 1: Process parameters for AIV screening

Device	Set pressure (barg)	P1, inlet pressure (bara)	Mass flow rate (kg/hr)[a]	Inlet temp (K)	P2,back-pressure (bara)	Molecular weight (g/mol)	Sound power Level (dB)
RV-102	110	118.6	218444	359.6	13.7	19.08	174.39
RV-103	113	118.2	118819	359.5	9.6	19.08	168.38
RV-104	115	118.9	80795	359.7	10.5	19.08	165.12

[a] Mass flow rate from each relief valve is different due to staggered opening (viz., set pressures are set differently for each device) of the relief valve during overpressure scenario.

All 3 relief valves have failed the AIV screening, with PWL exceeding 155dB. The recommendation by EI guidelines is to perform further AIV analysis to determine the LOF at each discontinuity. The next section will discuss on the result using the dynamic stress analysis method as a detailed assessment.

2.2 Dynamic stress analysis method

For this further analysis via dynamic stress analysis method, each discontinuity at the downstream of the relief valve downstream pipe is identified in-terms of connection type (viz., weldolet, tee joint), material type and branch diameter size. The sound power level is then translated to pipe wall vibration so as to obtain the dynamic stress (σ_{dyn}) generated [4, 5].

The dynamic stress is then compared against ASME VIII Div. 2 fatigue curve limits [6] to calculate the pipe discontinuity fatigue life. These limits determined based on the device operating duration and pipe discontinuity material type.

An instrument protective function study was carried out to determine the relief valve operating duration. This involves checking the robustness of the available safeguarding system of the compressor in an overpressure scenario, before having to activate the relief valves *(Relief devices are generally the last line of defence for overpressure events).* From the study outcome, the relief valve demand was determined to be once in 20 years, which is well within of a typical plant design life of 20 years. Such event is classified as short & rare event and the accepted dynamic stress limits are as follows:

Table 2: Dynamic stress allowable limits for short & rare operating duration.

Operating duration	Carbon steel	Austenitic Stainless Steel
Short & rare event	σ_{dyn}< 105MPa	σ_{dyn}< 148MPa

A result comparison of EI LOF calculation method vs dynamic stress method is shown in table 3. The dynamic stress method now enabled us to focus on only one failure case (viz., Configuration A) instead of all the four failure cases (viz., Configuration A to D).

Table 3: AIV result comparison between energy institute guideline vs. dynamic stress method

Configuration	Discontinuity (Branch Diameter Size (mm) x Connection type)	Main Pipe Outside diameter (mm)	Pipe wall thickness (mm)	Pipe Material	Energy Institute		Dynamic Stress Method	
					LOF	**Pipe Redesign Mandatory? (Mandatory / Recommended / Not mandatory)**	**Dynamic Stress Levels (MPa)**	**AIV analysis (Pass / Fail)**
A	DN20 weldolet	323.8	14.27	Carbon Steel	1.00	Mandatory	125	Fail
B	DN300 tee	406.4	9.53	Austenitic Stainless Steel	0.77	Recommended	127	Pass
C	DN50 weldolet	406.4	9.53	Austenitic Stainless Steel	1.00	Mandatory	127	Pass
D	DN80 weldolet	406.4	9.53	Austenitic Stainless Steel	0.94	Recommended	110	Pass

The main difference in results between EI and dynamic stress methods is due to fatigue life evaluation. The EI method uses fatigue life curves in accordance with BS 7608 or PD5500 while the dynamic stress method uses ASME Div. II fatigue life curves. BS 7608 or PD5500 provides comprehensive fatigue life for weld classes, whereas EI penalizes for using fillet weld type such as weldolet connections. As opposed to EI Guidelines, the dynamic stress method was developed based on ASME Div. II [6] has comprehensive fatigue life curves for different materials such as carbon steel and SS300 series high-alloy steel.

Furthermore, while LOF=1.0 at the discontinuities in EI Guidelines mandates redesign, corrective actions and/or detailed analysis of AIV, for a LOF less than 1.0, EI provides vague guidance which only covers recommendations (i.e. optional) for further analysis [1] (table 4). An LOF below 0.5 is considered as a pass for AIV analysis. With dynamic stress analysis method, the AIV analysis provides better quantitative approach, where AIV result is a pass or fail based on the material and duration of the operating device (viz. sound power source).

Table 4: LOF interpretation as per EI guidelines [1]

Score	Action
LOF ≥ 1.0	The main line shall be redesigned, resupported or a detailed analysis of the main line shall be conducted, and vibration monitoring of the main line shall be undertaken (***Note 1***) Corrective actions shall be examined and applied as necessary Small bore connections on the main line shall be assessed. A visual survey shall be undertaken to check for poor construction and/or geometry and/or support transmission to neighbouring pipework.
1.0 > LOF ≥ 0.5	The main line should be redesigned, resupported or a detailed analysis of the main line should be conducted, or vibration monitoring of the main line should be undertaken (***Note 1***) Correction actions should be examined and applied as necessary Small bore connections on the main line shall be assessed. A visual survey shall be undertaken to check for poor construction and/or geometry and/or support transmission to neighbouring pipework.

2.3 Finite Element Analysis Method

Since the discontinuity at DN20 weldolet (Configuration A) failed in the dynamic stress analysis study, a more advanced AIV study via finite element analysis (FEA) was utilized to further optimize the accuracy of the dynamic stress calculation. FEA was used to simulate the piping vibration excited by pipe internal sound dynamic pressure and calculate the piping dynamic stress resulting from the vibration.

Additional information such as the actual discontinuity geometry and site visual condition was taken into FEA to model the discontinuity accurately and reflect the as-built conditions of piping. FEA modelling was done using Nastran FEMAP (version 11.2) software. The drain connection was modelled using solid elements for detailed stress prediction; Plate elements were used for other areas of the model (i.e. to simulate constraints and for accurate mode shape prediction); The boundary conditions are shown in the screenshot of the model in figure 1.

The calculated dynamic stress from FEA modelling was calculated to be 66.1MPa (figure 2) which passed the AIV analysis as the dynamic stress is less than 105MPa.

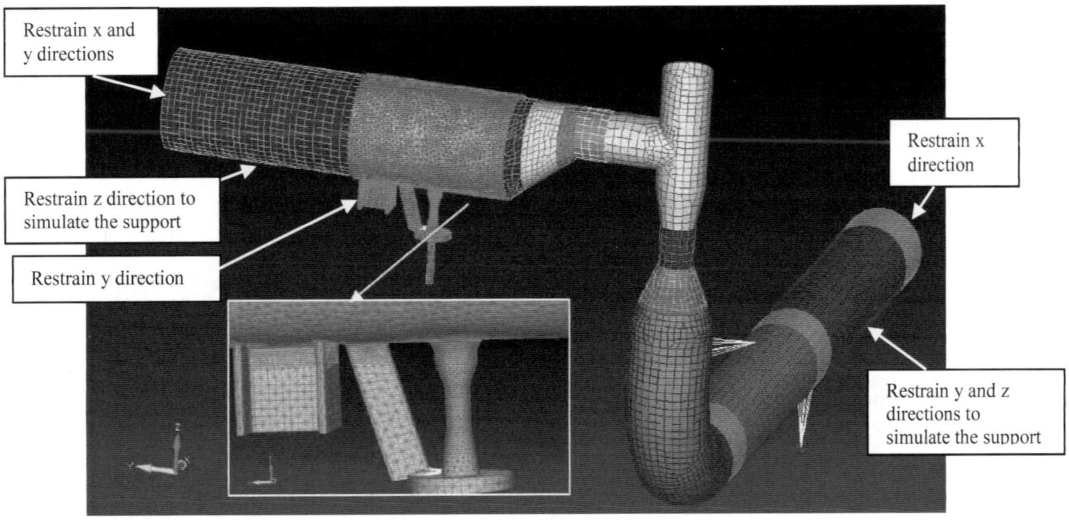

Figure 1: FEA modelling of DN20 x weldolet discontinuity

Figure 2: FEA model showing the location of maximum dynamic stress at the pipe discontinuity.

3. Conclusion

The compressor relief line has undergone an AIV study which comprise of (1) AIV screening using the EI method, (2) intermediate study was conducted using 2 methods which are EI guidelines and dynamic stress analysis method and (3) advance study was done using the FEA method. The discontinuity at the relief lines were found to be of no AIV concern and no physical line modification is required to mitigate AIV issues.

In addition to the EI Guidelines, the dynamic stress analysis provided a more refined fatigue estimation and the FEA method was able to fine tune the stress values to optimize the pipe's fatigue life which can be used to evaluate the AIV risk of existing process pipe. Through calculation of pipe fatigue life, this can help plant operators to better quantify the AIV risk to decide the extent of the required site physical modifications to mitigate AIV issues. In this case, it helped to avoid pipe modification at site. Such avoidance of physical pipe modification minimised shut-down/turn-around time required, plant production loss and site HSSE risk exposure, so it helped complete the project safer, faster and cheaper.

References

[1] Energy Institute 2008 *Guidelines for the Avoidance of Vibration Induced Fatigue in Process Pipework* (Energy Institute, London)
[2] Carucci, V A and Mueller R T 1982 *Acoustically Induced Piping Vibration in High Capacity Pressure Reducing Sytems* (ASME 82-WA/PVP-81982)
[3] F L Eisinger 1997 *Designing piping systems against Acoustically Induced Structural Fatigue* (Foster Wheeler Energy Corporation, Clinton, New Jersey)
[4] Norton, M P and Karczub, D G 2005 *Fundamentals of Noise and Vibration Analysis for Engineers* (Cambridge University Press)
[5] Karczub D G, Fagerlund A C 2005 *Dynamic stress predictions of acoustic induced pipe vibration failures* (24th International Conference on Offshore Mechanics and Artic Engineering, Halkidiki, Greece)
[6] ASME BPVC VIII Div. 2 2005 *Rules for Construction of Pressure Vessels Division 2 Alternative Rules* (The American Society of Mechanical Engineers, Two Park Avenue, New York)

Experimental study of efficiency of solar panel by phase change material cooling

Nicholas Tan Jian Wei, Wong Jian Nan, Cheng Guiping*
Faculty of Engineering, Nilai University, 71800 Nilai

E-mail: cheng@nilai.edu.my (Cheng Guiping)

Abstract. The dependence of efficiency of photovoltaic panels on their temperature during operation is a major concern for developers and users. In this paper, a phase change material (PCM) cooling system was designed for a 60W mono-crystalline solar panel. Tealights candle was selected as the cooling medium. The solar irradiance was recorded using Kipp & Zonen CMP3 pyranometer and Meteon data logger. Temperature distribution on the surface of solar panel, output voltage and output current of solar panel were measured. The average irradiance throughout data collection was found to be 705W/m^2 and highest irradiance was 1100 W/m^2. The average solar panel temperature was 43.6°C and a maximum temperature of 53°C was at the center of solar panel. Results showed that average power output and efficiency of the solar panel were 44.4W and 15%, respectively. It was found that the higher the solar irradiance, the lower the efficiency of solar panel and the higher the temperature and power output of solar panel. This is due to the fact that high irradiance results in high power input and high solar panel temperature. But high PV panel temperature reduces its power output. Therefore, the increase of power input outweighs that of power output, which leads to the decrease of efficiency of solar panel with the increase of solar irradiance. Compared with solar panel without cooling, the power output and efficiency of solar panel did not increase with PCM cooling. It indicates that Tealights candle as PCM cooling is not efficient in improving the efficiency of solar panel in this study.

1. Introduction

Solar energy is one of renewable energy sources favorable in Malaysia for its climatic conditions. According to Azhari et al [1], solar radiation for each day on an annual average is 4.21-5.56 kWh/m^2. The highest estimated solar radiation is as high as 6.8kWh/m^2 during August and November and lowest 0.61kWh/m^2 during December. The high radiation is very suitable to use photovoltaic (PV). However, the abundant irradiance will heat up the PV which will reduce the efficiency of the panels [2].

To increase the efficiency of the solar panels, several cooling techniques have been tried. Generally, there are two types of cooling: active cooling, which consumes energy (pump, fan, etc.) and passive cooling, which uses natural convection/conduction to enable heat extraction [3]. Most cooling methods are based on active water and air cooling, as these are the simplest techniques. However, passive cooling methods are self-operable, more reliable and cost effective than active cooling methods because of the absence of moving parts [4]. It can be divided into three main groups: air or water passive cooling, and conductive cooling. Phase change material (PCM) cooling is a special type of passive conductive cooling. Much research on PV/PCM cooling has been done by numerical simulations [5], laboratory tests and outdoor studies in Pakistan [6], Ireland [7] and India [8] in building integrated photovoltaics (BIPV). However, there is little research done in Malaysia. And not much research on PCM cooling of non BIPV system was conducted. It is known that the power drop and temperature rise of the solar panel largely depend on the climate of the sites [9]. In this paper, a PCM cooling system was designed and tested outdoor with a free standing solar panel. The effects of the phase change material cooling system on the output of the solar panel were studied.

2. Experimental Setup

The first step of fabricating the cooling system is the selection of the PCM. Paraffin and paraffin wax are frequently used as candidates of PCMs due to their non-toxic and inert properties [7]. Tealights candle was used as PCM in this study.

The proposed PCM cooling system is illustrated in Figure1. Aluminum tubes were fitted into a stainless steel container. The bottom of container was covered with wooden plank, which acts as an insulation layer. The cavities between the aluminum tubes as shown in Fig. 1(a) were filled with PCM. A layer of aluminum sheet was used to cover PCM, which prevents direct contact between PCM and the back of solar panel.

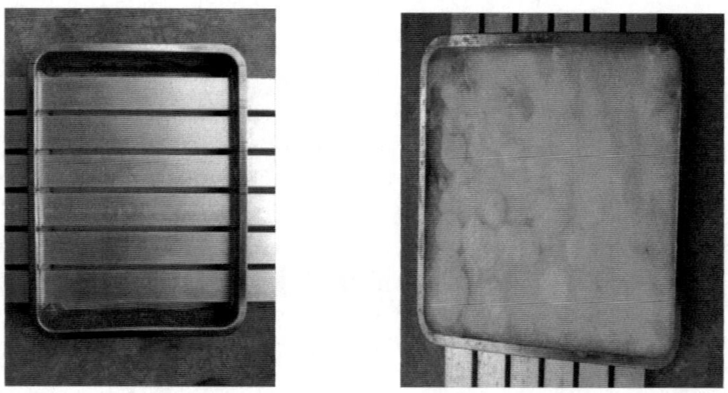

(a) (b)

Figure 1. (a) Back view without PCM, (b) Gaps between tubes filled with PCM

Finally, the cooling system was attached to a 60W monocrystalline solar panel and set at the desired direction and elevation with the help of a rack for outdoor experiments, as shown in Figure 2. Solar irradiance was measured using Kipp & Zonen CMP 3 Pyranometer, Meteon data logger and a laptop. Data collection was started at 11:00 and ended at 15:00. Readings were recorded through the data logger in the computer at intervals of one minute.

The surface temperatures at 9 positions shown in Figure2 and output of the solar panel were measured using K-type thermocouples and multimeter at intervals of 30 minutes, respectively.

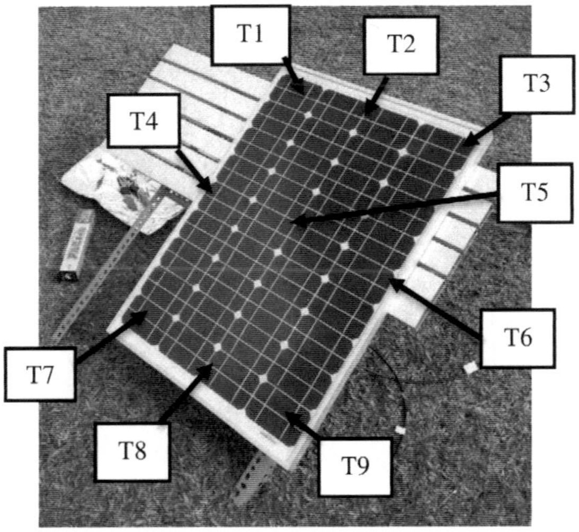

Figure 2. Complete set-up of PCM cooling system with thermocouple positions.

3. Results and Discussion

3.1 Irradiance intensity

Figure 3 shows the profile of irradiance on one day of data collection. It started from 780 W/m^2 at 11:00, reached a peak at 1026 W/m^2 at around 13:00 and then dropped to below 400 W/m^2 due to weather condition.

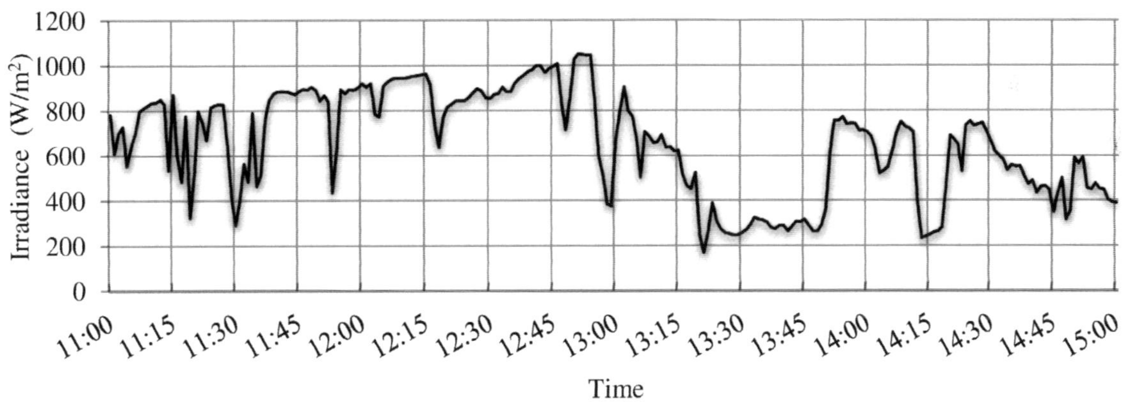

Figure 3. Irradiance *vs* time.

Distribution of irradiance throughout the data collection period of four days was tabulated in Table 1. It can be seen that most of the data points fell in the range of 700 to 800 W/m^2 in Day1, 700 to 900 W/m^2 in Day 2, 800 to 1000 W/m^2 in Day 3 and 4. The average irradiance throughout the data collection was 705W/m^2 and the highest irradiance was 1100 W/m^2 at 11:30 on Day 3. Since the power input to solar panel is a product of irradiance and area of the panel, the higher the irradiance the more the power input.

Table 1. Counts of irradiance.

Range (W/m^2)	Day 1	Day 2	Day 3	Day 4
0~199	0	1	16	0
200~299	21	27	9	12
300~399	19	19	5	14
400~499	13	23	11	1
500~599	10	23	15	4
600~699	33	29	19	10
700~799	114	36	40	28
800~899	31	48	65	89
900~999	0	29	55	83
1000~1099	0	6	6	0

3.2 Efficiency of Solar Panel

The efficiency of the tested solar panel was calculated by using the following equation [10]:

$$\eta = \frac{P_{out}}{P_{in}} = \frac{V_m \cdot I_m}{I.S} \quad (1)$$

Where: V_m – maximum voltage (V), I_m – maximum current (A), I – intensity of radiation (W/m^2), S – arear of panel (m^2), in this study S =0.41m^2.

Figure 4 shows that the efficiency of the solar panel decreased with the increase of irradiance. The highest efficiency was 35% at 11:30 at an irradiance of 391 W/m^2. The average efficiency of the solar panel during the four days' data collection was found to be 15%. Efficiency of a solar panel is the ability of the solar panel to convert light energy to electrical energy. It is reported that the majority of the soral radiation is converted to heat energy [11]. When radiation intensity increases, both the input and output power of the solar panel will increase. On the other hand, the temperature rise of solar panel will reduce the output power. Therefore, output power did not change much. This results in the inverse proportion of efficiency of solar panel to the irradiance.

Due to the fact that solar panel can only harness a limited amount of energy from the irradiation of the sun, the excessive solar radiation will only contribute to heat up the solar panel which in turn causes the drop of efficiency of the solar panel. This can be clearly seen in Figure 5, which shows the relationship between average temperature and efficiency of the solar panel. When panel temperature is higher, the efficiency is lower. The highest efficiency is 35% at 11:30 at an average panel temperature of 32 °C.

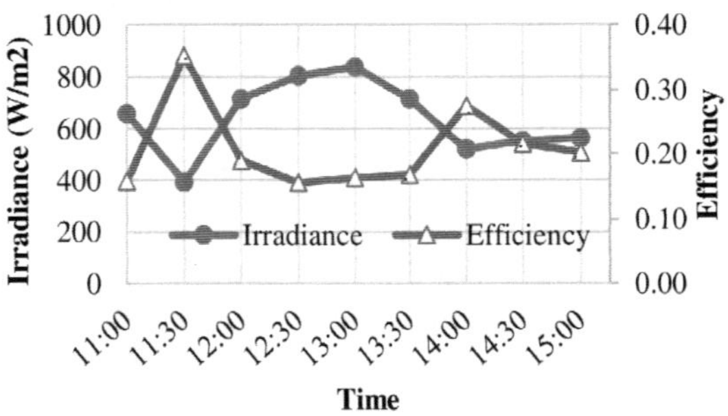

Figure 4. Irradiance *vs* efficiency.

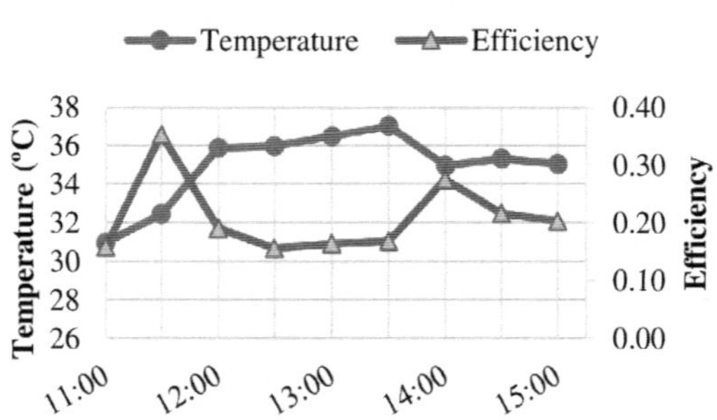

Figure 5. Temperature *vs* efficiency.

3.3 Temperature of Solar Panel

Table 2 shows the temperature distribution on the front surface of the solar panel with PCM cooling. It can be seen that T5 has the highest value among the 9 points, from 41 to 46 °C. That means the center of solar panel is the hottest point. The average solar panel temperature of the four days' data collection was found to be 43.6°C and a maximum temperature of 53°C was recorded at the center of solar panel on Day 4.

Table 2. Temperature of points 1 to 9 with PCM cooling.

Temperature (°C)	11:00	12:00	13:00	14:00	15:00
T1	35	46	45	41	43
T2	35	46	44	43	44
T3	35	47	44	41	43
T4	39	40	44	42	41
T5	41	46	46	42	41
T6	39	43	42	41	40
T7	37	42	42	41	40
T8	37	43	42	40	39
T9	36	40	41	40	40

The comparison of the temperature at the center of PV surface (T5) between solar panels with and without PCM cooling is tabulated in Table 3. There is no significant difference of temperature between the panels with and without PCM cooling. The average efficiency of the solar panels with and without PCM cooling was 14.6 and 15.0%, respectively.

PCM can absorb/discharge a large amount of energy during phase change. The capacity of a PCM for temperature control depends on its properties, heat transfer methods and system configuration (Huang, et al, 2011). PCM used in this study has a melting temperature of 48°C – 50°C, while the solar panel operates around 45°C with a maximum temperature at 53 °C. This indicates that PCM may be partially melted, which results in little contribution of latent heat to heat removal from PV panel. The non-melt PCM only absorbs sensible heat. Aluminum tubes in this study were designed to increase the surface area of conductive heat transfer from PCM and act as ducts for natural convection by wind induced air flow.But non-melt PCM has low thermal conductivity and the wind speed during the experiments was below 2 m/s. These lead to uneffective heat transfer from the solar panel. In order to fully utilize the advantages of phase change, PCM with lower melting point is recommended for further study.

Table 3. Surface temperature of the center of PV with and without PCM cooling.

Temperature (°C)	11:00	12:00	13:00	14:00	15:00
With PCM cooling	41	46	46	42	41
Without PCM cooling	39	45	44	41	41

4. Conclusion

PCM Cooling System was designed to study its effects on the efficiency and temperature reduction of solar panel in Malaysia climate. The measured average irradiance was 705W/m^2, and the maximum irradiance was up to 1100 W/m^2. The average solar panel temperature was 43.6°C with the highest value of 53°C and lowest 35°C. Comparison between solar panel with and without PCM cooling showed that Tealights candle as PCM in this study was not effective in reducing temperature and improving efficiency of solar panel. The reason is that the melting point of tealights candle is around the operating temperature of solar panel, which limits the contribution of heat removal from the panel by phase change. A PCM with a lower melting point than 40ºC is recommended for Malaysia climate.

References

[1] Azhari AW, Sopian K, Zaharim A, Al Ghoul M. A new approach for predicting solar radiation in tropical environment using satellite images-case study of Malaysia. *WSEAS Transactions on Environment and Development*. 2008 Apr;**4**(4):373-8.

[2] Skoplaki E, Palyvos JA. On the temperature dependence of photovoltaic module electrical performance: A review of efficiency/power correlations. *Solar energy*. 2009 May 31;**83**(5):614-24.

[3] Jakhar S, Soni MS, Gakkhar N. Parametric modeling and simulation of photovoltaic panels with earth water heat exchanger cooling. *Geothermal Energy*. 2016 Dec 1;**4**(1):10.

[4] Tan L, Singh B, Date A, Akbarzadeh A. Sustainable thermoelectric power system using concentrated solar energy and latent heat storage. InPower and Energy (PECon), *2012 IEEE International Conference* on 2012 Dec 2 (pp. 105-109). IEEE.

[5] Smith CJ, Forster PM, Crook R. Global analysis of photovoltaic energy output enhanced by phase change material cooling. *Applied Energy*. 2014 Aug 1;**126**:21-8.

[6] Hasan A, McCormack SJ, Huang MJ, Norton B. *Phase change materials for thermal regulation of building integrated photovoltaics* (Doctoral dissertation, Doctoral Thesis. Dublin Institute of Technology).

[7] Huang MJ, Eames PC, Norton B, Hewitt NJ. Natural convection in an internally finned phase change material heat sink for the thermal management of photovoltaics. *Solar Energy Materials and Solar Cells.* 2011 Jul 31;**95**(7):1598-603.

[8] Maiti S, Banerjee S, Vyas K, Patel P, Ghosh PK. Self regulation of photovoltaic module temperature in V-trough using a metal–wax composite phase change matrix. *Solar energy*. 2011 Sep 30;**85**(9):1805-16.

[9] Kibria MA, Saidur R, Al-Sulaiman FA, Aziz MM. Development of a thermal model for a hybrid photovoltaic module and phase change materials storage integrated in buildings. *Solar Energy*. 2016 Feb 29;**124**:114-23.

[10] *Measuring PV efficiency* (2010) Available at: http://www.pvpower.com/assets/Measuring-PV-Efficiency -Solar-Panels.pdf (Accessed: 6 December 2016).

[11] Wu S, Xiong C. Passive cooling technology for photovoltaic panels for domestic houses. *International Journal of Low-Carbon Technologies*. 2014 Jun 1;**9**(2):118-26.

Acknowledgements

This project was supported by Nilai University internal grant: Hybrid cooling system for PV panels under Malaysia climate conditions.

IOP Conf. Series: Materials Science and Engineering **217** (2017) 012012 doi:10.1088/1757-899X/217/1/012012

Fuel Characterization of Newly Discovered Nigerian Coals

Bemgba Bevan Nyakuma[1,2*], Olagoke Oladokun[1,2], Aliyu Jauro[3], and Denen Damian Nyakuma[4]

[1]Centre of Hydrogen Energy, Institute of Future Energy, Universiti Teknologi Malaysia, 81310 Skudai, Johor Baru, Malaysia.

[2]Department of Chemical Engineering, Faculty of Chemical & Energy Engineering, Universiti Teknologi Malaysia, 81310 Skudai, Johor Baru, Malaysia.

[3]National Centre for Petroleum Research and Development (NCPRD), Abubakar Tafawa Balewa University, P.M.B 0248, Bauchi, Bauchi State 740272, Nigeria.

[4] Directorate of ICT, University of Calabar, P.M.B 1115, Calabar, Cross River State 540271, Nigeria.

E-mail: bnbevan2@live.utm.my, bbnyax1@gmail.com

Abstract. This study seeks to characterize and highlight the fuel properties, rank, and classification of coals from *Ihioma* (IHM) and *Ogboligbo* (OGB) in Imo and Kogi states of Nigeria, respectively. The fuel properties were examined based on ultimate, proximate, and bomb calorific analyses. The results indicated that IHM coal contains comparatively higher *C* and *H* but lower *O*, *N*, and *S* content than OGB. In addition, the nitrogen (*N*) and sulphur (*S*) content for both coal samples were above 0.7 wt.% and 1.5 wt.%, respectively, which indicates high potential for pollutant emissions. Furthermore, the coal proximate properties were below 5 wt.% for *Moisture*; *Volatiles* (70 wt.%); *Fixed Carbon* (45 wt.%) and Ash (2.5 wt.%) on average. IHM coal has an HHV of 19.40 MJ/kg whereas OGB is 15.55 MJ/kg. This is due to the low carbon (*C*), hydrogen (*H*) and high oxygen (*O*) content in OGB whereas IHM contains higher *VM* and *HHV*. Furthermore, OGB presents better handling, storage, and transport potential. Furthermore, OGB has a higher fuel ratio and value index due to lower moisture, ash content, and volatiles. Based on the ASTM D388 standard, the coals were classified as Lignite (Brown) Low-Rank Coals (LRCs) with potential for energy recovery.

1. Introduction

Coal utilization presents significant potential for the energy recovery and power generation necessary for socio-economic growth and sustainable development [1]. The availability, accessibility, and acceptability of coal resources are key dynamics in the global quest for low-cost, secure energy supplies for the future [2, 3]. As a result, the discovery of new deposits in developing countries like Nigeria has rekindled interest in coal energy economy [4, 5]. However, lack of comprehensive data on the physicochemical, thermal kinetic, and thermodynamic coal properties is currently hampering transition to coal energy [6]. Coal data is vital to process design, optimization and scale-up of future power plants and conversion systems for efficient utilization [7, 8]. In addition, it can provide important information required by national policy makers and engineering professionals for evaluating the socioeconomic prospects and environmental impacts of future coal energy systems.

Currently, coal-fired power plants account for 41% of global electric power generation which is derived primarily from lignite (brown) coal utilization [9, 10]. Furthermore, lignite accounts for 80% of coal-based electricity generation in Germany, China, Russia, Poland and Australia [11], although IEA (International Energy Agency) reports that production declined from 834–807.4 Mt/yr from 2013 to 2016 [12].

Despite the global decline, the affordability, reliability, and versatility of lignite coal can stimulate socioeconomic growth, sustainable development, and poverty eradication in developing countries like Nigeria with large new deposits. However, coal utilization for energy recovery, fuel synthesis, and chemicals production is non-existent in Nigeria – Africa's largest economy and most populous nation [13, 14]. This is partially due to resistance from policy, academic and environmental stakeholders to the development, adoption, and implementation of coal despite Nigeria's perennial energy crises. In addition, it is due to concerns about pollutant emissions, global warming, and climate change arising from potential coal utilization [15]. However, the comprehensive characterization of coal fuel properties and implementation of clean coal technologies (CCT) can address the socioeconomic and environmental challenges [16] which currently hamper coal-based electric power generation [17-19]. Therefore, it is pertinent to identify, examine, and highlight coal fuel properties particularly in developing countries like Nigeria with vast new deposits. This will avail scientists, engineers, and policymakers with critical information required to foster the sustainable transition to future coal energy production.

Consequently, this study seeks to characterize the fuel properties, rank classification and potential applications of *Ihioma* (IHM) and *Ogboligbo* (OGB) coal from Imo and Kogi states in Nigeria, respectively. The elemental and calorific fuel properties of the coals based on ultimate, proximate, and bomb calorific analyses will be examined. It is envisaged that the results of this study will provide requisite design and optimization data for future thermochemical coal conversion.

2. Experimental

The coal samples from *Ihioma* (IHM) in Imo State and *Ogboligbo* (OGB) from Kogi State in Nigeria were acquired from the National Metallurgical Research and Development Centre (Jos, Nigeria). Next, the samples were pulverized and sifted using the Retsch^{TM} analytic sieve of mesh size 60 to acquire 250 μm sized particles prior to characterization.

The pulverized coal samples were subsequently characterized by ultimate, proximate, and bomb calorific analyses to examine the fuel characteristics required for rank classification. The ultimate analysis was determined using a CHNS elemental analyzer (Model: vario MICRO Cube^{TM}, Germany). The analysis was carried out in duplicate, according to the ASTM D5291 standard and the results were analyzed using the CHNS analysis software (*version* 3.1.1). The proximate analysis was examined by thermogravimetric (TG) analysis based on the procedure described in the literature [20].

The calorific (higher heating) value was determined using a bomb calorimeter (Model: IKA C2000, USA) according to ASTM D2015 whereas the lower heating value (LHV) were calculated as described by Eq 1 from literature [21]. The mineral matter (*Mm*) content of the fuel was calculated from the *Parr* formula described in the literature [10].

$$LHV = HHV - (22.604 \times H) - (2.581 \times MC) \quad (1)$$

$$Mm = (1.08 \times AC) + (0.55 \times S) \quad (2)$$

The terms in Eqs 1 and 2 represent *LHV* – Lower Heating Value (MJ/kg); *HHV* – Higher Heating Value (MJ/kg); *H* – Hydrogen content; *MC* – Moisture content; *Mm* – Mineral Matter; *AC* – Ash and *S* – Sulphur content. Lastly, the rank classification of the coals was examined according to the procedures of the ASTM D388 standard [22].

Based on the ultimate and proximate analyses of the IHM and OGB coal samples, the atomic and fuel property ratios were determined based on relations;

Lower Heating Value $$LHV = HHV - (22.604 \times H) - (2.581 \times MC) \quad (3)$$

Mineral Matter $$Mm = (1.08 \times AC) + (0.55 \times S) \quad (4)$$

Combustible to Pollutant Ratio $$\frac{C+H}{N+S} \quad (5)$$

Heating Value to Pollutant Ratio $$\frac{HHV}{N+S} \quad (6)$$

Heating Value to Combustible $$\frac{HHV}{C+H} \quad (7)$$

Fuel Ratio $$\frac{FC}{VM} \quad (8)$$

Fuel Value Index $$\frac{HHV}{AC+MC} \quad (9)$$

The atomic and fuel value ratios for the IHM and OGB coal samples are presented in Tables 2 and 4 as deduced from the Ultimate and Proximate analyses in Tables 1 and 3, respectively.

3. Results & Discussion

3.1. Elemental Analysis

The elemental composition of IHM and OGB coal is presented in Table 1 on a comparative basis with other coals in the literature [23]. As observed in Table 1, the *H, N, S* values for the coal samples examined in this study are in good agreement with values in literature. However, the *C* and *O* values are markedly different due to the typical variation in the fuel properties of different ranks of coals [10].

Table 1. Ultimate Analysis of IHM and OGB coal samples

Fuel Property	Symbol	IHM Coal (wt.%)	OGB Coal (wt.%)	Coal Values [23]
Carbon	C	46.87	37.33	62.9–86.9
Hydrogen	H	5.33	3.44	3.5–6.3
Nitrogen	N	0.66	0.81	0.5–2.9
Sulphur	S	1.50	2.27	0.2–9.8
Oxygen	O	45.64	56.15	4.4–29.9
Higher Heating Value	HHV (MJ/kg)	19.40	15.55	16.0–34.0
Lower Heating Value	LHV (MJ/kg)	18.32	14.85	**

As observed in Table 1, the *C* content in IHM is 46.87 wt. % whereas OGB is 37.33 wt.%. The *H* content is 5.33 wt. % in IHM and 3.44 wt.% in OGB coal. The *O* content in the coals is 45.64 wt.% for IHM while OGB is 56.15 wt.%. Lastly, the nitrogen (*N*) and sulphur (*S*) content for both coal samples is above 0.70 wt.% and 1.50 wt.%, respectively. The *N* and *S* content indicate that thermal conversion and utilization of IHM and OGB may be detrimental to the environment due to the potential risk of gaseous pollutant emissions. However, this can be potentially addressed by implementing clean coal technologies such as carbon capture and storage (CCS) or co-firing in biomass fuels for sustainable energy recovery and reduction of carbon footprint [24]. Comparatively, IHM contains higher proportions of *C* and *H* but lower *O*, *N*, and *S* content than OGB.

Likewise, the calorific (HHV) value of IHM (19.40 MJ/kg) is higher than OGB coal (15.55 MJ/kg). This is due to the higher proportions of the key combustible elements; *C* and *H* in IHM. The calorific value of coal is considered one of the most important factors for assessing energy recovery potential for electricity generation in power plants [25]. Furthermore, it is an important dynamic for evaluating the operational thermal efficiency [26] and engineering economics of coal-fired power plants [27]. Typically, the heating value requirement of coal in power plants ranges from 9.50 – 27.00 MJ/kg [25].

Therefore, IHM and OGB are potentially viable feedstock for energy recovery and electricity generation in future coal-fired power plants. The results also reveal that the thermal conversion of IHM coal will potentially result in higher net energy production which indicates it is of higher quality compared to OGB. Based on the ultimate analysis in Table 1, the atomic ratios of the coal samples were calculated as presented in Table 2.

Table 2. Atomic Ratios for IHM and OGB coal samples

Fuel Property	**Symbol**	**IHM Coal (wt.%)**	**OGB Coal (wt.%)**
Hydrogen/Carbon	H/C	0.11	0.09
Oxygen/Carbon	O/C	0.97	1.50
Nitrogen/Carbon	N/C	0.01	0.02
Combustible to Pollutants Ratio	C+H/N+S	24.17	13.24
Heating Value to Pollutants Ratio	HHV/N+S	8.98	5.05
Heating Value to Combustibles Ratio	HHV/C+H	0.37	0.38

The findings in Table 2, indicate that the atomic ratios H/C, O/C and NC were between 0.09 to 0.11, 0.97 to 1.50, and 0.01 to 0.02 for the coal samples. The lower ratios of IHM account for its higher calorific value (MJ/kg) compared to OGB based on the *van Krevelen* relation [28].

Furthermore, the factors; Combustible to Pollutant Elements (CPE) Ratio, Heating Value to Pollutant Elements Ratio and Heating Value to Combustible Elements Ratio of the coal samples were deduced from the ultimate analysis. The CPE ratio is a measure of combustible (carbon, hydrogen) to pollutant elements (nitrogen and sulphur) in the coal samples. However, the Heating Value to Pollutant Elements (HVP) Ratio is a measure of the calorific heating value of the coal sample to the nitrogen and sulphur.

Lastly, the Heating Value to Combustible Elements (HCE) ratio is a measure of the calorific to the combustible elements ratio. As observed, the CPE and HVP values of IHM are higher than OGB coal. However, the HCE of OGB is marginally higher than IHM coal. Overall, the results indicate that energy recovery from IHM coal will emit potentially lower pollutants.

3.2. Proximate Analysis

The proximate and calorific analyses of IHM and OGB are presented in Table 3 in terms of Moisture (MC), Volatiles (VM), Fixed Carbon (FC), and Ash (AC) content along with the calculated values for fuel ratio and mineral matter. As observed, the MC of IHM is 4.75 wt.% whereas OGB is 3.21 wt.%. The VM, FC and Ash content of IHM are 69.52, 23.30 and 2.43 wt. %, respectively. Conversely, the VM, FC and Ash content of OGB coal is 51.43, 44.41 and 1.03 wt.%, respectively.

The results indicate the average values were below 5 wt.% for MC; 70 wt.% for VM; 45 wt.% for FC and 2.5 wt.% for AC. Overall, the values are in good agreement with values typically reported in the literature [23]. In comparison, the MC, VM, AC and *Mm* of IHM are higher than OGB coal although the FC and fuel ratio of OGB is higher than IHM coal.

The Moisture Content (MC) is an important parameter for handling, storage, and transport (HST) of coal. In addition, it is an indicator of the heat energy potential or operational costs required to dry, co-fire, or convert coal in power plants [10]. Based on the results of the coals, the lower MC of OGB indicates its HST costs are potentially lower than IHM coal.

Table 3. Proximate Analysis and Calorific Values of IHM and OGB coals

Fuel Property	Symbol	IHM Coal (wt. %)	OGB Coal (wt. %)	Coal Values [23]
Moisture	MC	4.75	3.21	0.4–20.2
Volatiles	VM	69.52	51.43	12.2–44.5
Ash	AC	2.43	1.03	5.0–48.9
Fixed Carbon	FC	23.30	44.41	17.9–70.4
Mineral Matter	*Mm*	3.45	2.36	**

The Volatile Matter (VM) significantly influences the thermochemical reactivity, rank classification, and conversion efficiency of coal [28, 29]. In addition, it is a measure of condensable and non-condensable volatile compounds generated from the thermal decomposition of coal under given reaction conditions [10]. The lower VM of OGB coal also indicates better storage potential and low potential for spontaneous combustion compared to IHM. However, the higher VM of IHM is better suited for syngas production through pyrolysis and gasification technologies whereas OGB can undergo combustion for energy recovery.

The Ash Content (AC) of OGB was lower than IHM as observed in Table 3. Ash is the non-combustible residual or mineral matter produced from coal combustion. According to *Chukwu et al.*, [1], ash significantly affects waste handling, processing, and utilization equipment. In addition, the chemical composition of ash is a measure of the fouling, slagging and agglomeration potential of coal in conversion equipment [10]. Consequently, lower ash, as observed in OGB, is preferable from a cost and operational perspective.

The Fixed Carbon (FC) is the solid carbonaceous residue from drying and devolatilization typically used to estimate the coke (coking) potential and rank classification of coals [30]. Based on the findings of this study, OGB has a higher potential for coke formation and fuel ratio which is an important factor for metallurgical coal applications, energy recovery, and electric power generation.

Based on the proximate analysis, the Fuel Ratio and Fuel Value Index of the IHM and OGB coals were deduced as presented in Table 4.

Table 4. Fuel Ratios for IHM and OGB coal samples

Fuel Property	Symbol	IHM Coal (wt. %)	OGB Coal (wt. %)
Fuel Ratio	FC/VM	0.34	0.86
Fuel Value Index	HHV/AC+MC	2.70	3.67

The results indicate that OGB has a higher fuel ratio and value index compared to the IHM coal sample. This is due to the higher moisture, ash content and volatile matter content in IHM compared to OGB as presented in Table 3.

3.3. Coal Rank and Classification

The rank classification of the coals was examined according to ASTM D388 [22]. According to the standard, the coals are classified as *Lignite* based on HHVs which are less than 24 MJ/kg [10]. The sub-classification analysis indicates that IHM is *Lignitic* coal class A, whereas OGB is class B. In general, the coal samples are Brown or Low-Rank Coals (LRCs) with non-caking properties. This indicates the coals can be potentially utilized for electric power generation [10], coke blending or co-firing with biomass for energy recovery through pyrolysis, gasification, or combustion.

4. Conclusion

The paper investigated the fuel characteristics, rank classification, and potential applications of *Ihioma* (IHM) and *Ogboligbo* (OGB) coals from Nigeria. The characterization was based on ultimate, proximate, and calorific analyses. The results indicated the coal samples contain sufficient proportions of combustible elements for various energy recovery, power, and metallurgical applications. The IHM coal exhibited higher calorific properties compared to OGB. Furthermore, the fuel characterization indicated OGB has better handling, storage, and transport capabilities compared to IHM coal. However, IHM contains higher VM and HHV suitable for application in gasification and combustion applications. Furthermore, OGB has a higher fuel ratio and value index due its lower moisture, ash content and volatile matter. The rank classification indicates the coals are Lignite (Brown) Low-Rank Coals (LRCs) with potential for electricity generation or biomass co-firing for enhanced energy recovery and fuels synthesis. Lastly, the coal characteristics presented in this study can be utilized for the design, optimization, and scale-up of future conversion systems for energy recovery.

References

[1] Chukwu M, Folayan C, Pam G and Obada D. Characterization of Some Nigerian Coals for Power Generation. 2016 *Journal of Combustion*. (9728278):1-11.
[2] Höök M. Coal and Peat: global resources and future supply. Fossil Energy: Springer; 2013. p. 311-341.
[3] Bielowicz B. A new technological classification of low-rank coal on the basis of Polish deposits. 2012 *Fuel*. **96**:497-510.
[4] Ryemshak S A and Jauro A. Proximate analysis, Rheological properties and Technological applications of Nigerian coals. 2013 *International Journal of Industrial Chemistry*. **4**(1):1-7.
[5] Ohimain E I. Can Nigeria generate 30% of her Electricity from Coal by 2015. 2014 *International Journal of Energy and Power Engineering*. **3**(1):28-37.
[6] Nyakuma B. Physicochemical Characterization and Thermal Analysis of newly discovered Nigerian coals. 2016 *Bulgarian Chemical Communications*. **48**(4):746 – 52.
[7] Collot A-G. Matching Gasification Technologies to Coal properties. 2006 *International Journal of Coal Geology*. **65**(3-4):191-212.
[8] Bugge J, Kjær S and Blum R. High-efficiency coal-fired power plants development and perspectives. 2006 *Energy*. **31**(10):1437-45.
[9] OECD Working Paper. The Global Value of Coal 2012.
[10] Speight J G. The Chemistry and Technology of Coal. 3rd ed. Boca Raton, FL (USA): CRC Press - Taylor & Francis Group; 2012.
[11] Bielowicz B. New technological classification of Lignite as a basis for balanced energy management. 2010 *Mineral Resources Management*. **26**:25-39.
[12] IEA. Key Coal Trends: Statistics. IEA HQ Paris France; 2016.
[13] Leke A, Fiorini R, Dobbs R, Thompson F, Suleiman A and Wright D. Nigeria's renewal: Delivering inclusive growth in Africa's largest economy. 2014 *McKinsey Global Institute, McKinsey & Company*.
[14] Nyakuma B, Jauro A, Oladokun O, Uthman H and Abdullah T. Combustion Kinetics of Shankodi-Jangwa Coal. 2016 *Journal of Physical Science*. **27**(3):1-12.
[15] Pandey B, Agrawal M and Singh S. Assessment of air pollution around coal mining area: emphasizing on spatial distributions, seasonal variations, and heavy metals, using cluster and principal component analysis. 2014 *Atmospheric Pollution Research*. **5**(1):79-86.
[16] Franco A and Diaz A R. The future challenges for "Clean Coal Technologies": joining efficiency increase and pollutant emission control. 2009 *Energy*. **34**(3):348-54.
[17] Nyakuma B B and Jauro A. Chemical and Pyrolytic Thermogravimetric Characterization of Nigerian Bituminous Coals. 2016 *GeoScience Engineering*. **62**(3):1-5.
[18] Sambo A, editor Prospects of Coal for Power Generation in Nigeria. A paper presented at the International Workshop for the Promotion of Coal for Power Generation; 2009.
[19] Nyakuma B B and Jauro A. Physicochemical Characterization and Thermal Decomposition of Garin Maiganga Coal. 2016 *GeoScience Engineering*. **62**(3):6-11.

[20] Donahue C J and Rais E A. Proximate Analysis of Coal. 2009 *Journal of Chemical Education*. **86**(2):222-4.
[21] Basu P. Biomass Gasification and Pyrolysis: Practical Design and Theory. Burlington MA, USA: Academic Press (Elsevier); 2010.
[22] ASTM Standard D388. Standard classification of coals by rank. West Conshohocken, PA: ASTM International; 2015.
[23] Vassilev S V, Vassileva C G and Vassilev V S. Advantages and Disadvantages of Composition and Properties of Biomass in comparison with Coal: An overview. 2015 *Fuel*. **158**:330-50.
[24] Basu P, Butler J and Leon M A. Biomass co-firing options on the emission reduction and electricity generation costs in coal-fired power plants. 2011 *Renewable Energy*. **36**(1):282-8.
[25] Zactruba J. Burning Coals in Power Plants: Calorific Value and Moisture 2017 [Available from https://goo.gl/mk4pDS].
[26] Tanaka N and Wicks R. Power Generation from Coal Measuring and Reporting Efficiency Performance and CO2 Emissions. IEA, editor. Paris, France: OECD/IEA; 2010. 114 p.
[27] Burnard K and Bhattacharya S. Power generation from Coal: Ongoing Developments and Outlook. Paris, France: International Energy Agency (IEA); 2011. Report No.: 2079-2581.
[28] Speight J G. Handbook of Coal Analysis. Winefordner, J, editor. Hoboken, New Jersey, USA: John Wiley & Sons; 2015. 238 p.
[29] Smoot L D and Smith P J. Coal Combustion and Gasification: Springer Science & Business Media; 2013.
[30] Schweinfurth S P. An introduction to Coal Quality. Reston, Virginia 20192: U.S. Geological Survey, Reston, Virginia 20192; 2009.

Acknowledgment
The authors gratefully acknowledge the support of the National Metallurgical Research and Development Centre (Jos, Nigeria); the National Centre for Petroleum Research and Development (NCPRD) (Bauchi, Nigeria); and Universiti Teknologi Malaysia.

IOP Conf. Series: Materials Science and Engineering **217** (2017) 012013 doi:10.1088/1757-899X/217/1/012013

Evaluating the Energy Recovery Potential of Nigerian Coals under Non-Isothermal Thermogravimetry

Bemgba Bevan Nyakuma[1,2], Olagoke Oladokun[1,2], Aliyu Jauro[3], and Denen Damian Nyakuma[4]

[1]Centre of Hydrogen Energy, Institute of Future Energy, Universiti Teknologi Malaysia, 81310 Skudai, Johor Baru, Malaysia.

[2]Department of Chemical Engineering, Faculty of Chemical & Energy Engineering, Universiti Teknologi Malaysia, 81310 Skudai, Johor Baru, Malaysia.

[3]National Centre for Petroleum Research and Development (NCPRD), Abubakar Tafawa Balewa University, P.M.B 0248, Bauchi, Bauchi State 740272, Nigeria.

[4] Directorate of ICT, University of Calabar, P.M.B 1115, Calabar, Cross River State 540271, Nigeria.

Email: bnbevan2@live.utm.my, bbnyax1@gmail.com

Abstract. This study investigated the fuel properties and energy recovery potential of two coal samples from *Ihioma* (IHM) and *Ogboligbo* (OGB) environs in Nigeria. The ultimate, proximate, and bomb calorimetric analyses of the coal were examined. Next, the rank classification and potential application of the coals were evaluated according to the ASTM standard D388. Lastly, thermal decomposition behaviour was examined by non-isothermal thermogravimetry (TG) under pyrolysis conditions from 30 – 900 °C. The results indicated IHM and OGB contain high proportions of combustible elements for potential thermal conversion. The higher heating value (HHV) of IHM was 20.37 MJ/kg whereas OGB was 16.33 MJ/kg. TG analysis revealed 55% weight loss for OGB and 76% for IHM. The residual mass was 23% for IHM and 44% for OGB. Based on the temperature profile characteristics (TPCs); T_{on}, T_{max}, and T_{off}, IHM was more reactive than OGB due to its higher volatile matter (VM). Overall, results revealed the coals are Lignite (Brown) low-rank coals (LRCs) with potential for electric power generation.

1. Introduction

Carbonization (pyrolysis) is one of the oldest and most commonly utilized techniques for converting coal into solid, liquid, and gaseous fuels. It is also a practical approach for fuel synthesis and energy recovery processes. Furthermore, it is an important intermediate reaction in the gasification and combustion of carbonaceous fuels such as biomass and coal.

Coal is the most abundant and widely distributed fossil fuel around the globe. The carbonaceous nature of coal presents significant opportunities for conversion into energy, fuels, and chemicals. Currently, coal utilization plays an important role in the global energy mix, particularly for electricity generation. According to the IEA, global electricity production from coal accounts for 41% or 8,000 TWh although it is projected to rise to 12,000 TWh by 2035 [1]. As a result, coal utilization will significantly contribute to socioeconomic growth and development for the foreseeable future [2]. Furthermore, coal is a widely available, accessible and acceptable source of future energy with potential for poverty alleviation [3]. This is particularly important for energy exploration and exploitation in developing nations like Nigeria with vast new coal deposits [4, 5].

The coal resources in Nigeria are predominantly located in the upper, middle and lower Benue trough which extends from the SW to NE direction across the nation's sedimentary basin [6]. The sedimentary basin spans over 13 states and over 20 localities in the country. Currently, Nigeria's proven resources and reserves amount to 640 million tons and 2.8 billion tons, respectively [5, 6]. This mainly consists of Lignite (12%), Sub-Bituminous (49%) and Bituminous (39%) [7]. Consequently, coal is strategically positioned as an ideal feedstock to address the nation's energy crises [8, 9]. Furthermore, the current low demand for low-rank coals (LRC) like lignite suggests prices may remain low. Hence, the utilization of lignite for electric power generation will potentially result in cheap, reliable, and constant electric power supply in the country.

However, there is limited knowledge on the physicochemical, thermal kinetic, and thermodynamic properties of lignite coal in Nigeria. Previous studies have investigated the geochemistry [10], mineralogical [11], rheological [4], trace element [12], chemical products [13], rank and petrographic [13] properties of Nigerian coals. As a result, the reported findings in the surveyed literature only present insights into the geological, geochemical and mineralogical properties but not the thermochemical fuel properties of lignite coals required for energy recovery in Nigeria. Hence, there is an urgent need to critically examine the essential fuel properties of newly discovered lignite coals as required for application in power generation.

Therefore, the main objective of this study is to examine the physicochemical and thermal fuel properties of coals from *Ihioma* (IHM) and *Ogboligbo* (OGB) environs in Nigeria for potential energy recovery. This will be examined by ultimate, proximate, and bomb calorimetric analyses. Furthermore, the thermal decomposition behavior of the coal samples will be examined by non-isothermal thermogravimetry (TG) under pyrolysis conditions. Lastly, coal rank classification and the potential applications of the coal samples will be evaluated. It is envisioned that the coal properties will assist in the design and development of future energy recovery systems required to boost electric power generation and sustainable development in Nigeria.

2. Experimental

The rock samples of *Ihioma* (IHM) and *Ogboligbo* (OGB) coals from in Imo and Kogi States were acquired from the National Metallurgical Research and Development Centre (NMRDC) in Jos, Nigeria. The samples were subsequently pulverized and sifted using the Retsch™ analytic sieve of mesh size 60 to acquire 250 μm sized particles [14].

Next, the pulverised coals were characterized by elemental, proximate, and bomb calorific analyses. The elemental properties were examined by CHNS analysis using the vario MICRO Cube™ Elemental Analyzer (Germany). The proximate analysis was performed by thermogravimetric analysis (TGA) based on the procedures described in the literature [15].

The calorific (higher heating) value was examined by bomb calorimetry using the IKA C2000 bomb calorimeter (USA) according to the ASTM standard D2015. All tests were repeated twice to ensure the repeatability and reliability of the results. The coal rank classification was estimated based on the ASTM standard D388 as described in the literature [16].

Lastly, the thermal degradation behaviour of the coals was investigated through non-isothermal thermogravimetric (TG) analysis under inert (pyrolysis) conditions. During each run, approximately 15 mg of sample was loaded onto the Perkin Elmer TGA 4000 and heated in an alumina crucible at 10 °C/min from 30 °C to 900 °C. The gaseous species evolved during thermal decomposition were purged by nitrogen gas (N_2) at a flow rate of 50 ml/min. The procedure was aimed at simulating pyrolytic (pyrolysis) decomposition under non-isothermal heating during TGA.

3. Results and Discussion

3.1. Coal Fuel Characteristics

The ultimate, proximate, and bomb calorific analyses of IHM and OGB coals are presented in Table 1. The ultimate analysis is presented in dry ash free (*daf*) basis while the proximate analysis and calorific heating values are presented on dry basis (*db*) for comparison with coal data in the literature [17].

Table 1. Coal Fuel Characteristics

Element	Symbol	IHM Coal	OGB Coal
Carbon	C (wt.% *daf*)	50.49	38.94
Hydrogen	H (wt.% *daf*)	5.74	3.59
Nitrogen	N (wt.% *daf*)	0.71	0.84
Sulphur	S (wt.% *daf*)	1.62	2.37
Oxygen	O (wt.% *daf*)	41.44	54.25
Moisture	MC	4.75	3.12
Volatiles	VM (wt.% *db*)	72.99	53.09
Ash	AC (wt.% *db*)	2.55	1.06
Fixed Carbon	FC (wt.% *db*)	24.46	45.85
Fuel Ratio	FC/VM	0.34	0.86
Higher Heating Value	HHV (MJ/kg)	20.37	16.33
Lower Heating Value	LHV (MJ/kg)	19.20	15.60

The ultimate analysis revealed that IHM and OGB contain sufficient proportions of the combustible fuel elements required for potential energy recovery. As observed, the *C* and *H* content of IHM are significantly higher than OGB coal. Typically, the higher the *C* and *H* content, the higher the calorific value of the coal [5]. Hence, the fuel quality of IHM coal is superior to OGB in terms of potential energy recovery from thermal conversion. In comparison, the values of *H* are in agreement with values 3.50 – 6.30 wt.% for other coals in literature. However, the *C* was lower than the reported values 62.90 – 86.90 wt.% due to the maturity and rank of the coals examined [18].

Furthermore, the content of pollutant precursor elements; nitrogen (*N*) and sulphur (*S*) of IHM are lower than OGB coal. In comparison, the values were found to be in agreement with typical values of *N* (0.50 – 2.90) wt.% and *S* (0.20 – 9.80) wt.% for coals in the literature [18]. According to *Speight* [19], the presence of *N* and *S* is influential for determining potentially damaging atmospheric emissions resulting from coal energy utilization [19]. This indicates that thermal conversion of IHM coal could potentially result in lower atmospheric gaseous emissions compared to OGB.

The oxygen content of IHM is also lower than OGB. The presence of oxygen in the form of carboxyl, phenolic, or heteroatoms in the coal structure plays an important role in chemical reactivity, rank classification, and potential products from conversion [18, 19]. In comparison, the *O* content for IHM and OGB were higher than reported values (4.40 – 29.90) wt.% in the literature [18]. As stated previously, this may be attributed to the low maturity and rank of the coals. Hence, it can be surmised that IHM and OGB are low-rank coals (LRC).

The proximate analysis of the coals was also determined as presented in Table 1. The results are in good agreement with reported values for MC (0.40 – 20.20) wt.%; AC (5.00 – 48.90); FC (17.90 – 70.40) wt.% except for VM (12.20 – 44.50) wt.% [17]. This is a confirmation of the LRC status of the coal samples based on their high organic volatile matter content. Furthermore, this suggests that the coals will exhibit high reactivity comparable to biomass fuels. Hence, the coals examined in this study are potentially good fuel feedstock for thermal applications.

Lastly, the heating values of the coals are 20.37 MJ/kg for IHM and 16.33 MJ/kg for OGB. The values are in good agreement with the range 16.00 – 34.00 MJ/kg reported for coals [6, 18] and 9.50 – 27 MJ/kg required for coal-fired power plants [5, 20]. Furthermore, the HHVs are below 24 MJ/kg [3] indicating the coals can be categorized as low-rank Lignite (Brown) coal as defined by ASTM D388 [16]. Therefore, the coals can be potentially utilized for coal-fired power generation [3, 19], co-firing with biomass [21], pyrolysis or gasification into energy or chemicals [22, 23].

3.2. Thermal Degradation Characteristics

The thermogravimetric (TG) analysis of IHM and OGB coals was examined from 30 – 900 °C by heating the samples at 10 °C/min under inert atmosphere. The resulting weight loss (TGA) and derivative weight loss (DTG) curves are presented in Figures 1 and 2, respectively. The TG curves exhibit the downward sloping weight loss curves for thermally decomposition coals reported in the literature [24-28]. This indicates temperature significantly influenced the devolatilization of the coal samples under the conditions investigated.

As observed in Figure 1, the thermal decomposition mainly occurred from 30 – 550 °C. However, at temperatures above 550 °C, thermal decomposition plateaued indicating low devolatilization despite the relatively high volatile matter (VM) content. Similarly, Haykiri-Acma and Yaman [14], Sonobe *et. al.,* [21] reported low devolatilization rates for lignite coals under pyrolysis conditions. The observed trends are ascribed to the complete devolatilization of organic VM in the coal samples. In addition, the tailing observed in Figure 1 may be due to low reactivity of the coals resulting fixed carbon and ash content potential coke yield [29, 30] from coal devolatilization.

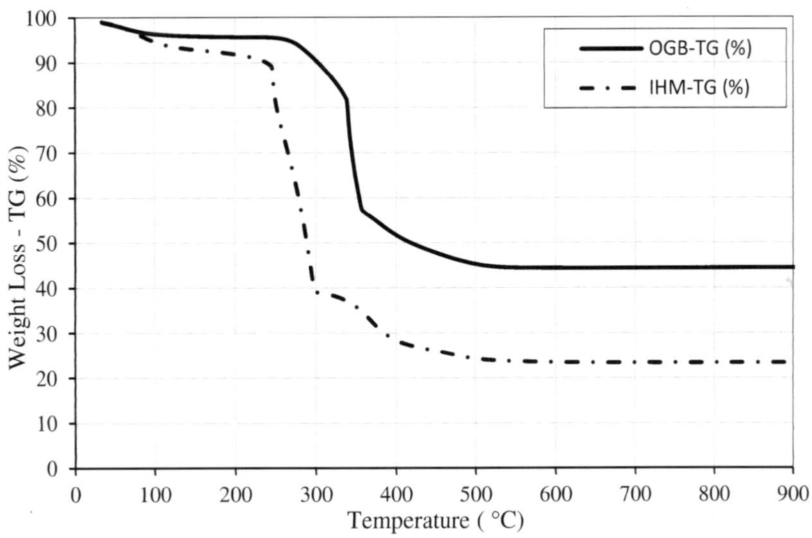

Figure 1. TG Plots for IHM and OGB coals

Likewise, the thermal decomposition behaviour of the coals was examined by derivative weight loss (DTG) analysis as presented in Figure 2. As observed in Figure 2, the DTG curves for IHM revealed three peaks (two symmetric and one asymmetric) during devolatilization. The first was a small symmetric peak from 30 – 150 °C denoting drying or loss of low molecular weight volatiles. The second was a large, sharp, asymmetric tailing peak from 205 – 300 °C denoting the first stage of devolatilization due to loss of organic volatile matter. This reveals the devolatilization of IHM began at 205 °C which is in fairly good agreement with the onset temperature, T_{on} (250 °C) for the lignite decomposition reported by *Sonobe et. al.,* [21].

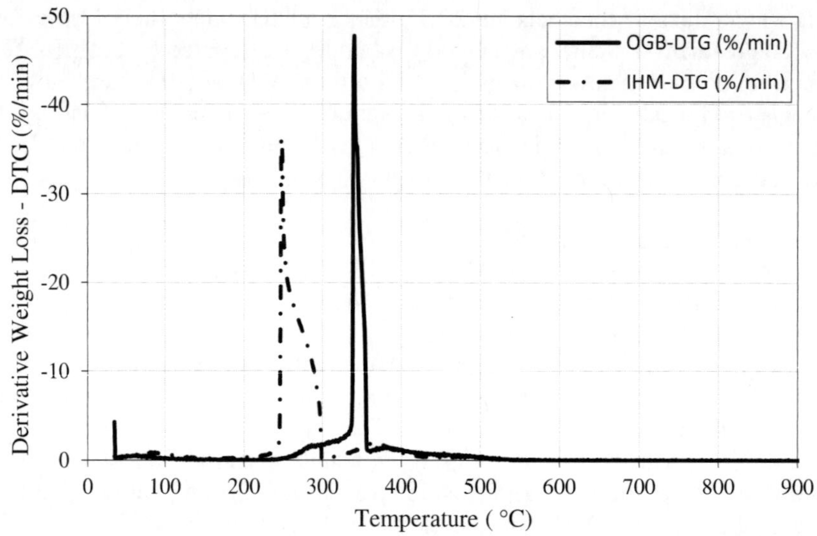

Figure 2. DTG Plots for IHM and OGB coals.

The maximum peak temperature, T_{max} for IHM was at 246.99 °C. The third and last peak was observed from 300 – 550 °C denoting the second stage of devolatilization as characterized by the offset temperature T_{off} of 550 °C. Similar results have been reported for lignite coals in the literature [17, 21]. In contrast, the DTG curves for OGB coal devolatilization revealed two major peaks (one small symmetric and a larger asymmetric peak). The first peak was from 30 – 130 °C denoting drying whereas the second larger peak was from 240 – 530 °C. The second peak can be ascribed to devolatilization or loss of organic volatile matter during TG analysis. Based on this, the onset temperature T_{on} for OGB coal was 240 °C whereas the peak decomposition temperature T_{max} was 340.24 °C and offset temperature T_{off} of 530 °C.

Based on the temperature profile characteristics T_{on}, T_{max} and T_{off}, it can be concluded that IHM is more reactive than OGB. This can be ascribed to the higher content of volatile matter (*VM*), carbon (*C*) and hydrogen (*H*) in IHM compared to OGB coal. Overall, the devolatilization of the coals indicated that thermal decomposition under pyrolysis conditions from 30 – 900 °C resulted in 54.58% weight loss (conversion) for OGB whereas IHM experienced 75.73%. The residual mass, which indicates the coke (char) potential, was 23.30% for IHM and 44.41 % for OGB, respectively, which confirms IHM is more reactive than OGB.

4. Conclusion

The study examined the energy recovery potential of newly discovered Nigerian Lignite (Brown) coals using non-isothermal thermogravimetry under pyrolysis conditions. Therefore, the physicochemical, thermal, and calorific fuel properties of IHM and OGB coals from Nigeria were characterized. The results revealed that IHM and OGB contain satisfactory combustible elements, low moisture, and ash for potential energy recovery from thermochemical conversion. Furthermore, the coal samples displayed sufficiently high heating value (HHV) from 16 – 20 MJ/kg. Based on this, the coals were categorized as Lignite (Brown) coals with potential application in coal-fired power plants or co-firing with biomass. The results for thermal decomposition indicated that coal pyrolysis resulted in 55% weight loss for OGB while IHM experienced 76% which indicates IHM is more reactive than OGB. Based on the fuel properties, the coals were classified as low-rank coals (LRC) with potential application in metallurgical or power generation applications. Overall, the study presents novel data on the physicochemical and thermal fuel properties of the newly discovered Nigerian Lignite coals. This will be vital to the design, optimization, and scale-up of future coal energy recovery applications.

References

[1] OECD Working Paper. The Global Value of Coal 2012.

[2] Sambo A, editor Prospects of coal for power generation in Nigeria. A paper presented at the International Workshop for the Promotion of Coal for Power Generation; 2009.

[3] Bielowicz B. New technological classification of Lignite as a basis for balanced energy management. 2010 *Mineral Resources Management*. **26**:25-39.

[4] Ryemshak S A and Jauro A. Proximate analysis, Rheological properties and Technological applications of some Nigerian coals. 2013 *International Journal of Industrial Chemistry*. **4**(1):1-7.

[5] Chukwu M, Folayan C, Pam G and Obada D. Characterization of Some Nigerian Coals for Power Generation. 2016 *Journal of Combustion*. (9728278):1-11.

[6] Ogala J, Siavalas G and Christanis K. Coal Petrography, Mineralogy and Geochemistry of Lignite samples from the Ogwashi–Asaba Formation, Nigeria. 2012 *Journal of African Earth Sciences*. **66**:35-45.

[7] Energy Commission of Nigeria (ECN). National Energy Policy of Nigeria (NEP). In: Policy, N, editor. Abuja, Nigeria: Energy Commission of Nigeria; 2003. p. 1-89.

[8] Nyakuma B, Jauro A, Oladokun O, Uthman H and Abdullah T. Combustion Kinetics of Shankodi-Jangwa Coal. 2016 *Journal of Physical Science*. **27**(3):1-12.

[9] Ohimain E I. Can Nigeria generate 30% of her Electricity from Coal by 2015. 2014 *International Journal of Energy and Power Engineering*. **3**(1):28-37.

[10] Ogala J E. The Geochemistry of Lignite from the Neogene Ogwashi-Asaba formation, Niger Delta basin, Southern Nigeria. 2012 *Earth Sciences Research Journal*. **16**(2):151-64.

[11] Adedosu T, Adedosu H and Adebiyi F. Geochemical and Mineralogical significance of trace metals in Benue Trough Coal, Nigeria. 2007 *Journal of Applied Sciences*. **7**:3101-5.

[12] Olobaniyi S and Ogala J, editors. Major and Trace element Characteristics of the Tertiary lignite series within the Ogwashi-Asaba Formation, Southern Nigeria. Proc 23rd Colloquium of African Geology/14th Conference of the Geological Society of Africa (Johannesburg, South Africa; 2011.

[13] Nwadinigwe C. Wax and resin characteristics of Nigeria's Lignites and Sub-Bituminous coals. 1992 *Journal of Mining and Geology*. **28**(1):75-80.

[14] Haykiri-Acma H and Yaman S. Synergy in devolatilization characteristics of Lignite and Hazelnut shell during Co-Pyrolysis. 2007 *Fuel*. **86**(3):373-80.

[15] Donahue C J and Rais E A. Proximate Analysis of Coal. 2009 *Journal of Chemical Education*. **86**(2):222-4.

[16] ASTM Standard D388. Standard classification of coals by rank. West Conshohocken, PA: ASTM International; 2015.

[17] Vassilev S V, Vassileva C G and Vassilev V S. Advantages and Disadvantages of Composition and Properties of Biomass in comparison with Coal: An overview. 2015 *Fuel*. **158**:330-50.

[18] Vassilev S V, Baxter D, Andersen L K and Vassileva C G. An overview of the Chemical Composition of Biomass. 2010 *Fuel*. **89**(5):913-33.

[19] Speight J G. The Chemistry and Technology of Coal: CRC Press; 2012.

[20] Zactruba J. Burning Coals in Power Plants: Calorific Value and Moisture World Wide Web (W3)2017 [Available from https://goo.gl/mk4pDS.

[21] Sonobe T, Worasuwannarak N and Pipatmanomai S. Synergies in Co-Pyrolysis of Thai Lignite and Corncob. 2008 *Fuel Processing Technology*. **89**(12):1371-8.

[22] Idris S S, Rahman N A, Ismail K, Alias A B, Rashid Z A and Aris M J. Investigation on the thermochemical behaviour of low-rank Malaysian coal, Oil Palm Biomass, and their blends during Pyrolysis via Thermogravimetric Analysis (TGA). 2010 *Bioresource Technology*. **101**(12):4584-92.

[23] Idris S S, Rahman N A and Ismail K. Combustion characteristics of Malaysian Oil Palm Biomass, Sub-Bituminous Coal and their respective blends via Thermogravimetric Analysis (TGA). 2012 *Bioresource Technology*. **123**:581-91.

[24] Nyakuma B. Physicochemical Characterization and Thermal Analysis of newly discovered Nigerian coals. 2016 *Bulgarian Chemical Communications*. **48**(4):746 – 52.

[25] Sonibare O, Ehinola O, Egashira R and KeanGiap L. An investigation into the Thermal Decomposition of Nigerian Coal. 2005 *Journal of Applied Sciences*. **5**(1):104-7.

[26] Nyakuma B B and Jauro A. Physicochemical Characterization and Thermal Decomposition of Garin Maiganga Coal. 2016 *GeoScience Engineering*. **62**(3):6-11.

[27] Varol M, Atimtay A, Bay B and Olgun H. Investigation of Co-Combustion characteristics of low-quality Lignite coals and Biomass with Thermogravimetric Analysis. 2010 *Thermochimica Acta*. **510**(1):195-201.

[28] Nyakuma B B and Jauro A. Chemical and Pyrolytic Thermogravimetric Characterization of Nigerian Bituminous Coals. 2016 *GeoScience Engineering*. **62**(3):1-5.

[29] Zimmerman R E. Evaluating and Testing the Coking Properties of Coal. San Francisco, California USA: Miller Freeman Publications, Inc.; 1979.

[30] Speight J G. Handbook of Coal Analysis. Winefordner, J, editor. Hoboken, New Jersey, USA: John Wiley & Sons; 2015. 238 p.

Acknowledgments

The authors gratefully acknowledge the support of the National Metallurgical Research and Development Centre (Jos, Nigeria); the National Centre for Petroleum Research and Development (NCPRD) (Bauchi, Nigeria); and Universiti Teknologi Malaysia.

Experimental Design and Analysis of Pump as Turbine for Microhydro System

M.A. Ismail, W.K. Muzammil, M. Rahman, M.W.K. Ibrahim, S. Misran
Department of Mechanical Engineering, Faculty of Engineering, University Malaysia Sabah, Malaysia

E-mail: lanz_mr@ums.edu.my

Abstract - Pump as Turbine (PAT) is typically used as electromechanical components in microhydro systems, especially by rural communities in developing countries to reduce initial capital cost. The technology is readily available and easily accessible compared to commercially available turbines. The aim of this paper is to present the experimental design and analysis of PAT for microhydro systems over a range of rotational speeds. An end suction centrifugal pump was tested by inversing the flow across the pump. The rotational speed of the impeller was controlled by manipulating the braking force applied to the output shaft. The corresponding flow rate, pressure, and torque were recorded and presented. The experiment results show that the centrifugal pump can operate in turbine modes without any modification on mechanical components with the highest efficiency of 65.04%; however, at off-design operation, the efficiency decreases significantly due to unmatched flow velocity with the wall boundaries inside the pump.

1. Introduction

Rural electrification plays a significant role in enhancing the quality of life for rural communities by promoting access to modern energy services. This leads to increased economic strength and improves productivity, consequently reducing inequality. The traditional way to supply electricity to rural areas is to extend the national grid into these areas; however, after taking into account financial viability, this approach is usually found to be uneconomical. Using microhydro technologies for off-grid electricity generation is the most suitable method whenever there is an accessible potential site. The use of appropriate technologies to suit local conditions is one of the critical factors affecting the success of an off-grid microhydro electrification system [1, 2]. Economic feasibility is one of the main considerations associated with microhydro systems. Hence, the application of low-cost equipment for the microhydro system has always been the central focus for rural electrification.

The use of Pump as Turbine (PAT) in place of commercial electromechanical components has proven to be a feasible, low-cost solution. Some successful microhydro projects have been reported for rural area sites, especially in developing countries [3-5]. There has been growing interest on utilising the PAT as a substitute for the commercial microhydro turbine, in particular for generating power between the ranges of 10 kW to 25 kW. Even though the PAT has lower efficiency than commercial turbines, its availability covers a wide variety of flows and heads, which makes microhydro projects more economical and practical. Also, the PAT offers benefits such as simple construction, easily-attainable spare parts, readily-available maintenance services, and installation that can be carried out by local

IOP Conf. Series: Materials Science and Engineering **217** (2017) 012014 doi:10.1088/1757-899X/217/1/012014

laymen. These desirable characteristics of the PAT are indeed important, as they reduce dependence on third-party professional services, which can be very expensive.

Even though there are many ways to predict the behaviour of PATs, such as simulation analysis, theoretical frameworks and prediction models, the best approach is to test the pump at a predefined operation range of flow rate and pressure. Each PAT is unique to its application; thus, it needs to be tested in its working environment. There are many prediction and simulation models; however, they are limited to certain applications and cannot definitively predict the actual performance. Therefore, the experimental results will give specific indicators of its operational performance. The outcome of the experimental work will accurately present real operational conditions from the perspective of microhydro applications [6]. Pump manufacturers normally supply performance curves of pump modes but seldom provide the turbine mode operation. The testing procedure to acquire PAT performance is similar to pump test standard, ISO 3555:1977 [7]. The characteristic of PAT is highly dependent on the shape and operation characteristic. The efficiency of PAT for low specific speeds varies between 60% to 80% [8, 9]. The effectiveness of the turbine mode is usually inferior to the pump mode.

This paper presents an experiment work which involved the design and construction of a hydraulic test rig and the turbine performance experiment procedure. The hydraulic test rig was designed and fabricated on a laboratory scale as a modular system to allow for a high degree of flexibility and adaptability. The main purpose of this study was to test the PAT to evaluate its operational abilities in a laboratory environment. Establishing the practical operation of PAT was critical to investigate its actual performance.

2. Methodology

Figure 1 shows the schematic of the hydraulic test rig that was adopted for the PAT test. The main components consist of a feed pump, PAT, test bed, PVC piping network and control valves. Bends and tees were used as the pipe connectors and fittings. All the pipes and fittings were fitted with flanged joints and connected with bolts and nuts. Two ball valves were installed in-line with the piping system as the regulation mechanism.

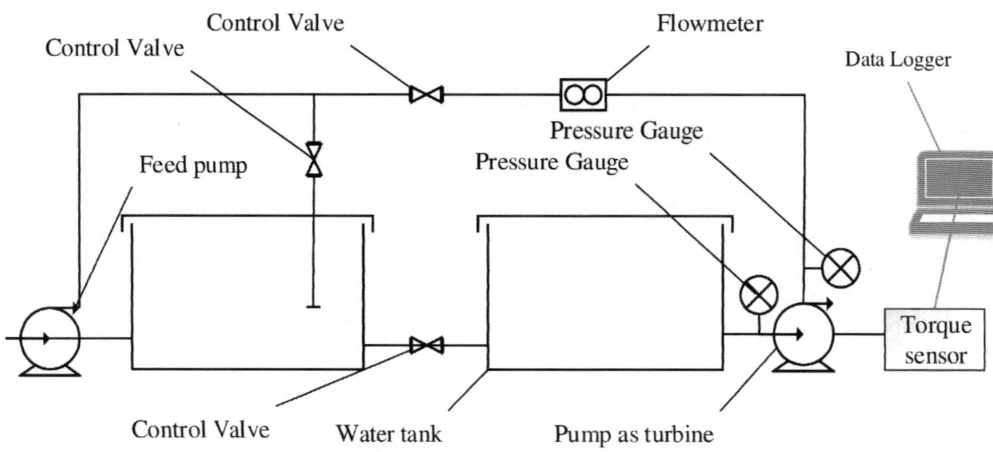

Figure 1. Schematic view of the hydraulic test rig

The operating parameters were recorded by the flowmeter, pressure gauge, and torque sensor. The flowmeter was installed in-line with the pipe, measuring the water flow rate before it entered the PAT. Two pressure gauges were positioned at each end of the PAT to measure pressure difference across the device.

The driving force behind the flow of water across the PAT was recorded as a pressure drop at the corresponding flow rate. The torque sensor was mounted in-line with the rotating shaft and measured the rotating speed and torque produced by the rotating impeller. This type of testing involved recording the torque required to maintain a constant speed.

The hydraulic test rig was an open-loop system. The inlet and outlet piping systems were joined to two water tanks that acted as a reservoir. The open-loop system allowed a continuous water supply to the piping system. The inlet of the feed pump was positioned at the lower water tank line to ensure sufficient suction pressure. This served to prevent flow separation and reduce air suction to the feed pump. The pressurised water flowed through the piping network to the PAT before being returned to the water tank.

2.1 Test Rig Main Components and Sensors

The hydraulic test rig consists of eight main components and sensors as describe below. Table 1 indicates the specifications of the feed PAT.

Table 1. Feed pump and centrifugal pump specifications

Parameters	Feed Pump	PAT
Maximum flow rate, Q (l/s)	20.0	8.0
Maximum pressure, H (m)	18.5	14.0
Suction Head (m)	7.0	4.0
Specific Speed	1453.0	1075.0

Feed pump: The feed pump that was used to supply pressurised water to the PAT was a monoblock pump with a power rating of 2.2 kW and 2900 rpm maximum speed. The pump used an induction generator with a single-phase power input. The feed pump was mounted on a large fixed frame bed and secured by a bolt connection. This kept the pump attached and stabilized thus reducing the vibration generated during the experiment. This pump will produce a maximum water pressure of 18 m at a flow rate of 20.0 l/s. The additional pressure at the pump inlet will be added to the total pressure generated at the pump outlet.

Pump as Turbine: The centrifugal pump (Euroflo brand) used as the Pump as Turbine was a low specific speed pump that delivered high pressure at a low flow rate. Its dimensions and performance standards were in accordance with the EN733:1995/DIN24255. It was manufactured locally in Malaysia. The impeller had a diameter of 214.0 mm, and the flange diameter was 50.0 mm at the inlet and 65.0 mm at the outlet. The pump's power rating was 2.2 kW, and the rotational speed was 1450 rpm. The best efficient point was at 65.0% with a corresponding pressure and flow rate of 14.0 m and 8.0 l/s, respectively. The non-dimensional specific speed (N_s) of the pump was 1075.0 units. The pump had a bare shaft coupled with a torque sensor that rotated in-line with the shaft axis.

Piping system: The pipes were made from polyvinyl chloride, uPVC (Class D, (PN12)), and complied with MS 628:1991 standards. The wall thickness was recorded at 3.0 mm, with the permissible working pressure of 12 bars at 20° Celsius. Reducers and expanders were used to connect the pipes to the pump that had a different diameter. The uPVC pipe was fitted with a set of connectors and fittings including tees, bends, and a valve with flange ends.

Water tank: Two water tanks were connected and combined by a pipe at the side wall of the tanks. One control valve was used to adjust water flow between the tanks. The combined tanks had a total volume capacity of 1800 liters. The water level was kept at a defined height to generate sufficient suction pressure at the feed pump inlet. A wooden pallet held the tank foundation with a steel frame for support.

Control valve: The feed pump did not have a control mechanism to regulate the flow and pressure generated at the pump outlet. Two control valves controlled the flow rate. The flow rate was adjusted by manipulating the opening of the control valve and diverting a portion of the flowing water from the piping system to the water tank, by passing the PAT.

Pressure gauge: The pressure gauges were placed at the inlet and outlet of the PAT. The pressure difference between the two points gave the pressure difference across the PAT. The digital pressure gauges were an Ashcroft® D1005PS with a measuring range of 0 psi to 30 psi.
Flowmeter: The real-time flow rate of the water in the PAT was recorded using a digital flowmeter. It was a propeller type flowmeter that had an NPT connector at both ends with model number TM200-N. The flow range was 76.0 to 760.0 l/min. The maximum pressure rating was 225 psi at 23^{0}C. The diameter of the flow rate was 50.0 mm, similar to the diameter of the uPVC piping system.

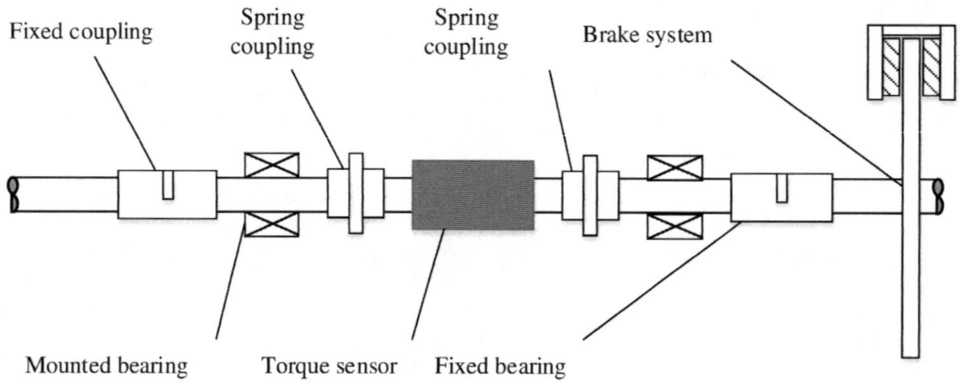

Figure 2. Schematic drawing of braking system with torque sensor and fittings

Braking system: The braking system used mechanical brakes that applied a frictional force to a rotating disc brake. A motorcycle hydraulic brake system was adapted and fixed at the end of the torque sensor using a sprocket brake hub. The brake system applied a torsional force, and the torque sensor recorded the measurement. Figure 2 illustrates the braking system with torque sensors and fittings.

2.2 Accuracy of the Performance Data

The manufacturer's data sheets give the following performance data. The percentage of error for the flowmeter is rated to be no more than +/- 3.0%. The percentage terminal point of accuracy pressure at the PAT inlet was rated to be no more than +/- 0.25%. The percentage of combined error for the torque sensor was rated to be no more than +/- 0.1%, based on the manufacturer's calibration.

2.3 General Assembly and Operation Procedure

Figure 3 shows the completed hydraulic test rig that was used to test the PAT while Figure 4 illustrates the torque sensor with the brake system installed in line with the PAT shaft. The arrangement of rig components was adapted to suit the existing hydraulic lab facilities. The rig was designed to have modular components, which would allow for a simple system that could be upgraded for future research. The hydraulic test rig was fabricated and installed in the Mechanical Laboratory of the University Malaysia Sarawak in mid-2015. The experimental analysis results were collected the following year. Intensive preliminary experiments were performed before the rigorous testing to make certain that the data collected was reliable.

The feed pump supplied a continuous flow to the piping system. The water was contained in the reservoir and was open to atmospheric pressure. A tee with a ball valve was used for flow rate control between the PAT and feed pump. The flow was adjusted by regulating the opening of the two valves by hand.

The water flow rate measurement was collected before the water entered the PAT using a propeller type flowmeter that was installed in-line with the pipe. As the water entered the PAT, the impeller rotated and produced rotational power. The impeller shaft was coupled with a torque sensor and braking system. The desired rotational speed was set by applying the brake to the shaft. During this process, the torque sensor recorded the torque and rotational speed produced by the rotating shaft. Meanwhile, the corresponding pressure across the PAT was recorded with a pressure gauge positioned at the inlet and outlet of the PAT.

Each experimental process gave a set of data consisting of flow rate Q, pressure P, torque τ and rotational speed N. Measuring the flow rate and pressure were quite complicated since pressure pulsation from the impeller caused readings to fluctuate. There was no data logger for the flow rate and pressure, so data was acquired manually. Because of the fluctuations in data readings, each experiment's results were averaged over a range of several seconds, and standard deviations were calculated and presented.

The results were presented in accordance with the given objective of investigating the performance of the PAT. Given that the sensitive parameters were interdependent, the rotational speed was the only parameter that was manipulated. The remaining parameters of pressure, torque and flow rate corresponded accordingly. The purpose of the experimental work on the PAT was to determine the efficiency, flow rate and torque over the operation rotational speed.

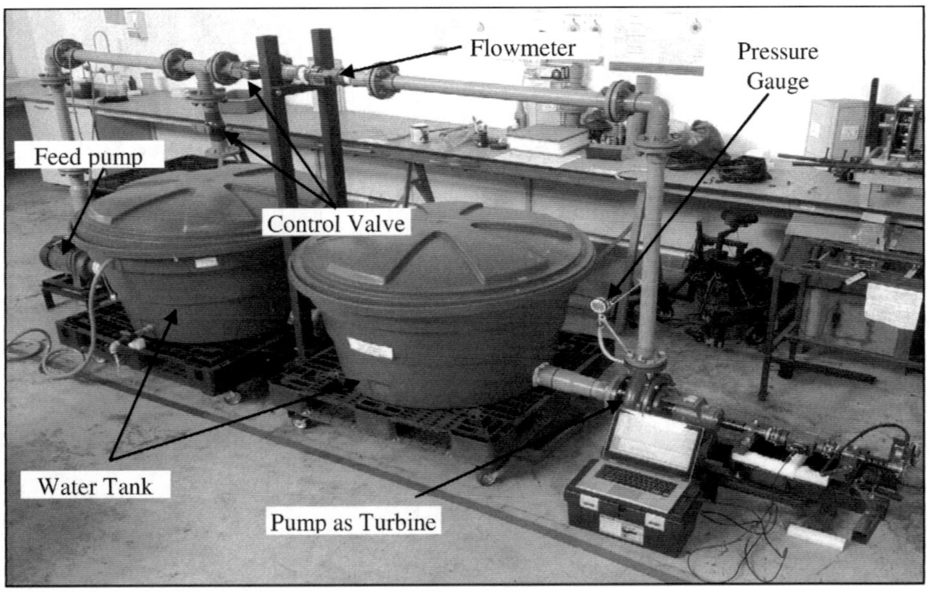

Figure 3. Complete hydraulic test rig

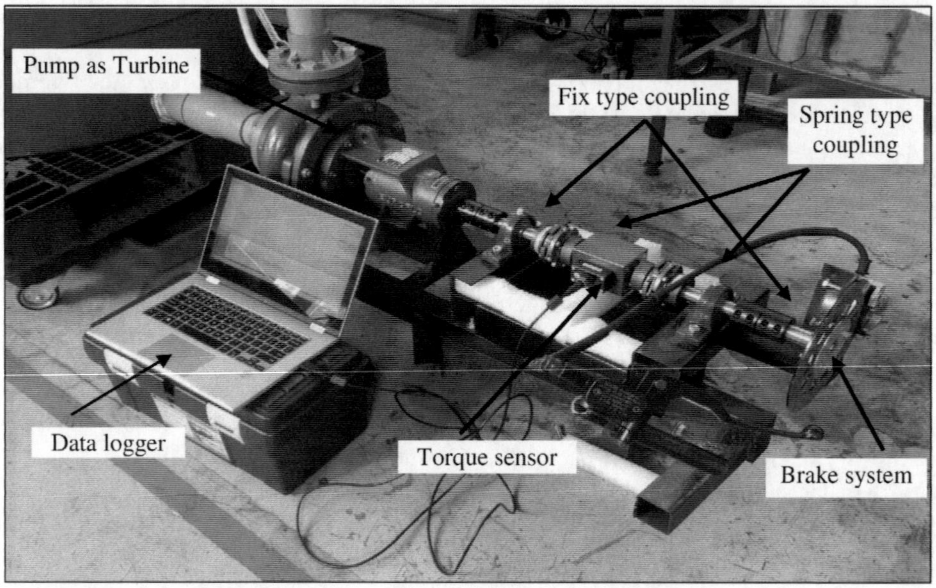

Figure 4. Hydraulic test rig brake system and data logger

2.4 Experimental Expressions

The PAT is a mechanical device that converts energy from the flowing, pressurised fluid to the impeller. This conversion of energy can be expressed in mathematical equations. Considering that the PAT impeller generates torque τ, with a rotational speed of N (rev/min), the power produced from the shaft can be expressed by Equation (2).

Mechanical power = (torque) (angular velocity)

$$P_{mechanical} = \tau \times \omega \text{ (Watt)} \tag{1}$$

$$P_{mechanical} = \frac{2\pi N \tau}{60} \text{(Watt)} \tag{2}$$

The pressure difference across the PAT is denoted by *H*. Therefore, the available hydraulic power input to the PAT is found in equation (3).

Hydraulic power = (water density) (gravitational acceleration) (flow rate) (pressure)

$$P_{hydraulic} = \rho g Q H \text{ (Watt)} \tag{3}$$

The conversion of energy consists of power losses. These power losses can be categorised as mechanical losses and hydraulic losses. The mechanical loss includes bearing and shaft seal losses, while hydraulic loss consists of disc friction loss, flow separation loss, recirculation loss, incidence loss, shock loss, mixing loss and leakage loss. The expression of total energy conversion is

Total hydraulic power input = (mechanical losses) + (hydraulic losses) + (mechanical power)

Where,

Mechanical loss = (bearing loss) + (seal loss)

Hydraulic loss = (flow friction loss) + (mixing loss) + (recirculation loss) + (incidence loss) + (disc friction loss) + (leakage loss)

The efficiency of the PAT is therefore expressed in Equation (4).

$$\eta = \frac{P_{mechanical}}{P_{hydraulic}} \times 100\,(\%) \qquad (4)$$

Likewise, the equation for efficiency can be written as shown in Equation (5).

$$\eta = \frac{\frac{2\pi N\tau}{60}}{\rho g Q H} \qquad (5)$$

3. Results and Discussions

The PAT was allowed to run continuously for 15 minutes before readings from the sensors were recorded. This allowed the piping to be filled with water and ensured that the sensors were recording steady readings. The initial information that was collected consisted of visual observations, i.e. leakages and vibrations. The results showed that the centrifugal pump was able to run in turbine mode without any mechanical failure. A small amount of vibration was observed, but was at a minimum level and was, therefore, acceptable. The vibration may have been caused by hydraulic oscillations (pressure pulsations/recirculation/rotor-stator interaction).

The rotational speed of the PAT was adjusted to between 800 rpm and 1400 rpm by adjusting the braking force applied to the rotating shaft. The corresponding pressure, flow rate, and torque were measured and documented. The mechanical power, hydraulic power, and efficiency were determined based on the observed parameters and mathematical expression of energy conversion from Equations (2), (3), and (4).

All measurements collected from the sensors were subjected to some uncertainty due to the hydraulic test rig setup limitations, simplifications of the experimental procedure and uncontrolled changes to the test environment. In the area of experimental uncertainties, the value of standard deviations from the data was presented to demonstrate how widely the measured values spread by the average of min values. The statistical analysis to determine the experimental uncertainties can illustrate the degree of accuracy and precision for the experimental results.

Tables 2, 3 and 4 illustrate the min values of flow rate, pressure, and torque with their respective standard deviations. These values were collected simultaneously at their respective operation rotational speeds.

Table 2. Mean and standard deviation values for flow rate

Rotational speed, N (rev/min)	Flow rate, Q (gal/s)	Standard deviations, σ
1400	99.52	1.52
1300	119.81	1.60
1200	125372	1.48
1100	126.62	1.82
1000	128.40	1.39
900	131.50	1.22
800	133.40	1.80

Table 3. Mean and standard deviation values for pressure

Rotational speed, N (rev/min)	Pressure, H (Psi)	Standard deviation, σ
1400	17.00	0.21
1300	14.30	0.22
1200	14.11	0.19
1100	13.90	0.25

1000	12.90	0.31
900	12.30	0.19
800	12.09	0.24

Table 4. Mean and standard deviation values for torque

Rotational speed, N (rev/min)	Torque. τ (Nm)	Standard deviations, σ
1400	-	-
1300	3.50	0.56
1200	3.70	0.63
1100	4.70	0.39
1000	4.80	0.55
900	5.10	0.66
800	6.40	0.77

Figures 5, 6 and 7 illustrate the mean and standard deviations of the respective parameters drawn from the 100 to 1000 values from the data logger for each rotational speed.

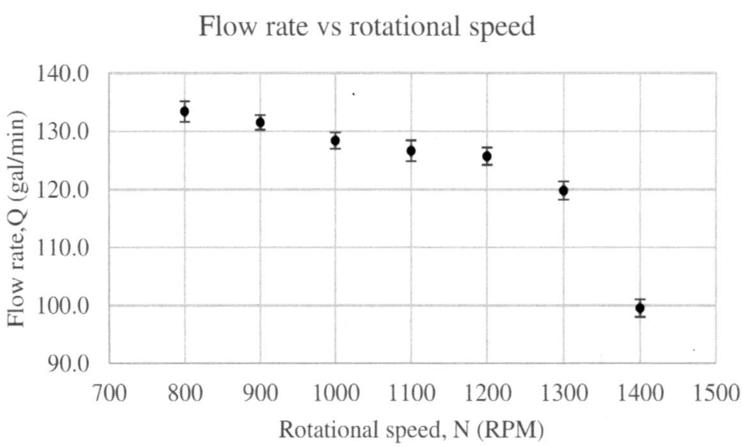

Figure 5. Mean and standard deviations of flow rate vs. rotational speed

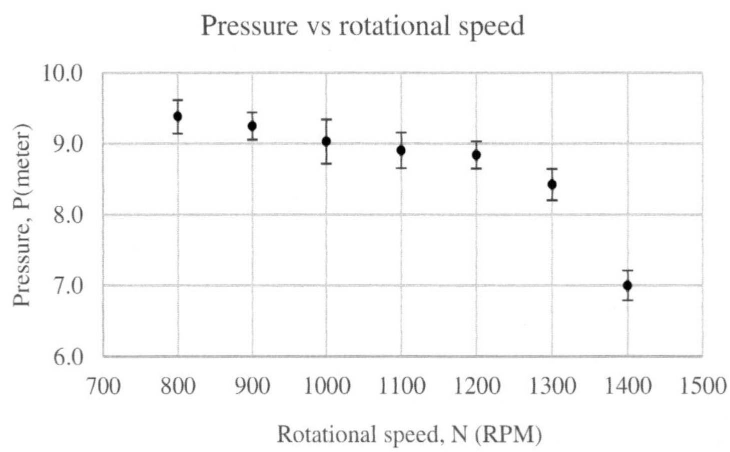

Figure 6. Mean and standard deviations of pressure vs. rotational speed

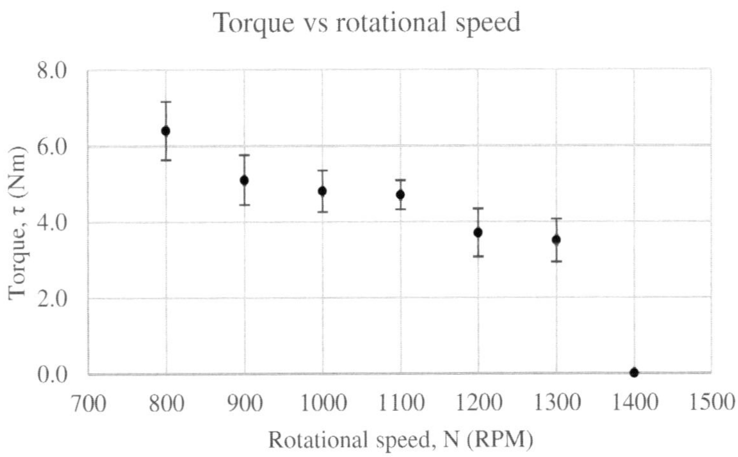

Figure 7. Mean and standard deviations of torque vs. rotational speed

Table 5. Experiment results

Rotational speed, N (rev/min)	Torque. τ (Nm)	Mechanical power, $P_{mechanical}$ (watt)	Flow rate, Q (l/s)	Pressure, H (meters)	Hydraulic Power, $P_{hydraulic}$ (watt)	Efficiency, η (%)
800	6.40	539.65	10.11	9.38	929.96	58.03
900	5.10	483.79	9.96	9.25	903.66	53.54
1000	4.80	505.92	9.73	9.03	861.56	58.72
1100	4.70	544.92	9.59	8.90	837.83	65.04
1200	3.70	467.98	9.30	8.84	806.41	58.03
1300	3.50	479.57	9.25	8.42	764.37	62.74
1400	0.00	0.00	7.54	7.00	517.58	0.00

It can be observed that the pressure and flow rate that enters the impeller steadily decreasing between 800 to 1300 RPM. The steady decline of flow rate and pressure across the PAT is because, at higher rotational speed, the water flows faster. The torque profile exerted by the impeller shows that higher torque was generated at lower rotational speed. The torque steadily reduces as the rotational speed increases. It should be noted that this is not an ideal experimental setup to achieve complete PAT performance. However, it was sufficient to test the Pump as Turbine at a predefined operation range.

Table 5 shows the experiment results. The highest rotational speed was recorded at 1400 RPM at free-running speed, which was the highest speed recorded by the torque sensor when no brake force was applied to the shaft. The lowest rotational speed has been registered at 800 rpm, generating the highest torque at 6.40 Nm.

The corresponding operational flow rate for the experimental set-up was recorded between 7.0 l/s and 10.0 l/s, similar to pump operation mode at a flow rate of 8.0 l/s. The highest efficiency (65.04%) was attained at 1100 rpm. The highest power was generated by the PAT at 800 rpm, with mechanical power of 929.96 watts and a corresponding efficiency of 58.03%. From these results, it can be determined that the torque increase as the rotational speed decrease. The efficiency varied, but not prohibitively so.

It should be noted that highest efficiency of the pump was 65.0% at 8.0 l/s and 14.0 m of pressure. Results from the experimental work showed that the efficiency was higher in turbine mode than in pump mode. However, the induction generator performance was excluded from the analysis. The overall performances normally take into account the complete component systems when considering efficiency in pump mode. Nevertheless, the mechanical efficiency of 65% is reasonable if we examine the savings from the lower acquisition cost of the turbine system.

It should be highlighted that these results are unique to the pump that was selected for the analysis. The results should be applied to other pumps with caution, and comparisons should be made considering the same specific speed, but for different sizes, shapes, and operation ratings.

4. Conclusion

This paper outlined the experimental testing of PAT performance using a hydraulic test rig. The rotational speed of the PAT was modified by adjusting the torque applied to the shaft. The corresponding pressure, torque and flow rate of the PAT system were measured using sensors and a data logger. Meanwhile, the vibration and leakages on the seal packing were observed. From the experimental results, it can be concluded that the centrifugal pump could run in turbine mode with satisfactory operational performance.

The ideal flow rate for the pump in turbine mode was higher than in pump mode; however, due to the constraints of the hydraulic test rig, the control parameters were limited to the designed test rig set-up. For future research, it is recommended that some of the system components, such as feed pump capacity and diameter of the pipe, are upgraded. This will allow a higher flow rate and greater pressure to be transferred to the pump; consequently, a higher rotational speed that matches the synchronous speed of the induction generator could be achieved, allowing full performance characteristics to be acquired.

REFERENCES

[1] C. P. Jawahar and P. A. Michael, "A Review on Turbines for Micro Hydro Power Plant," *Renewable and Sustainable Energy Reviews,* vol. 72, pp. 882-887, 2017.

[2] B. S. Solanki and A. C. Tiwari, "Review of various Aspects of using Centrifugal Pump in Turbine Modefor Small Hydro Power Generation," *International Journal of Current Engineering and Technology,* vol. 3, 2016.

[3] S. Chuenchooklin, "Development of Pico-hydropower Plant for Farming Village in Upstream Watershed, Thailand," in *Agricultural Development within the Rural-Urban Continum*, 2006.

[4] M. Arriaga, "Pump as Turbine – A Pico-Hydro Alternative in Lao People's Democratic Republic," *Renewable Energy,* vol. 35, pp. 1109-1115, 2010.

[5] M. A. Ismail, A. K. Othman, S. Islam, and H. Zen, "End Suction Centrifugal Pump Operating in Turbine Mode for Microhydro Applications," *Advances in Mechanical Engineering,* vol. 2014, p. 7, 2014.

[6] N. Patel, M. Chauhan, V. J. Sanjay, and R. N. Patel, "Experimental Investigations on Pump Running in Turbine Mode with Different Draft Tubes," in *Fluid Mechanics and Fluid Power – Contemporary Research: Proceedings of the 5th International and 41st National Conference on FMFP 2014*, A. K. Saha, D. Das, R. Srivastava, P. K. Panigrahi, and K. Muralidhar, Eds., ed New Delhi: Springer India, 2017, pp. 881-890.

[7] J. Fernández, E. Blanco, J. Parrondo, M. Stickland, and T. Scanlon, "Performance of a Centrifugal Pump Running in Inverse Mode," presented at the Proceedings of the Institution of Mechanical Engineers, 2004.

[8] H. Nautiyal, Varun, A. Kumar, and S. Yadav, "Experimental Investigation of Centrifugal Pump Working as Turbine for Small Hydropower System," *Energy Science and Technology,* vol. 1, 18 February 2011 2011.

[9] H. Nautiyal, Varun, and A. Kumar, "Reverse Running Pumps Analytical, Experimental and Computational Study: A Review," *Renewable and Sustainable Energy Reviews,* vol. 14, pp. 2059-2067, 2010.

IOP Conf. Series: Materials Science and Engineering **217** (2017) 012014 doi:10.1088/1757-899X/217/1/012014

Acknowledgement

Special thanks are given for the financial support for this work provided by Centre of Research and Innovation (UMS/PPI 1.3.2/800-4/8/303), University of Malaysia, Sabah.

IOP Conf. Series: Materials Science and Engineering **217** (2017) 012015 doi:10.1088/1757-899X/217/1/012015

Numerical investigations on axial and radial blade rubs in turbo-machinery

Ahmed M Abdelrhman[1], Eric Sang Sung Tang[1], M Salman Leong[2], Haidar F Al-Qrimli[3], G Rajamohan[1].

[1]Curtin University Malaysia, CDT 250 98009, Miri, Sarawak, Malaysia
[2]Institute of Noise and Vibration, University Technology Malaysia, Malaysia
[3]Heriot-Watt University, 62200 Putrajaya, Malaysia

Email: ahmed.mohammed@curtin.edu.my

Abstract. In the recent years, the clearance between the rotor blades and stator/casing had been getting smaller and smaller prior improving the aerodynamic efficiency of the turbomachines as demand in the engineering field. Due to the clearance reduction between the blade tip and the rotor casing and between rotor blades and stator blades, axial and radial blade rubbing could be occurred, especially at high speed resulting into complex nonlinear vibrations. The primary aim of this study is to address the blade axial rubbing phenomenon using numerical analysis of rotor system. A comparison between rubbing caused impacts of axial and radial blade rubbing and rubbing forces are also aims of this study. Tow rotor models (rotor-stator and rotor casing models) has been designed and sketched using SOILDSWORKS software. ANSYS software has been used for the simulation and the numerical analysis. The rubbing conditions were simulated at speed range of 1000rpm, 1500rpm and 2000rpm. Analysis results for axial blade rubbing showed the appearance of blade passing frequency and its multiple frequencies (1x, 2x 3x etc.) and these frequencies will more excited with increasing the rotational speed. Also, it has been observed that when the rotating speed increased, the rubbing force and the harmonics frequencies in x, y and z-direction become higher and severe. The comparison study showed that axial blade rub is more dangerous and would generate a higher vibration impacts and higher blade rubbing force than radial blade rub.

1. Introduction

In order to improve performance and aerodynamics efficiency of the turbomachines, the clearance between the rotor and stator and between the rotor and the casing has to be minimized to become smaller. Due to the smaller clearance, the occurrence of blade rubs when the rotor is contacted with stator or the casing at a certain operating speed and that will probably result in high nonlinear vibration and finally could lead to severe machine damage. Blade rubbing is known to be a very serious malfunction in turbomachinery and it's reported to be 42% from the total malfunctions in rotating machinery [1, 2]. The clearance between the rotating blades and the inner surface of the rotor casing and the stator blades has to be designed to be as smaller for better efficiency. However, this will obviously lead to blade rubbing and also the excessive vibration which could lead to the failures of blades in turbine engines. The complicated vibration generated due to the rubbing also could cause a reduction of performance in the system and shorten lives of the blades. The occurrence of the rub-impact could also lead to the high vibrational level, high temperature, light wearing rate and finally could result in blade loss. The instantaneous huge unbalance that caused by the blade loss could also lead to severe blade rubs. Therefore, the rubbing is frequently occurred in the turbine engines during the operation due to the centrifugal effects.

As reported by Zhang and Chen. [3], many researchers investigated the rotor-stator interactions using different methods such as an analytical, numerical and experimental method. For example, Zhang and Chen. [3], studied on the jump phenomena and full annular rub of Jeffcott rotor by analytical and numerical methods. They also tended to investigate the feature of the rotor response which will be affected by rotational speed and the system rotor parameter based on Routh-Hurwitz criteria. Lu and Chu. [4], proposed an improved model of a vertical Jeffcott rotor with the polar equation in order to investigate the nonlinear dynamics characteristics on radial and torsional vibration of a rotor under full annular rub, partial rub and no-rub conditions. Tai, Ma, Liu, Liu and Wen. [5], presented a lumped mass model of a single rub-impact rotor system for analysing the nonlinear steady-state response of the system depended on the size of the gap, the stiffness of stator and unbalance. Choy and Padovan. [6], investigated the characteristics of the non-linear dynamics of rubbing between rotor and casing and also the relevance of the variety of rubbing excitation parameters such as rub force and energy levels to rubbing duration and incidence separation angles. In order to understand the mechanisms of blade rubbing, Laverty. [7], studied the mechanics of rubbing between compressor blade tip seals and rotor casing. They found that the incursion rate of the rubbing is mainly contributed to the overall rubbing energy. Cao, Yang, Chen, Wang, Jiang, Li, Wei and Zhao. [8] suggested a hysteretic contact force model (HCFM) under hard coating and soft coating of the equations Yuan, Chu and Hao. [9], employed an unbalance and axial rub-impact rotor model with the derivation of Lagrangian equation to all six degrees of freedom and the ODE solvers in Matlab for numerical simulation. Zhang and Ding. [10], conducted a numerical calculation on the lateral and torsional vibration of a rotor system with axial-rub between disk and stator based on the lateral coupled equations, the torsional motion of the rotor and lateral motion of disk. Ma, Yin, Tai, Wang and Wen. [11], investigated the dynamic behavior between rotating blade and casing under rubbing condition with a constant rotational speed at the run-up process. They were noticed that the steady process was occurred at 10000rpm by considering the effects of the static misalignment, stiffness of casing, deformation of the casing and the vibration response due to the intermittent rubs. Ma, Yin, Wu, Tai and Wen. [12], simulated a bladed rotor system with blade tip rubbing. They observed the appearance of the blade passing frequency (BPF) and its multiple frequencies are shown very obvious when four blades were rubbing with the rotor casing compared to single blade rubbing condition. The current author also investigated experimentally the blade radial rub-impact in multi-stage rotor system [13-15].

The investigations on the axial dynamic behaviour of axial rub-impact seem to be relatively few. This study is intended to investigate the axial dynamic behaviour of axial rub-impact a rotor system with a single stage at different operating speeds via numerical simulation using ANSYS software. Consequently, a comparison between the axial rub-impact and radial rub-impact is also discussed in this study.

2. Simulation procedures

2.1. Models design specifications:

To study the effects of blade axial rubbing impact, vibration phenomena and the rubbing force, two rotor models (rotor-stator model and rotor casing model) were designed using 3D modeling software (SolidWorks), see figure (1). The blading system is identical in both models. For rotor-stator model, the rotor blade is designed based on the geometric properties while the stator blade is designed as the same geometric properties of a rotor blade with the additional width of 9mm and thickness of 10mm. The penetration depth of the rotor blade and stator blade and between the rotor blades tip and rotor inner casing is 6mm. The geometric properties of the rotor blade, stator blade, and rotor casing are tabulated in Table 1).

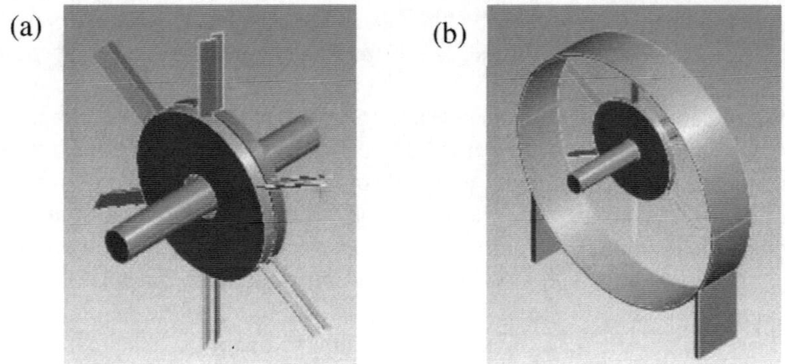

Figure 1. (a) Rotor-Stator Model, (b) Rotor-Casing Model

Table 1. Material and Geometric Parameters

Part	Material Parameter	Geometric Parameter
Shaft	Density: 7850 kg/m^3, Shear modulus: 75GPa	Length: 0.6m, Radius: 0.04m
Disk	Density: 7850 kg/m^3, Young's modulus: 200GPa, Poisson's ratio: 0.3	Radius: 0.2m, Thickness: 0.03m
Blade, n=6	Density: 7850 kg/m^3. Young's modulus: 200GPa	Length: 0.2m, Width: 0.03m, Thickness: 0.004m, Stagger angle, θ: 30^0
Casing	Density: 7850 kg/m^3, Young's modulus: 200GPa, Poisson's ratio: 0.3	Inner Radius: 0.5m, Outer Radius: 0.51m, Length: 0.2m, Length of support: 0.3m

2.2. Meshing Independence

Meshing is an important part in any computer simulation. Meshing is creating of some grid-points "nodes" connected to another nodes "element". In order to acquire good solution results, the more number of elements or nodes created the higher accuracy of the solution could be achieved in ANSYS. In this study, meshing independence is applied by altering the element sizes for the validation with Yang's model by comparing the lowest absolute error between the present model and Yang's model. Therefore, 0.3m, 0.002m and 0.006m of element sizes are adopted, as shown in Figure 2.

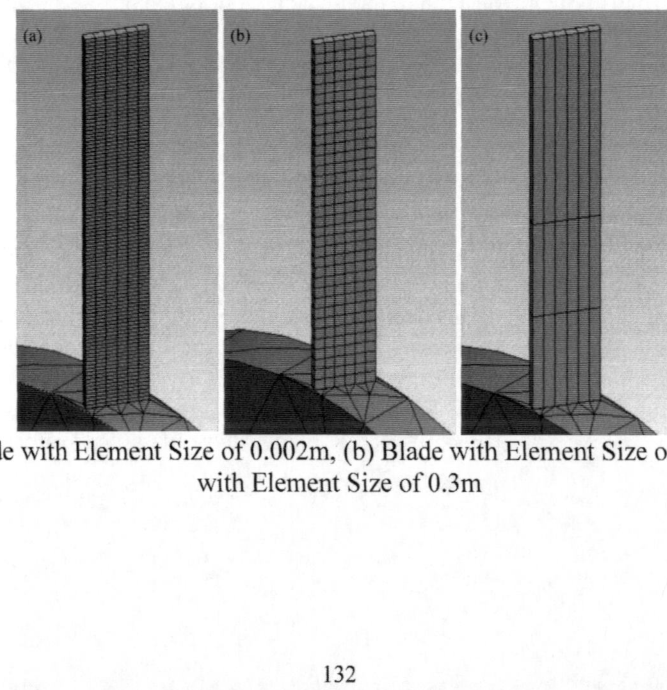

Figure 2, (a) Blade with Element Size of 0.002m, (b) Blade with Element Size of 0.006m, (c) Blade with Element Size of 0.3m

IOP Conf. Series: Materials Science and Engineering **217** (2017) 012015 doi:10.1088/1757-899X/217/1/012015

2.3. Boundary Condition
In the analysis of rotor-stator model by ANSYS, the stationary part (stator) was set to be fixed at all the directions which the translation in X, Y and Z directions were arrested in order to prevent the rotation in all degrees of freedom. A rotational speed of 1000rpm, 1500rpm and 2000rpm was applied respectively on the z-axis of the shaft of the rotor system. The occurrence of the axial rubbing between the rotor blade and stator blade is occurring when the rotor starts rotating. Meanwhile, for the rotor-casing model, the frictional coefficient is assumed as 0.246 and a rotational speed of 1000rpm was applied on the z-axis of the shaft. The occurrence of the radial rubbing between the blade tips and the rotor inner casing is simulated and investigated and its impacts will be compared with blade axial rub impacts. The total simulation time id set to be 1.2 second, the run-up process starts from t = 0 to 0.2s and run-down process start from t= 1s to 1.2s while the constant rotational speed is from t = 0.2s to t = 1s.

2.4. Contact analysis
The contact region has to establish the relationship between the surfaces of the two contacted objects in order to avoid them just passing through each other without any effect. In this study, both models are simulated using transient structural analysis in ANSYS 15 Workbench. The type of contact applied between the rotor blade and stator blade is frictionless contact while the contact between the blade tip and the inner casing is frictional contact type. The frictionless contact is applied for the rotor-stator model as the contact between the rotor blade and stator blade is far open with the penetration depth. Meanwhile, the frictional contact is applied for the rotor-casing model as the contact between the inner casing and the tip of the rotor blade is closed with penetration depth.

2.5. Unstressed Modal and Pre-stressed Modal Analysis
Modal Analysis was conducted in ANSYS Workbench 15.0 for both unstressed and pre-stressed models. For unstressed modal analysis, the stress does not taken into account for the analysis while for the pre-stressed modal analysis, there is the stress initialization due to the centrifugal force during the rotation of the rotor blade. A constant centrifugal force will be experienced by an object when it is rotating with a constant angular velocity given by equation:

$$F = m\omega^2 r \qquad (1)$$

where F is the centrifugal force, m is the mass of the spinning object, ω is the angular velocity and r is the radial distance from the axis of rotation. In this study, the rotational speed of the rotor is set to be 1000rpm and 1500rpm and 2000rpm respectively. It is believed that the frequency response of the system can be altered by the stress stiffening where the natural frequencies changed with the speed of the rotor. Presence of unexpected resonances is expected when the natural frequencies changed with the high speed of the rotor system.

Table 2. Unstressed Modal Analysis

Unstressed Modal Analysis				
Mode	**Stator**	**Vibrational Mode**	**Rotor**	**Vibrational Mode**
1	78.016	First Flap-wise (blade)	81.065	First Flap-wise (blade)
2	78.032	First Flap-wise (blade)	81.369	First Flap-wise (blade)
3	78.045	First Flap-wise (blade)	81.384	First Flap-wise (blade)
4	78.061	First Flap-wise (blade)	81.405	First Flap-wise (blade)
5	78.083	First Flap-wise (blade)	81.427	First Flap-wise (blade)
6	78.101	First Flap-wise (blade)	81.449	First Flap-wise (blade)
7	492.93	Second Flap-wise (blade)	283.36	First torsional (disk) + First flap-wise (blade)
8	493.04	Second Flap-wise (blade)	460.4	First Bending (shaft) + Second flap-wise (blade)
9	493.09	Second Flap-wise (blade)	461.5	First Bending (shaft) + Second flap-wise (blade)
10	493.4	Second Flap-wise (blade)	481.65	First Bending (shaft) + Third flap-wise (blade)
11	493.53	Second Flap-wise (blade)	482.58	First Bending (shaft) + Third flap-wise (blade)
12	493.9	Second Flap-wise (blade)	507.74	Forth Flap-wise (blade)
13	572.21	First Edgewise (blade)	507.94	Forth Flap-wise (blade)
14	573.46	First Edgewise (blade)	508.79	Forth Flap-wise (blade)
15	578.61	First Edgewise (blade)	508.83	Forth Flap-wise (blade)
16	581.01	First Edgewise (blade)	519.12	Fifth Flap-wise (blade)
17	582.91	First Edgewise (blade)	519.41	Fifth Flap-wise (blade)
18	596.7	First Edgewise (blade)	545.38	First Edgewise (blade)
19	764.84	First torsional (blade)	546.17	Frist Edgewise (blade)

Table 3. Pre-stressed Modal Analysis

Pre-stressed Modal Analysis				
Mode	**Natural Frequency of Rotor (Hz)**			**Vibrational Mode**
	1000rpm	**1500rpm**	**2000rpm**	
1	86.562	94.741	104.74	First Flap-wise (blade)
2	86.816	95.071	105.19	First Flap-wise (blade)
3	86.937	95.189	105.31	First Flap-wise (blade)
4	87.038	95.298	105.43	First Flap-wise (blade)
5	87.078	95.34	105.47	First Flap-wise (blade)
6	87.343	95.603	105.73	First Flap-wise (blade)
7	277.42	277.81	278.33	First torsional (rotor) + First Flap wise (blade)
8	438.43	438.66	438.93	First Bending (shaft) + First Edgewise (blade) + Second Flap-wise (blade)
9	440.06	440.30	440.56	First Shaft Bending + First Edgewise + Second Flap-wise
10	467.66	469.35	471.17	First Shaft Bending + Third Flap-wise
11	469.64	471.38	473.24	First Shaft Bending + Third Flap-wise
12	504.6	513.69	524.42	Third Flap-wise (blade)
13	504.71	514.86	527.78	Third Flap-wise (blade)
14	505.79	515.83	528.48	Third Flap-wise (blade)
15	506.8	517.05	531.17	Third Flap-wise (blade)
16	513.68	522.95	536.06	Forth Flap-wise (blade)
17	514.87	524.13	537.21	Forth Flap-wise (blade)
18	538.00	539.59	542.72	First Edgewise + First Flap-wise (blade)
19	538.29	539.94	543.28	First Edgewise + First Flap-wise (blade)

The mode shapes and natural frequencies of both unstressed and pre-stressed models are tabulated in Table 2) and Table 3) respectively, with the description for each mode shape of the structure at 1000rpm 1500rpm and 2000rpm.

3. Results and Discussions

3.1. Observations on Dynamic Behavior of rotor-stator model at 1000rpm

Figure *3*3(a), (b) and (c) showed a presence of smaller vibration excitation when the collision (rub) between stator blade and rotor blade occurred during the run-up process where t = 0 to 0.2 seconds. After t = 0.2 seconds, the displacement pattern changed where the rotor was whirling at a constant rotational speed of 1000rpm. The amplitude of the vibration is much higher compare to the vibration at run-up process. It was observed that the blade has undergone resonance in a flap-wise vibrational mode in the x-direction, bending vibration mode in the y-direction and edgewise vibrational mode in the z-direction. It is possible that the occurrence of resonance could lead to large deflection on the blade when the collision (rub) excite more vibration. The rubbing force in Figure 3 (d), when the collision between the rotor blade and stator blade occurred and when the 6 rotor blades with the stator blades. During the run-up process at t = 0 to 0.2 seconds, one of the six blades was experienced a higher level of rubbing level of 2548.5N and 3042.5N. The reason is that the stator blade was rubbing with rotor blade when the penetration depth was changed to a higher depth more than the initial penetration depth of 6mm due to the vibration of the blade. The rotor rubs against the stator would excite the vibration amplitude. The highest level of rubbing force was found to be 4128N during the run-down process.

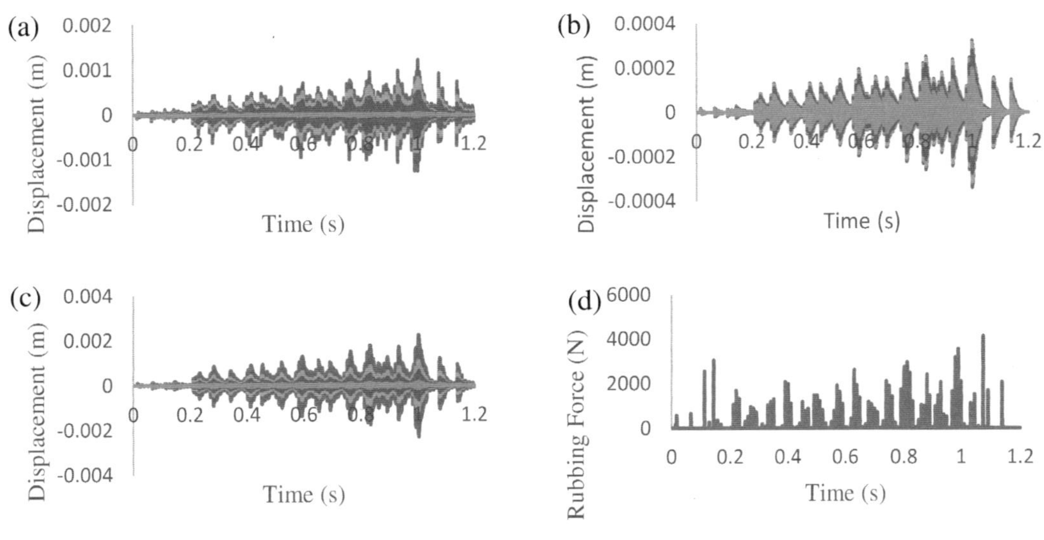

Figure 3. (a) Displacement of Stator Blade against Time (1000rpm, x-direction), (b) Displacement of Stator Blade against Time (1000rpm, y-direction), (c) Displacement of Stator Blade against Time (1000rpm, z-direction), (d) Rubbing Force at 1000rpm

3.2. Observations on Dynamic Behavior of rotor-stator model at 1500rpm

At running speed of 1500 rpm, during the run-up process, the stator blade has undergoes to a steady-state vibration when the rubbing between the stator blade and 6 rotor blades occurred. There is no presence of higher amplitude of excited vibrations showed to coincide with any of the natural frequencies as shown in Figure 4(a), (b) and (c). The pattern of the vibration has changed after t = 0.2 seconds and the rotor was whiling at a constant rotational speed of 1500rpm. It was observed that the vibration's features of the stator blade were identical with the vibration's features of the stator blades at the rotational speed of 1500rpm. During the run-down process (t= 1.0 to 1.2 seconds) the rotor vibration was gradually decreased the whirling speed, the excited vibration of the stator blade was gradually decreased and there was no any resonance occurred by coinciding with excitation resonant frequencies during rubbing. By comparing the features of vibrations in all directions, the stator blade at 1500rpm was experienced a higher vibration (displacement) than the stator blade at 1000rpm during axial rubbing condition. Therefore, it could be concluded that the vibration of the stator blade will become severe when the rotational speed increased. (Ma, Yin, et al 2016).Besides, less rubbing forces were recorded during the run-up process and the run-down process according to Figure 4(d). This proved that the penetration depth did not alter much compare to the initial penetration depth of 6mm and only the occurrence of higher rubbing levels at constant rotational speed. It was possible that the larger vibration of the blade was due to the occurrence of resonance at t = 0.2 seconds to t = 1 seconds and also caused the alteration of penetration depth between the stator blade and rotor blade where a stator blade was rub with six rotor blade during whirling motion of the rotor. During rotational speed of 1500rpm, some of the blades were experienced the intermittent rubs but some did not due to the changed of penetration depth.

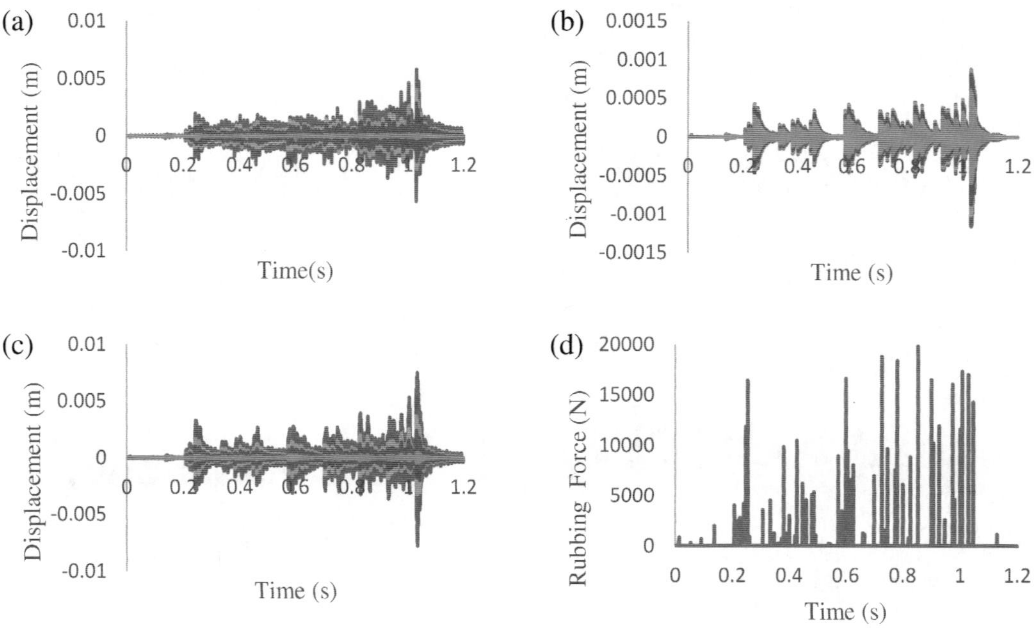

Figure 4. (a) Displacement of Stator Blade against Time (1500rpm, x-direction), (b) Displacement of Stator Blade against Time (1500rpm, y-direction), (c) Displacement of Stator Blade against Time (1500rpm, z-direction), (d) Rubbing Force under 1500rpm

3.3. Frequency Response of vibration at 1000rpm

The frequency response of the rotor system is obtained by Fast Fourier Transform (FFT) analysis to study the characteristic behaviour of the rotor system when rubbing event occurred in the rotor system. Figure 55(a) shows the healthy data of the measured vibration of the blade where no rubbing events. The fluctuation of the complicated non-linear frequency was not occurred due to no rubs events and there was only the appearance of the rotating speed frequency (16.67Hz) and its multiple frequencies

(6fr = 100Hz) which are also known as the blade passing frequency (BPF = 100). The blade passing frequency only occurs in real condition for healthy data of the rotor-stator model. Based on Figure **5**5(b), (c) and (d), which shows the frequency response of the rotor system at x, y and z directions. Figure 5(b, c and d) presenting the axial rubbing conditions at x, y and z-direction, respectively. That the appearance of the amplitudes of blade passing frequency (6fr, BPF) and its multiple frequency in x, y and z-direction such as BPF, 2BPF, 3BPF, 4BPF, 5BPF, 6BPF, 7BPF, 8BPF, 9BPF, 10BPF and 11BPF, i.e., 6fr, 12fr, 18fr, 24fr, 30fr, 36fr, 42fr, 48fr, 54fr, 60fr and 66fr. The vibration of the blade was excited due to the intermittent rubs. One of the multiple frequency (30fr, 5BPF) in x, y and z-direction was coincided with the Third flap-wise natural frequency of the rotor blade at 1000rpm based on the pre-stressed modal analysis in Table 3 where the natural frequency of 504.6Hz in mode 12 is nearly near to the blade passing frequency (30fr, 5BPF). The coincidence between natural frequency of rotor blade and blade passing frequency was showed to be more significant in the z-direction (axial rubbing condition) which was obtained the highest amplitude of 0.29mm. Meanwhile, the frequency of 77.5 Hz with the amplitudes of 54μm and 29.8μm appeared in x and z-direction respectively but it was observed that it did not coincide with any of natural frequency.

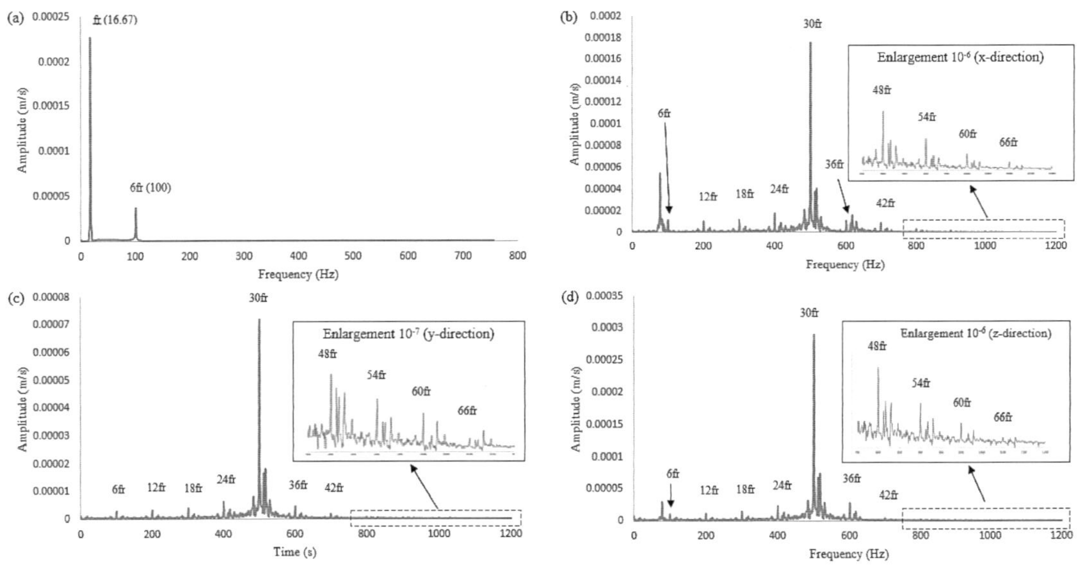

Figure 5. (a) Frequency Response at 1000rpm (Healthy Data), (b) Frequency Response at 1000rpm (x-direction), (c) Frequency Response at 1000rpm (y-direction), (d)Frequency Response at 1000rpm (z-direction).

3.4. Frequency Response of vibration at 1500rpm

Figure 6 illustrate the frequency response of the rotor system at 1500rpm with axial rub induced. Figure **6**6(b) and (d) shows that the appearance of the amplitudes of blade passing frequency (6fr, BPF) and its multiple frequencies in x, and z-direction such as BPF, 2BPF, 3BPF, 4BPF, 5BPF, 6BPF, 7BPF, 8BPF, 9BPF, 10BPF and 11BPF, i.e., 6fr, 12fr, 18fr, 24fr, 30fr, 36fr, 42fr and 48fr. One of the multiple frequencies (78Hz) was coincided with the First flap-wise natural frequency of the stator blade based on the unstressed modal analysis where the natural frequency of 78.016Hz is nearly near to the multiple frequencies (78Hz). Meanwhile, the frequency of 513 Hz with the amplitudes of 151μm but it was observed that it did not coincide with any of natural frequencies. Consequently, the coincidence was occurred between the one of the multiple frequencies (513Hz) and the natural frequency of rotor blade at 1500rpm with high amplitude of 63.3μm in y direction in Figure *66*(c) based on the pre-stressed modal analysis with the effect of centrifugal where the Third Flap-wise natural frequency of 513.69 Hz in mode 12 is nearly closed to the frequency of 513Hz. Subsequently, the severe resonance was occurred in the frequency response in the z-direction in Figure *66*(d) due to the two high amplitudes of 169μm and 256μm under rubbing condition. The frequency of 78Hz and 513Hz were coincided with the First Flap wise and Forth Flap-wise natural frequency respectively based on the modal analysis in Table 2 and Table 3.

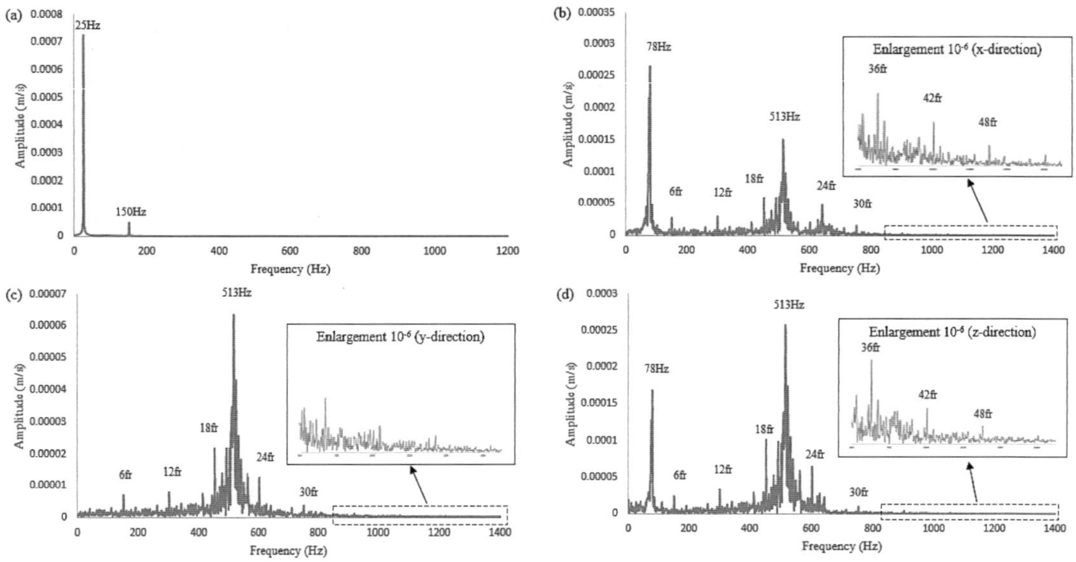

Figure 6. (a) Frequency Response at 1500rpm (Healthy Data), (b) Frequency Response at 1500rpm (x-direction), (c) Frequency Response at 1500rpm (y-direction), (d)Frequency Response at 1500rpm (z-direction).

3.5. Frequency Response of Vibration at 2000rpm

Figure *77* shows the frequency responses of the rotor system at a rotational speed of 2000rpm. It showed to be much complex compared to the frequency response at 1000rpm and 1500rpm. The blade passing frequency of 200Hz at 2000rpm and its multiple frequencies appeared and it was strongly proved that the amplitude of blade passing frequency of 200Hz was much higher due to the number of rotate cycles for the intermittent rubs between the stator blade and the six rotor blades.

Besides, the frequency of 84Hz with the amplitudes of 392μm and 155μm in x and z-direction respectively were coincided with the natural frequency of rotor blade where the frequency of 84Hz was nearly closed to the First Flap-wise natural frequency of 81 Hz according to the unstressed modal analysis in Table 2. Meanwhile, it was clearly that the frequency of 547Hz with the amplitudes of 52.5μm and 213μm in y and z-direction respectively were coincided with the natural frequency of rotor blade as the frequency of 547Hz was nearly closed to the First Edgewise natural frequency of 546.17 Hz at 2000rpm based on the unstressed modal analysis. Furthermore, the blade passing frequency in all direction at 2000rpm will be excited by increasing the rotational speed compared to the blade passing frequency at 1000rpm and 1500rpm.

Figure *77* also clearly illustrated that the occurrence of severe resonance due to the appearance of high amplitudes which is more than the resonance happened at the rotational speed of 1000rpm and

1500rpm. The more the number of rotate cycles, the more the interaction between the rotor blade and stator blade.

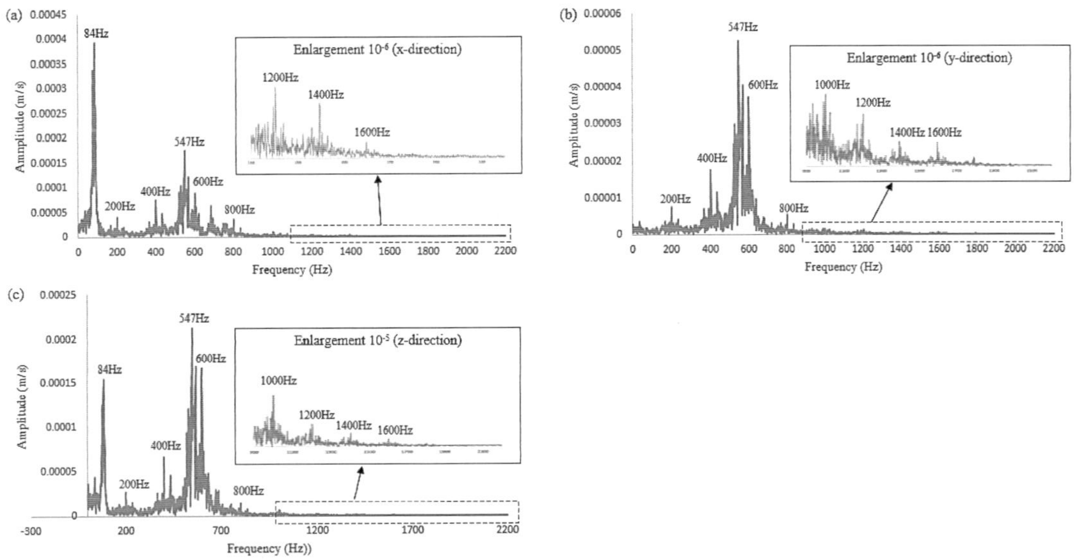

Figure 7. (a) Frequency Response at 2000rpm (x-direction), (c) Frequency Response at 2000rpm (y-direction), and (d) Frequency Response at 2000 rpm (z-direction)

3.6. Comparison between Axial and radial Rub Frequency's Response

Figure 8 shows the frequency response of the rotor system with radial blade rub induced fault at 1000rpm. It is clear that the blade passing frequency of 100Hz with the amplitude of 0.154μm and its multiple frequencies of 300Hz and 400Hz significantly appeared in y direction among x and z directions. This is possibly due to the fact that the occurrence of radial rub-impact between the tip of the rotor blade and inner casing happens in the y direction. Meanwhile, rotating frequencies of 16Hz, 40Hz and 31Hz with high amplitudes of 0.814μm, 2.96μm and 0.361μm respectively were coincided with the natural frequency of the casing due to the rotating frequencies during rubbing events were nearly closed to the natural frequency of casing based on the modal analysis of casing in Table 3. By comparing the frequency response of radial rub-impact in all direction figure (7) with the frequency response of axial rub-impact in (Figure 55) at rotating speed of 1000rpm, the appearance of blade passing frequency of 100Hz with the amplitude of 0.00238μm and 0.154μm in x and y-direction respectively from the casing when the six rotor blade were contacted with the inner casing. But the blade passing frequency did not appear in the frequency response in the z direction. This is because the appearance of blade passing frequency was more significantly in y-direction when the tip of the rotor blade rub with casing in the y direction. However, it was observed that the amplitudes of blade passing frequencies for radial rub-impact were relatively smaller than amplitudes of the blade passing frequencies for axial rub-impact at the same operating speed and simulation conditions. This due to the effect of the casing which might cause less vibration. Some of the vibration caused by the rotor blade by collision (rub) might be neutralized by the overall of the casing that made up of steel material. Besides, by comparing the rubbing force in Figure 8 for radial rub-impact with axial rub-impact in Figure 55(d), it was observed that the highest rubbing force occurred is only 4.75N only between the rotor blade and inner upper casing compared to the axial rub-impact. For the axial rub-impact, the highest rubbing force of 4128N is recorded between stator blade and rotor blade.

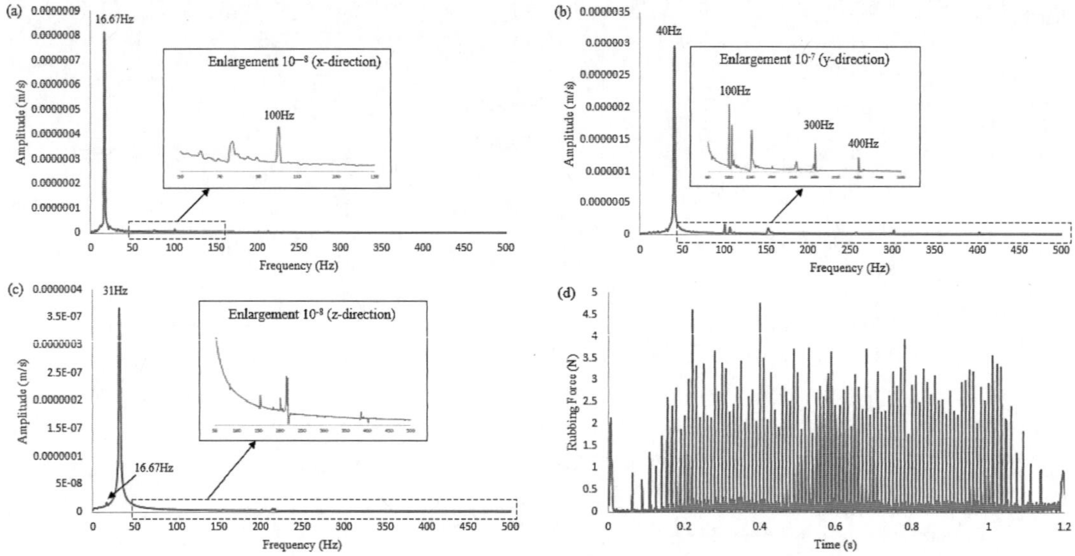

Figure 8. (a) Frequency Response at 1000rpm in Radial Rub Impact (x-direction), (b) Frequency Response at 1000rpm in Radial Rub Impact (y-direction), (c) Frequency Response at 1000rpm in Radial Rub Impact (z-direction), (d)Rubbing Force.

4. Conclusion

In this study, vibration response of axial and radial blade rubbing conditions in a rotor system were investigated using numerical simulation in ANSYS Software. The vibration response in all direction of axial rubbing was investigated at the rotational speed of 1000rpm, 1500rpm and 2000rpm. The rotor frequency response was obtained using Fast Fourier Transform (FFT) for the further investigation of the vibration's characteristics during the rubbing events. The blade passing frequency (BPF) and its multiple harmonic frequencies obviously appeared in x, y, and z-direction when the 6 rotor blades were rubbing with a stator blade 1 in the axial direction at the rotational speed of 1000rpm, 1500rpm and 2000rpm. With increasing the intermittent rubs by increasing the rotational speed, the blade passing frequency would be excited and also increased with high amplitudes. By comparing the frequency response at 1000rpm and 1500rpm with 2000rpm for blade axial rub it could be concluded that the more severe resonances with high amplitudes occurred at the higher rotational speed (2000rpm). Furthermore, it was clear that there was a difference between the axial rub-impact and radial rub-impact. The rub between blade to blade in the axial direction is showed to be more severe than the rub between the blade tips to the inner surface of the casing after the vibration response of the rotor-stator model and rotor-casing model was investigated and compared. The rubbing forces generated due to the blades axial rubs and blades radial rubs was also investigated and evaluated in this study and it was found that the blades axial rubs would generate a higher rubbing force than blades axial rubs, which may lead to fast machine failure compared to radial blade rubbing.

References

[1] Abdelrhman A M, Hee L M, Leong M and Al-Obaidi S. 2014. Condition Monitoring of Blade in Turbomachinery: A Review. *Advances in Mechanical Engineering*.**6**. 210717

[2] Abdelrhman A M, Leong M S, Saeed S A M and Al Obiadi S M. 2012 A review of vibration monitoring as a diagnostic tool for turbine blade faults. In: *Applied Mechanics and Materials,* Trans Tech Publ). pp. 1459-63

[3] Zhang H and Chen Y. 2011. Bifurcation analysis on full annular rub of a nonlinear rotor system. *Science China Technological Sciences*.**54**. 1977-85

[4] Lu W and Chu F. 2014. Radial and torsional vibration characteristics of a rub rotor. *Nonlinear Dynamics*.**76**. 529-49

[5] Tai X, Ma H, Liu F, Liu Y and Wen B. 2015. Stability and steady-state response analysis of a single rub-impact rotor system. *Archive of Applied Mechanics*.**85**. 133-48

[6] Choy F and Padovan J. 1987. Non-linear transient analysis of rotor-casing rub events. *Journal of Sound and Vibration*.**113**. 529-45

[7] Laverty W. 1982. Rub energetics of compressor blade tip seals. *Wear*.**75**. 1-20

[8] Cao D, Yang Y, Chen H, Wang D, Jiang G, Li C, Wei J and Zhao K. 2016. A novel contact force model for the impact analysis of structures with coating and its experimental verification. *Mechanical Systems and Signal Processing*.**70**. 1056-72

[9] Yuan Z, Chu F and Hao R. 2007. Simulation of rotor's axial rub-impact in full degrees of freedom. *Mechanism and Machine Theory*.**42**. 763-75

[10] Zhang K and Ding Q. 2009. Lateral and torsional vibrations of a two-disk rotor-stator system with axial contact/rubs. *International Journal of Applied Mechanics*.**1**. 305-26

[11] Ma H, Yin F, Tai X, Wang D and Wen B. 2016. Vibration response analysis caused by rubbing between rotating blade and casing. *Journal of mechanical science and technology*.**30**. 1983-95

[12] Ma H, Yin F, Wu Z, Tai X and Wen B. 2016. Nonlinear vibration response analysis of a rotor-blade system with blade-tip rubbing. *Nonlinear Dynamics*.**84**. 1225-58

[13] Abdelrhman A M, Leong M S, Hee L M and Ngui W K. 2013 Application of wavelet analysis in blade faults diagnosis for multi-stages rotor system. In: *Applied Mechanics and Materials,* Trans Tech Publ). pp. 959-64

[14] Abdelrhman A M, Leong M S, Hamdan Y M and Hui K H. 2015 Time Frequency Analysis for Blade Rub Detection in Multi Stage Rotor System. In: *Applied Mechanics and Materials,* Trans Tech Publ). pp. 95-9

[15] Abdelrhman A M, Leong M S, Hee L M and Hui K H. 2014 Vibration analysis of multi stages rotor for blade faults diagnosis. In: *Advanced Materials Research,* Trans Tech Publ). pp. 133-7

A techno-economic assessment of grid connected photovoltaic system for hospital building in Malaysia

Normazlina Mat Isa[1, 2], Chee Wei Tan[1], AHM Yatim[2],

[1]Department of Electrical Engineering, Polytechnic Merlimau, KB1031, Merlimau, Melaka
[2]Department of Electrical Power Engineering, Faculty of Electrical Engineering, Universiti Teknologi Malaysia, 81310 Skudai, Johor
E-mail: mzzlina@gmail.com,

Abstract. Conventionally, electricity in hospital building are supplied by the utility grid which uses mix fuel including coal and gas. Due to enhancement in renewable technology, many building shall moving forward to install their own PV panel along with the grid to employ the advantages of the renewable energy. This paper present an analysis of grid connected photovoltaic (GCPV) system for hospital building in Malaysia. A discussion is emphasized on the economic analysis based on Levelized Cost of Energy (LCOE) and total Net Present Post (TNPC) in regards with the annual interest rate. The analysis is performed using Hybrid Optimization Model for Electric Renewables (HOMER) software which give optimization and sensitivity analysis result. An optimization result followed by the sensitivity analysis also being discuss in this article thus the impact of the grid connected PV system has be evaluated. In addition, the benefit from Net Metering (NeM) mechanism also discussed.

Keywords: optimization, grid connected, photovoltaic, levelized cost of energy, net present cost, net metering

1 Introduction

Recently, the tremendous increase in fossil fuel prices and emission problems that caused by traditional power generation has contributed to a transformation of latest technology in generating electricity by using clean renewable energy source such as solar and wind. A report by U.S Environmental Protection Agency (USEP) [1] shows that 32% of the total greenhouse gas emission(GHG) was comes from the electric power industry and another report of "Trends in global CO2 emission,2015" [2] indicate that almost 61% of total global CO2 emissions country or region are China (30%), the United States (15%), the European Union (EU-28) (10%) and India (6.5%). The statistic has present the serious challenges to the stake holder, engineers and researchers to find the solutions to reduce the pollution and at the same time full fill the energy requirement. Thus, a distributed generation(DG) especially photovoltaic(PV) technology systems are being utilized means of mitigating this challenges by generating electricity directly from sunlight[3].

Grid connected photovoltaic (GCPV) system is one of the PV technology where an electricity generating by solar panel is connected to the utility grid. A GCPV system consist of a numbers of solar panels, one or several unit of inverters, a power conditioning unit (PCU) and a grid connection equipment. The system size can range from a small residential and commercial rooftop (2 – 10kW) to a large utility scale solar farm stations (1-10MW). In comparison to stand alone photovoltaic (SAPV) system, GCPV system is more reliable power source and considered low cost since they do not need a storage system[4]. Also, GCPV allow the power trading since it will supply power back to the grid if there are excess electricity generated by PV as well as GCPV system will help in reducing greenhouse gas emission. This is done with the displacing power needed to the connected or local load and providing additional electricity to grid. In addition, during the maximum solar irradiance hours, only fewer generations' plants are required. Thus, it will reduce a transmission and distribution (T&D) losses. Installing the GCPV also will help to reduce the stress on the utility grid. This is because when demand goes up in urban areas, the utility has to upgrade its distribution network capacity.

Due to numerous benefits provided by GCPV system, many articles in literature have been discuss about this system that cover the area like modelling [5]–[8], sizing [9], [10] and control of GCPV [9], [11]–[13]. A review of trends and challenges of grid-connected photovoltaic systems including the standards framework for the system is discussed in article [14]. The technique for sizing grid-connected photovoltaic system using evolutionary program is presented in reference [9]. Aris in [15] presents the application of Particle Swarm Algorithm (PSO) method to optimize the photovoltaic grid-connected systems. In other work, Masoud [16] focuses on optimization of a grid-connected PV-based power plant. In this article, the portion of the purchased power from the utility grid and the area of the installed PV system are optimized in order to have a cost-effective and reliable energy system. A design of a grid-connected solar PV system for Uganda using HOMER energy software tool is presented in [17].

In Malaysia, a government is committed to decrease the GHG emission by 40% in year of 2035 ,compared to 2005 level [18]. Hence, various programmes throughout several policies [19] is introduced to promote renewable energy particularly solar energy [20]. Also, the launch of feed-in-tariff (FiT) scheme for renewable energy electricity has started and supervised by Sustainable Energy Development Authority (SEDA)[21]. The updated status of grid-connected PV installed in Malaysia are presented in references [22] and [23]. There are 125 sites with total PV capacity power is approximately 1137.21 kWh of Grid Connected Building Integrated Photovoltaic (GCBIPV) systems until the end of June, 2011. In article [24], an economic evaluation of GCPV system for residential houses in Malaysia is conducted. Ali in [25] discussed the Malaysian Energy Centre (PTM)'s Suria1000 project which have several problems associated with rooftop PV systems. Due to support from the government in PV technology, many research papers have been published discuss a current research and prospectus scenarios for the solar energy development in Malaysia, including the solar potential [26], incentives for solar energy [27],[28] and solar installation either in off grid or grid connected system. The potential of PV system in low voltage (LV) distribution network is reviewed in [29] while an economic analysis for grid connected hybrid PV-wind is conducted in article [30]. The articles in literature reviews presented mostly highlight the research work and implementation of GCPV system through the development of BIPV in Malaysia. Yet, a development of GCPV system in a ground area for commercial building application like school, hotel and hospital are still limited. Therefore, this paper present an analysis of GCPV system to be installed in a commercial building. The selected area is a ground area of public hospital building in Malaysia. The evaluation is based on techno-economic point of view which employing the HOMER simulation software to carry out the technical and economic analysis. Net metering (NeM) mechanism also discussed to show the profits that hospital owner can obtained from the GCPV installation. There are six sections outlined in this paper. After this "Introduction" part, section 2 explain a system description and followed by section 3 which describe the inputs parameters required for the simulation. The operation strategy is explained in section 4 while results and analysis is elaborated in section 5. Finally, the conclusion is described in section 6.

2 System description

A grid connected PV system under this study that compose of PV panel and inverter is shown in Figure 1. In this design system, the battery system is excluded therefore the cost of the whole system is decreases by around 40-50%, depending on the type of the batteries used and the capacity required. Besides, since the designed system is supported by the grid, it is not only improve the reliability condition, but also it has the economic benefits by reducing a system cost. The following subsection will describe the techno-economic parameter setup for each element used in the system including PV panel, inverter and utility grid.

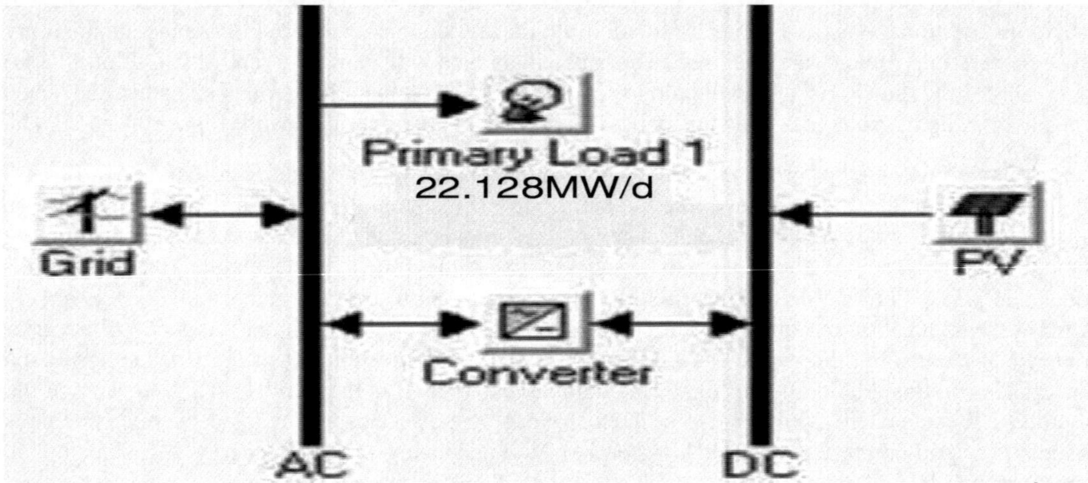

Figure 1: Grid connected PV system

2.1 Solar PV

In order to utilize the PV panel, the component size consideration is important whereby the size of PV arrays is dependent to solar radiation availability, load profile and the renewable fraction required. Renewable fraction is defined as the energy derived from renewable resources and used as part of supply to the load. Here, the renewable fraction is related to the PV production. Size of PV array can either be increased or decreased which depends to the amount of unmet electric load and renewable energy fraction set in the design. The cost data for PV panel is provided in Table 1. PV lifetime is set to be 20 years as long as PV panels typically have their best performance within 20 years and their output will continuously degrades after 20 years. A derating factor is assumed to be 90%. The initial capital, replacement and operational & maintenance (O&M) cost of PV considered in this study are RM2500 per kW, RM2500 per kW and 10 RM/kW/year respectively for 500 kW panel. The O&M cost includes the cost to remove dirt and dust on the PV panel and the PV panel direction adjustment.

2.2 Inverter

The PV arrays that produce direct current (DC) at a voltage which depends on the specific design and solar irradiation have to be converted to alternate current (AC) before it can be fed to the grid. Inverters commonly used to do a conversion process. The inverter size depends on maximum DC input power (i.e. size of the PV array in peak watt) as well as the maximum specified output power (i.e. the AC power provided to the grid). The appropriate inverter size is able to maximize the quantity of energy harvested from the PV arrays in addition to minimizing inverter cost. However, the overall system performance will not be affected or reduced. In this project, the average price of PV inverter is set at RM400 per kW and the O&M cost is estimated to be 5% of the total investment per year. The replacement cost for inverter is RM375 per kW. The efficiency of the inverter is 90% for 10 years lifetime. The inverter cost data is presented in Table 1.

2.3 Grid

When the PV system is connected to the grid and inter-connected with other power generating devices, the grid capital cost is equal to the interconnection charge. Otherwise, the grid capital cost is maintained at zero. In HOMER, the replacement cost of the grid is always zero. The grid O&M cost is equal to the annual cost of buying electricity from the grid (i.e. cost of energy plus demand cost) minus any income from the electricity sold to the grid. The grid O&M cost for the grid-connected PV systems also includes the standby charge.

Table 1 : Economical data of PV and inverter

1. PV Panel [31]	**Value**
Rated power (W)	315
Lifetime (years)	20
Derating factor	90%
Capital cost per kw (RM)	2500
Replacement cost per kw (RM)	2500
O&M cost (RM/kW/yr)	10
2. Inverter[32]	
Rated power (kW)	60
Lifetime (year)	10
Efficiency	90%
Capital cost per kw (RM)	400
Replacement cost per kw(RM)	375
O&M cost per kW per year(RM)	20

3 Input parameters for simulation

The hospital situated in East Peninsular Malaysia with coordinate of 3.4500°N and 102.4167°E is selected as a case study. The area of the hospital is 178,061.68 m^2 with 10 blocks of clinic and 24 wards. The area allocate 5975.90 m^2 for wards where the hospital have 650 beds and 13 units of operating theatre. The area of kitchen and canteen and public area are 2001.42m^2 and 16206.79m^2, respectively. However, in the analysis only use Block 1 and 2 load profiles for the simulation purpose.

3.1 Load profile

The load profile for the building is generated in HOMER software using monthly data and also the daily load data from the same source. The daily load profile for Block 1 and 2, HOSHAS is shown in Figure 2, and the variation of the monthly load profile is illustrated in Figure 3. The highest load consumption in this hospital is 922 kW daily in peak time for almost 24 hours. This is due to the activity of the hospital routine in the clinical and ward block. The average load consumption is 850 kWh/day distributed as peak load of 22.5 kW, with load factor of 0.464.

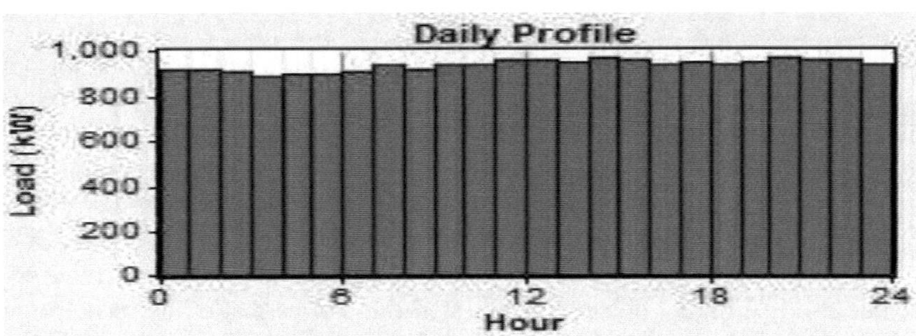

Figure 2: Hourly load profile of HOSHAS

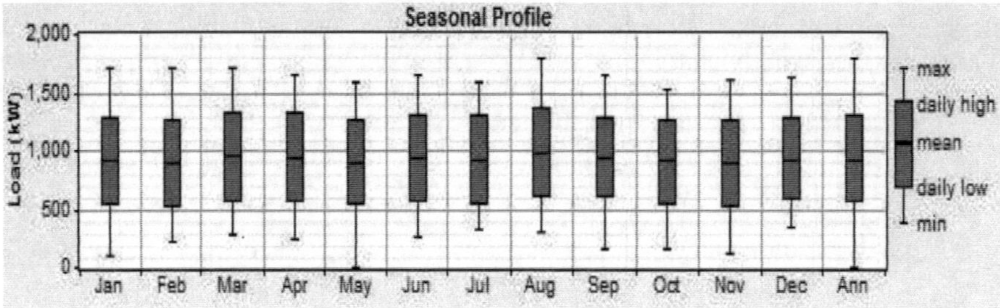

Figure 3: Monthly load profile of HOSHAS

3.2 Solar radiation

In HOMER, the solar resource input can be represented by either the solar radiation data or the clearness index. The solar radiation data was obtained from National Aeronautics and Space Administrative, NASA [33]. The data is 22-year average monthly solar radiation data of the location. The latitude and longitude, 3.4500°N and 102.4167°E were considered in the HOMER software, from which it received the radiation data. Figure 4 shows the solar radiation data for the studied location. The irradiance ranges from 5.056 kWh/m^2/day to 6.036 kWh/m^2/day and the annual average of solar irradiance is estimated to be 5.05 kWh/m^2/day. It is noticed that solar irradiance is high in February, March, April, September and October, and is low in May and November. The reason of low solar irradiation is because of the South-West Monsoon season during this period.

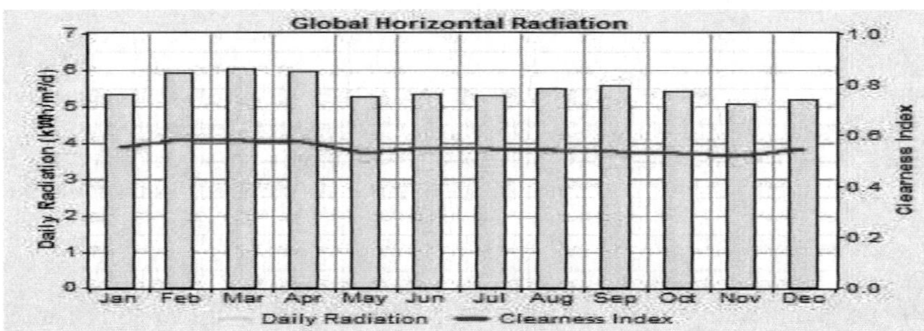

Figure 4: Solar irradiation for the studied area

3.3 Energy price

Main utility in Malaysia, Tenaga Nasional Berhad (TNB) charged the consumers not based on peak time usage but based on kWh of usage. TNB charged consumer based on kWh usage per month in step of 100kWh. As an example, for the first 100kWh , TNB charges 21.8 sen/kWh and it followed by 33.4 sen/kWh for another 100kWh. As this is the analysis on the hospital as a commercial building, a commercial tariff from TNB is take into consideration. Table 2 listed the electricity tariff in Malaysia for commercial customer like hospital[34]. In HOMER software, the term grid power price is refer to the electricity charged for energy purchased from the grid while a demand rate is referred to the price charges for the peak grid demand. Sellback rate is refer to the price that utility pays for power sold to the grid. The rate charges can be modelled in HOMER which is varies according to the time of the electricity used. However, based on Malaysia situation, HOMER cannot analysis the energy trading based on tariff data. Therefore, by using average method, it was assumed that tariff charged by the TNB is flat rated at 36.5 sen/kWh.

IOP Conf. Series: Materials Science and Engineering **217** (2017) 012016 doi:10.1088/1757-899X/217/1/012016

4 Operating strategy

The designed system operates based on grid connected system and Net metering (NeM) mechanism. Net metering is a billing mechanism that will give credits solar energy system owners for the electricity they add into the grid. NeM allow the residential and commercial customer who generate their own electricity from solar power to feed electricity they do not use back into the grid[35]. As an example, if the residential customer has a PV system's on their rooftop, they may generate more electricity than the home demands during daylight hours. If the home is net-metered, the electricity meter will run backwards to provide a credit against what electricity has been consumed at night or at other periods where the home's electricity use has exceeds the system's output. In NeM, customers are only billed for their 'net' energy use. In this study, the generated power from PV will be used to supply the load demand in the building. Here, electricity generated by such system will reduce the power bill and the producer may be able to sell surplus electricity produced to the local electricity supplier. When renewable resources and storage system are insufficient to satisfy the load demand, the electricity of the grid would supply the deficit power [36]. In a grid- connected PV-based power plant, the use of a storage device can truly enable the power system to fully meet the power demand and increase the reliability of the power system. The grid itself acts as a virtual store of electricity, when electricity produced by PV is sell back to the grid. In Malaysia, Sustainable Energy Development Authority (SEDA) will manage the registration and installation of any renewable energy [illegible] e meter arrangement in NeM mechanism is illustrated in Figure 5.

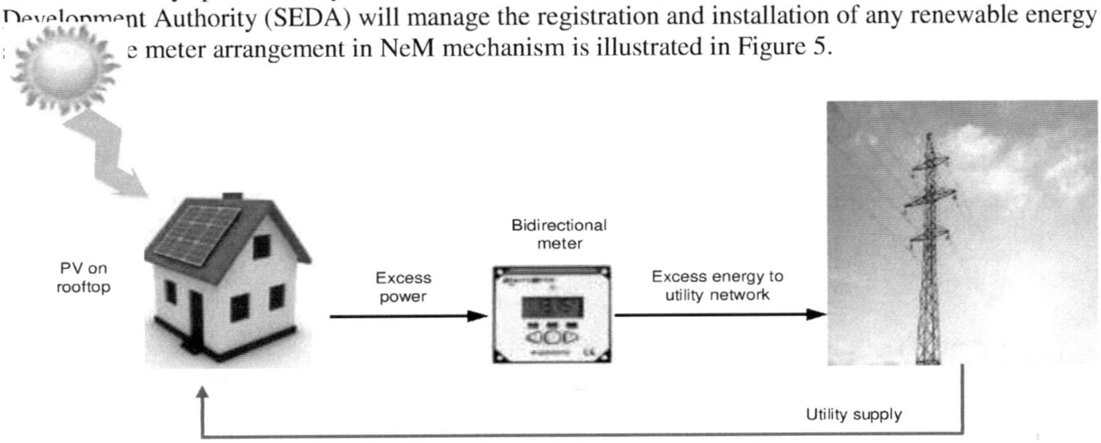

Figure 5: Net metering mechanism topology

Table 2: Electricity tariff in Malaysia for commercial customer

Tariff category	Current rates(1 January 2014)
Tariff C1 - Medium Voltage General Commercial Tariff	
For each kilowatt of maximum demand per month	30.3 MYR/kW
For all kWh	0.365 MYR/kWh
The minimum monthly charge is RM600.00	
Tariff C2 - Medium Voltage Peak/Off-Peak Commercial Tariff	
For each kilowatt of maximum demand per month during the peak period	45.1 MYR/kW
For all kWh during the peak period	0.36 MYR/kWh
For all kWh during the off-peak period	0.22 MYR/kWh
The minimum monthly charge is MYR600.00	

5 Result and analysis

Through HOMER software, the economic analysis will be performed base on net present cost (NPC), levelized cost of energy (LCOE) and annual interest rate. In case of this optimization results, total NPC will represent the life cycle cost of a system, while the revenue includes an income from selling power to the grid. The cost analysis is done by comparing the optimal configuration of load grid connected system and GCPV system. The annual interest rate is stated for sensitivity analysis and it is varies from 1% to 4% in duration of 20 years project lifetime.

5.1 Payback period

Payback period can be defined as the expected number of years required to recover the original investment. The economic analysis for the payback period for the PV module is called energy payback time (EPBT). Based on [37], the formula to calculate the EPBT is as in the (1):

$$EPBT = \frac{kWh_e}{E} \quad (1)$$

Where kWh_e is embodied energy and E is energy generation rate of the PV array. In this study, embodied energy is equal to 5600, PV array rated power at 50 kW and E taken from HOMER simulation equals to 60000 kWh/year. Hence, the EPBT is

Therefore, $EPBT = \frac{5600 \ X \ 50}{60000}$ (2)

EPBT = 4.67 ≈ 5 years.

So, the payback time for this system is 5 years.

5.2 Installation Area of PV panels

As this is the grounded grid connected PV system, a suitable and sufficient area to place the PV panels have to be satisfied. This is to ensure the PV panels can produce certain power needed by the load. Based on the [27], the formula to find the exact area is defined in (3):

$$A = \frac{P_{PV}(t)}{R(t) * \eta_{PV}} \quad (3)$$

Where R is solar insolation in kW/m^2, η_{PV} is efficiency of solar panels and $P_{PV}(t)$ is power generated by the PV plant. In this case, the sufficient area to installed the solar PV panel in the hospital area is 182m^2

5.3 Optimization result

The optimal size indicates for the GCPV designed consist of 1000kW grid, 60 000kW PV and 4000 kW inverter, to full fill the average load profile of 0.922MW. This optimal size made the PV panel produce 108,169,616 kWh/yr and only need to purchase 3,795,915 kWh/yr from the grid, which is help the building owner to decrease their energy bill. The electricity produced by solar PV is shown in Figure 6. The excess electricity is 98,866,640 kWh/yr which can be sell back to the grid. In this GCPV, the unmet load is 247,376 kWh and renewable fraction is 0.96. The emission caused of carbon dioxide, sulphur dioxide and nitrogen dioxide are 318,746 kg/yr, 1,382 kg/yr and 672 kg/yr , respectively. Compared to conventional power generation, the emission caused of this gasses reduced almost 50%. The optimal cost needed for this optimization design are $153,200 (RM 679,579.81) and $165,489 (RM 734,109.05) per year for initial capital cost and operating cost, respectively. The total net present cost (TNPC) and levelized cost of energy (LCOE) require $2,615,252 (RM11,601,257.11) and $0.022 (RM0.0975) per kWh , respectively. This system have the renewable fraction of 0.97 and the capacity shortage is only 0.022. Furthermore, the optimal result on economic expenditure for this GCPV is tabulated in Table 3.

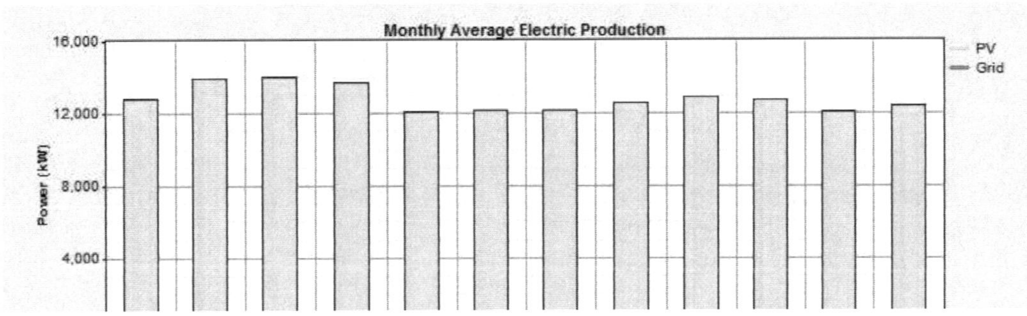

Table 3: Optimal result on economic view

Component	Capital ($)	Replacement ($)	O&M ($)	Fuel ($)	Total ($)
PV	150,000	0	8,926	0	158,927
Grid	0	0	2,448,513	0	2,448,513
Converter	3,200	2,232	2,380	0	7,813
System	153,200	2,232	2,459,820	0	2,615,253

Figure 6: Monthly average electricity produced

5.4 Sensitivity analysis

A sensitivity analysis is executed on the uncertain parameters that should give a significant effect on the TNPC. In HOMER software, the variable can be determined within certain value. The variable will be measure on how it will change the input variable can be specified. In this simulation analysis towards the economic view, an annual interest rate is selected as a sensitivity variable. The reason is because the inflation rates that fluctuate in every year may give the impact on the annual interest rate. According to Bank Negara Malaysia (BNM) [38] [39], the annual interest rate varies from 1% to 3.25% (current). Nevertheless, in this sensitivity analysis, the annual interest rate is assumed from 1% to 4% for 20 years of lifetime. The sensitivity parameters are set in order to observe the effect to the total NPC and LCOE. The increment in the annual interest rate does not affect the PV array capacity but it will give a significant impact to the LCOE, as shown in Figure 7. The result shows that although the annual interest rate is increase up to 4% in coming 20 years, the TNPC for GCPV is decrease. Means, the life cycle cost for GCPV is still considered cheap compared to the conventional system. However, the LCOE has shown a significant increment when there are changes in annual interest rate from 1% to 4% for 20 years lifetime, as shown in Figure 8.

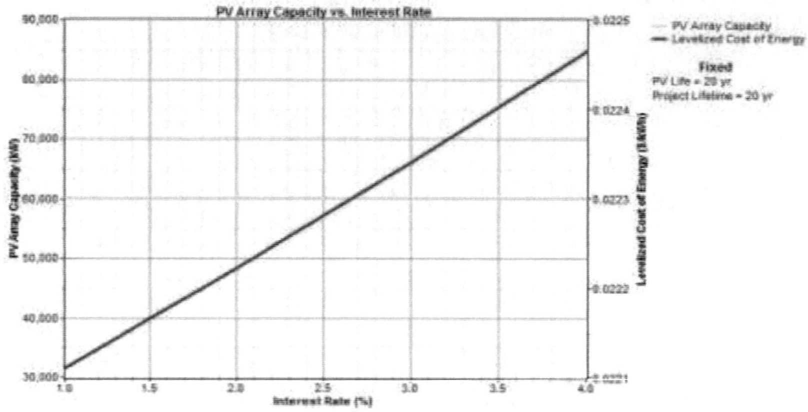

Figure 7: PV array capacity production versus interest rate in 20 years

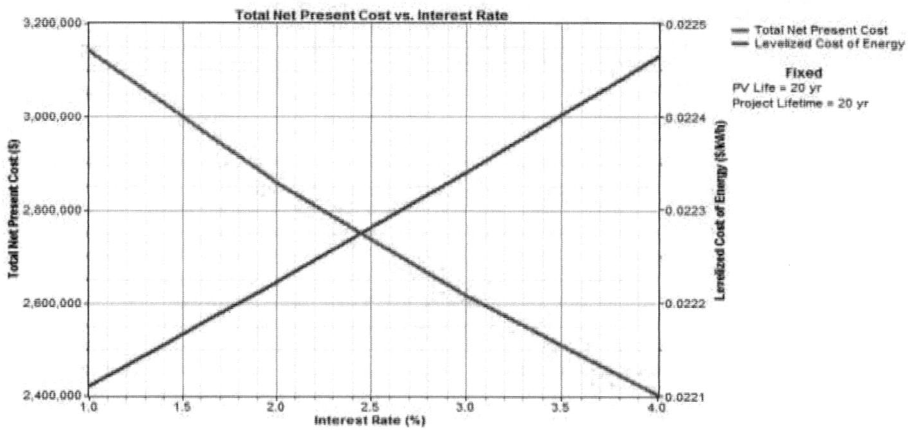

Figure 8: TNPC versus annual interest rate in 20 years lifetime

5.5 Benefits from NeM scheme

Previously, FiT mechanism is introduce and implemented in Malaysia where customer who generate the electricity have to sell all their electricity as the FIT are RE payments of electricity in kilowatt-hour (kWh). Around the globe, Germany is a country that successfully implementing FiT in order to promote renewable energy resource. In 1990, Germany started to implement FiT to avoided cost. At that time, the tariff payment is linked to electricity price until in 1999, the capacity of electricity generated by wind energy had reached 4400MW, and they started to revise back the FiT scheme. The new tariff calculation is based on generation of the technology. In Malaysia, FiT scheme is launch to attract more independent power producer to generate electricity as well as it act as a kick-start for the renewable energy generation. As for this analysis, if the FIT mechanism is considered, the hospital management will benefits of MYR1200 for each 40kW a month. However, FIT mechanism has been stopped due to limited quota. Then, NeM is introduce in Malaysia with more benefits where customers can avoid the dependence to TNB. Also, the implantation of NeM is aim to encourage job creation in the downstream solar PV market, provide further impetus to growth of clean technology in Malaysia and to reduce the level of carbon emissions. Throughout this GCPV installation, if the hospital is net-metered, the hospital owner will benefits the reduction in energy bill as well as the billing structure for NeM consumer are as shown in Table 4. NeM customer will only receive one bill every month. If NeM Rebate (RM) more than utility supply (RM) then customer is will pay the net amount to utility. If NeM Rebate (RM) more than utility supply (RM) then customer is given monetary credit in the next bill. Unused credit amount will carry up to end of calendar year. At the end of each calendar year, change in tariff category or upon the termination of NeM contract with utility, any remaining credit will be adjusted to zero. The credit rate for energy excess to utility classified into two categories based on utility displaced cost. For low voltage connection, credit rate is at RM0.31/kWh and medium voltage connection rate is RM0.238/kWh (with assumption RM1 = USD0.2257).

Table 4: Billing structure in NeM mechanism for customer in Malaysia

Item	RM	Total Payable
kWh import from utility X tariff	A	TNB supply : (A+B+C+D+E)
Good & Services Tax (GST) 1 : (A +E) X 6%	B	
RE fund: A X 1.6%	C	
Penalties	D	
Imbalance Cost Pass Thru	E	
kWh export X solar rate	F	NeM Rebate : (F+G)
GST 2: F X 6%	G	
Total Payable		(A + B+C+D+E)- (F+G)

6 Conclusion

An optimal design based on economic point of view and technical sizing for grid connected PV system is presented in this paper. The configuration of grid connected system is installed to hospital building which have average load of 0.922MW, located in Temerloh Pahang, Malaysia. In order to generate the sufficient PV energy, a size of PV panel and inverter should be optimized. Therefore in this simulation through HOMER software, a 60 000 kW PV panel and 4 000 kW inverter is required to be installed in the area of 182m2 of hospital area. The installation of GCPV system in the hospital area will give a reduction in energy cost whereas the amount for TNPC and LCOE in 20 years period are $2,615,253 (RM11,601,257.11) and $0.022 (RM0.0975) per kWh, respectively. The sensitivity analysis is performed to observe the impact to both TNPC and LCOE due to changes in annual interest rate. Since the hospital is health and hopefulness building, the installation of GCPV is expected to give cleaner environment for the community inside the hospital. The implementation of NeM in the hospital will give the profit to hospital owner where they can reduce their energy bill. As a conclusion, although the initial cost to develop the GCPV, it will give a worth profits to the policy makers, building owners and society in terms of decrement in pollutant to the environment.

References

[1] "Inventory of U . S . Greenhouse Gas Emissions and Sinks : 1990 - 2014," 2016.
[2] J. G. J. Olivier, G. J. Maenhout, M. Muntean, and J. A. H. W. Peters, "Trends in global CO2 emissions: 2015 report," 2015.
[3] J. O. Petinrin and M. Shaaban, "Renewable energy for continuous energy sustainability in Malaysia," *Renew. Sustain. Energy Rev.*, vol. 50, pp. 967–981, 2015.
[4] J. R. Perez-Gallardo, C. Azzaro-Pantel, S. Astier, S. Domenech, and A. Aguilar-Lasserre, "Ecodesign of photovoltaic grid-connected systems," *Renew. Energy*, vol. 64, pp. 82–97, 2014.
[5] M. S. Mahmoud, S. Azher Hussain, and M. a. Abido, "Modeling and control of microgrid: An overview," *J. Franklin Inst.*, vol. 351, no. 5, pp. 2822–2859, May 2014.
[6] G. Bayrak and M. Cebeci, "Grid connected fuel cell and PV hybrid power generating system design with Matlab Simulink," *Int. J. Hydrogen Energy*, vol. 39, no. 16, pp. 8803–8812, 2014.
[7] F. Ding, P. Li, B. Huang, F. Gao, C. Ding, and C. Wang, "Modeling and simulation of grid-

connected hybrid photovoltaic/battery distributed generation system," *2010 China Int. Conf. Electr. Distrib. CICED 2010*, vol. 3, no. 1, p. 10 pp.-pp., 2010.

[8] E. Benkhelil and A. Gherbi, "Modeling and simulation of grid-connected photovoltaic generation system," pp. 295–306, 2012.

[9] S. I. Sulaiman, T. K. A. Rahman, I. Musirin, S. Shaari, and K. Sopian, "An intelligent method for sizing optimization in grid-connected photovoltaic system," *Sol. Energy*, vol. 86, no. 7, pp. 2067–2082, 2012.

[10] M. A. M. Ramli, A. Hiendro, K. Sedraoui, and S. Twaha, "Optimal sizing of grid-connected photovoltaic energy system in Saudi Arabia," *Renew. Energy*, vol. 75, pp. 489–495, 2015.

[11] H. Shareef, A. Mohamed, and A. H. Mutlag, "A current control strategy for a grid connected PV system using fuzzy logic controller," *2014 IEEE Int. Conf. Ind. Technol.*, pp. 890–894, 2014.

[12] V. F. Pires, J. F. Martins, and C. Hao, "Controlling a Power Converter as a Photovoltaic Power Source and Compensator in Smart-Grids," pp. 173–178, 2012.

[13] D. Sun, B. Ge, D. Bi, and F. Z. Peng, "Electrical Power and Energy Systems Analysis and control of quasi-Z source inverter with battery for grid-connected PV system q," *Int. J. Electr. Power Energy Syst.*, vol. 46, no. 2009, pp. 234–240, 2013.

[14] M. Obi and R. Bass, "Trends and challenges of grid-connected photovoltaic systems – A review," *Renew. Sustain. Energy Rev.*, vol. 58, pp. 1082–1094, 2016.

[15] A. Kornelakis and Y. Marinakis, "Contribution for optimal sizing of grid-connected PV-systems using PSO," *Renew. Energy*, vol. 35, no. 6, pp. 1333–1341, 2010.

[16] M. Zebarjadi and A. Askarzadeh, "Optimization of a reliable grid-connected PV-based power plant with / without energy storage system by a heuristic approach," *Sol. Energy*, vol. 125, pp. 12–21, 2016.

[17] S. Twaha, M. Ha, M. Anwari, and A. Khairuddin, "Applying grid-connected photovoltaic system as alternative source of electricity to supplement hydro power instead of using diesel in Uganda," vol. 37, pp. 185–194, 2012.

[18] P. Y. Gan, R. Komiyama, and Z. Li, "A low carbon society outlook for Malaysia to 2035," *Renew. Sustain. Energy Rev.*, vol. 21, pp. 432–443, 2013.

[19] K. M. Poh and H. W. Kong, "Renewable energy in Malaysia: a policy analysis," *Energy Sustain. Dev.*, vol. 6, no. 3, pp. 31–39, 2002.

[20] A. H. Shamsuddin, "Development of Renewable Energy in Malaysia-Strategic Initiatives for Carbon Reduction in the Power Generation Sector," *Procedia Eng.*, vol. 49, pp. 384–391, Jan. 2012.

[21] "Feed in Tariff rates for solar PV," *20 October 2012*, 2012. .

[22] M. Z. Hussin, N. Hasliza, A. Yaacob, Z. M. Zain, and A. M. Omar, "A Development and Challenges of Grid-Connected Photovoltaic System in Malaysia," no. July 2005, pp. 191–196, 2012.

[23] M. Z. Hussin, A. Yaacob, Z. M. Zain, S. Shaari, and A. M. Omar, "Status of a Grid-connected MBIPV Project in Malaysia," in *international Symposium & Exhibition in Sustainable Energy & Environment*, 2011, no. June, pp. 1–3.

[24] K. Hasim, M. K. Almsafir, and V. Kumaran, "An Economic Evaluation of Grid- Connected Photovoltaic Generation System for Residential House in Malaysia," *J. Adv. Sci. Eng. Res.*, vol. 4, no. 1, pp. 47–60, 2014.

[25] I. Ali, A. Sher, and J. Pasupuleti, "Implementation of Photovoltaics in Malaysia," no. December, pp. 7–8, 2009.

[26] M. Z. A. Ab Kadir, Y. Rafeeu, and N. M. Adam, "Prospective scenarios for the full solar energy development in Malaysia," *Renew. Sustain. Energy Rev.*, vol. 14, no. 9, pp. 3023–3031, Dec. 2010.

[27] S. C. Chua, T. H. Oh, and W. W. Goh, "Feed-in tariff outlook in Malaysia," *Renew. Sustain. Energy Rev.*, vol. 15, no. 1, pp. 705–712, 2011.

[28] S. Ahmad, R. Mat, F. Muhammad-sukki, A. Bakar, and R. Abdul, "Role of feed-in tariff policy in promoting solar photovoltaic investments in Malaysia : A system dynamics approach," *Energy*, vol. 84, pp. 808–815, 2015.

[29] J. Wong, Y. S. Lim, J. H. Tang, and E. Morris, "Grid-connected photovoltaic system in Malaysia : A review on voltage issues," *Renew. Sustain. Energy Rev.*, vol. 29, pp. 535–545,

2014.
[30] R. N. S. R. Mukhtaruddin, H. A. Rahman, and M. Y. Hassan, "Economic Analysis of Grid-Connected Hybrid Photovoltaic-Wind System in Malaysia," pp. 577–583, 2013.
[31] N. M. Isa, H. S. Das, C. W. Tan, A. H. M. Yatim, and K. Y. Lau, "A techno-economic assessment of a combined heat and power photovoltaic/fuel cell/battery energy system in Malaysia hospital," *Energy*, vol. 112, pp. 75–90, 2016.
[32] "Inverter," 2014. .
[33] T. Givler and P. Lilienthal, "Using HOMER ® Software , NREL ' s Micropower Optimization Model , to Explore the Role of Gen-sets in Small Solar Power Systems Case Study : Sri Lanka Using HOMER ® Software , NREL ' s Micropower Optimization Model , to Explore the Role of Gen-sets in Smal," no. May, 2005.
[34] TNB, "Commercial Tariff," *[Online]*, 2014. .
[35] S. Chemud, "The Impact of Roof Top Solar Pv Under Net Energy Metering program on National Utility and Public in Malaysia," 2016.
[36] M. A. Eltawil and Z. Zhao, "Grid-connected photovoltaic power systems : Technical and potential problems — A review," vol. 14, pp. 112–129, 2009.
[37] B. Rismanchi, R. Saidur, H. H. Masjuki, and T. M. I. Mahlia, "Cost-benefit analysis of using cold thermal energy storage systems in building applications," 2012.
[38] "Malaysia interest rate." .
[39] Z. A. Karima and B. A. Karimb, "Interest rates targeting of monetary policy: An open economy SVAR study of Malaysia," *Gadjah Mada Int. J. Bus.*, vol. 16, no. 1, pp. 1–22, 2014.

IOP Conf. Series: Materials Science and Engineering **217** (2017) 012017 doi:10.1088/1757-899X/217/1/012017

Carbon materials derived from rice husks at low and high temperatures

G J H Melvin[1, *], Z Wang[2], N J Siambun[1], M M Rahman[1]

[1] Faculty of Engineering, Universiti Malaysia Sabah, Jalan UMS, 88400 Kota Kinabalu, Sabah, Malaysia.
[2] Institute of Carbon Science and Technology, Shinshu University, 4-17-1 Wakasato, Nagano 380-8553, Japan.

*E-mail: melvin.gan@ums.edu.my

Abstract. Rice husk (RH) can be classified as an agriculture residue, majorly produced from by-product of rice milling industries. However, RHs are only mainly utilized for low value energy resource. A great number of researches and innovations have shown that heat treated RHs can turn into valuable carbon materials. In this study, the RHs were carbonized at 800°C and 2500°C, respectively. Their structure, morphology, elemental composition, and quality were characterized using transmission electron microscopy, X-ray photoelectron spectroscopy, and Raman spectroscopy. The carbon materials obtained from low and high temperature carbonization processes showed different characteristics. High purity and crystallinity of carbon materials were obtained from RHs carbonized at 2500°C. Furthermore, from Raman results, RHs carbonized at 2500°C exhibited low D/G ratio. This further reveals that the RHs carbonized at 2500°C possess minimal defects. The unique characteristics of RHs carbonized at high temperature indicate that they could be a promising material to be utilized in particular or various applications.

1. Introduction

Carbon materials including fullerene, carbon black, carbon fiber, carbon nanotubes, graphene, porous carbon, and their other derived forms; due to their distinctive physical and chemical characteristics, have received considerable attention, from fundamental research to wide range of applications [1, 2]. However, these carbon materials can be considered as high-grade carbon materials, which are high in cost and rarely available in bulk quantity. In order to obtain carbon materials with low cost fabrication process and more environment friendly technique, carbonization of waste materials can be considered as a beneficial method. Since the waste materials are available in large scale, greener technology to produce bulk quantity of valuable carbon materials from waste materials can be promoted. Waste materials can be in the form of agricultural or organic wastes, synthetic compound wastes, and so on. Several conventional sources for carbon materials are methane, ethane, acetylene, benzene, and another hydrocarbon based fossil fuel [2]. In comparison with the waste materials as the source to obtain carbon materials, their lacks in abundancy and comparatively high cost have been the drawback for their usage, for particular cases. For instance, green nanocomposites [3], supercapacitors [4, 5], conductive materials [6], adsorbent materials [7], and so on, are some examples of usage of carbon materials derived from waste materials.

Rice husk (RH) can be categorized as an agriculture residue, mainly produced from by-product of rice milling industries [8, 9]. From a statistic, it was reported that the annual rice production is approximately 500 million tons in developing country, and the RH available to be utilized is approximately 100 million tons [8]. Commonly, RHs are exploited as low value energy resource or simply burnt at the field as they are considered as a waste, which will greatly affect the air quality and unfavorable to the environment [9]. Thus, it is important to conduct research on RHs and then utilize them comprehensively. RH contains organic materials such as lignin, hemicellulose, and cellulose, which is in favor to turn them into raw carbon materials [10]. A great number of researches and innovations have shown that heat treated RHs can turn into valuable carbon materials [8-10]. Based on the consideration regarding cost, environment issues, production quantity, and so on, RHs can be considered as a strong candidate to be fully utilized as the source of carbon materials.

Generally, in order to obtain high-grade carbon materials, most of the synthesis, chemical process or reactions are necessary. Alternatively, carbonization process, through heat treatment can be considered as an advantageous method. Diverse structure and morphology of carbon materials derived from waste sources can be obtained, depending on the parameters such as heat temperature control in the carbonization process [11, 12]. In this study, the RHs were carbonized at low temperature of 800°C and high temperature of 2500°C. Then, their structure, morphology, elemental composition and quality were characterized and compared.

2. Experimental

Dried RHs without further modification were used as the source to be carbonized. Two RH samples were prepared and carbonized at different temperatures of 800°C and 2500°C. First, the raw RHs were carbonized at 800°C, 1 h using furnace, under the presence of argon gas. Next, the carbonized RHs at 800°C were again loaded into graphite box, further heat treated at 2500°C, 0.5 h using graphite resistance furnace operating in a high purity argon gas atmosphere at 760 Torr. Both samples were cooled down naturally and then grounded into powder form before conducting further evaluation. RHs carbonized at 800°C will be denoted as LTRH (low temperature) and RHs carbonized at 2500°C will be denoted as HTRH (high temperature), hereafter.

The samples were observed using transmission electron microscopy (TEM; JEM-2100F, JEOL) operated at 200 kV, in order to investigate the structure and morphology. X-ray photoelectron spectroscopy (XPS; PHI Quantera II) was performed to characterize the elemental composition. Raman spectroscopy measurement was performed on a Raman spectrometer (Renishaw inVia Raman spectroscope) with 532 nm laser excitation, in order to determine the quality of carbon materials derived from RHs at low and high temperatures.

3. Results and discussions

TEM images are shown in figure 1. From figure 1(a), LTRH indicate that they can be considered as disordered carbon form or amorphous carbon and no clear edges were observed. Commonly, carbon materials derived from waste materials at low temperature, such as 800°C, will produce porous carbon [4, 5]. Conversely for HTRH, from figure 1(b), mixture of flattened graphene, wrinkled graphene, and partially carbon cage can be observed. Furthermore, they also exist in the nanometer size range, which includes few- and multi-layer graphene layers. Moreover, no amorphous carbon was observed on the graphene surface. Heat treatment at 2500°C produced clean surface, clear edges, and relatively high crystallinity of carbon materials. This can be attributed to the realignment or restructure of vapor carbon at high temperature [12]. The carbons vaporized at high temperature, and their deposition during the cool down process, contributed to the structure formed by HTRH.

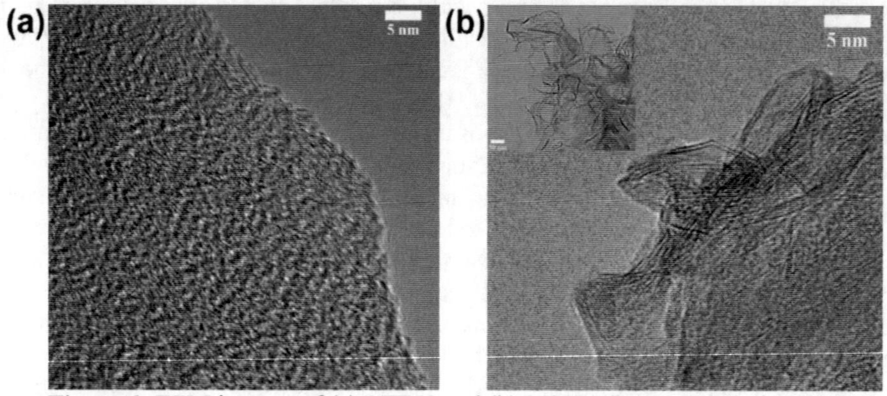

Figure 1. TEM images of (a) LTRH and (b) HTRH (inset scale bar 50 nm).

XPS evaluation was conducted to determine the elemental composition of the RHs heat treated at low and high temperatures, and the results are as depicted in figure 2(a). Both XPS wide survey spectra confirmed the presence of carbon (C 1s) peak at around 284 eV and oxygen (O 1s) peak at around 533 eV. Clearly, in contrast with HTRH, LTRH showed Si 2s and Si 2p (silicon) peaks. This might be attributed to the silica absorption of rice plants in the form of silicic acid from the soil [13]. Commonly, heat treated RHs at low temperatures (around 600~1000°C) will have main components of carbon and silica, depending on certain conditions [14, 15]. For HTRH, clear C 1s peak can be observed, with almost no other elements, and this implies that they are in the form of high purity carbon materials. However, extremely low oxygen (O 1s) peak was observed too, might be attributed to oxygen absorption when exposed to the environment. Interestingly, combination of TEM and XPS evaluation, RHs carbonized at 2500°C show graphene layer structure with high crystallinity and high purity of carbon.

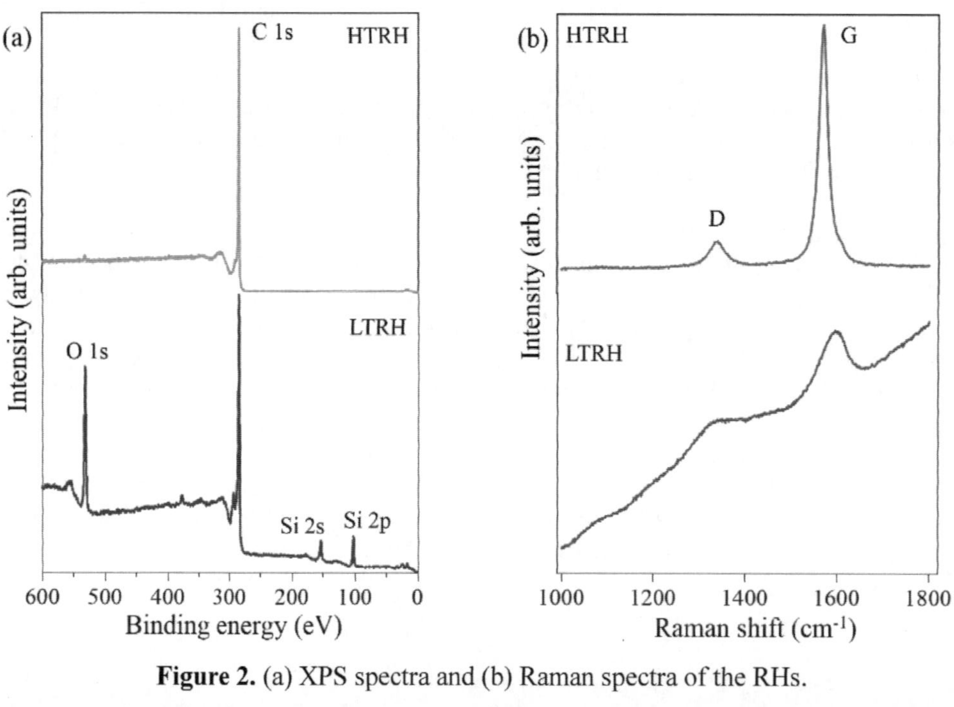

Figure 2. (a) XPS spectra and (b) Raman spectra of the RHs.

Raman spectroscopy is a nondestructive and an effective tool to probe structure at the atomic scale on the basis of vibrational symmetry. Generally, characteristic bands such as D band (disorder-induced band that is often associated with defective graphite structures) at around ~1350 cm^{-1} and G band (graphite band) at around ~1580 cm^{-1} can be observed for various carbon materials. Raman spectra for LTRH and HTRH are illustrated in figure 2(b). For HTRH, clear D and G peak can be observed. For LTRH, the trend of moving upward spectra can be attributed from the broadening of D peak or affected by SiC peak usually traced at lower wavenumber. Nevertheless, the spectra shows certain similarity with low sp^3 amorphous carbon [16]. The D and G peak position, full width at half maximum (FWHM) for D and G peak, and relative intensity of D and G peak (I_D/I_G) are tabulated in table 1. The position of D peak for LTRH and HTRH has high similarity. The position of G peak for LTRH shifted to higher wavenumber, caused by the broad D peak, as mentioned earlier. Even though direct comparison cannot be concluded for FWHM between LTRH and HTRH, for both D and G peaks, the FWHM values of HTRH showed that they are high crystalline quality with some defects introduced, as depicted in figure 2(b). The I_D/I_G is associated with the amount of structural defects of carbon materials, and thus provides direct information about the quality of carbonized RHs. The I_D/I_G for HTRH is calculated to be 0.23, which is comparatively lower than other carbon materials or RHs carbonized at lower temperature [17, 18]. However, based on the existence of D peak, defects are still detectable in both of our samples, regardless they are heat treated at low or high temperatures. One of the reasons, the defects might be introduced to the samples when they are grounded before Raman evaluation. It is worth to notice that D peak also can be found in most of graphitic carbons, which might be due to the defects on the edges, or in our case the carbon cage for HTRH. For comparison, Raman information for multi-walled carbon nanotubes (MWCNT, d = 40–70 nm) was also included in table 1. The position of D and G peak is in the typical range for MWCNT. The D and G peak of MWCNT have high similarity, thus contributed to similar FWHM. Noticeably, the I_D/I_G is also higher than HTRH, although no modification was conducted to the MWCNT. Commonly, modification will be conducted to improve the dispersion of MWCNT, which also will further introduce defects (higher I_D/I_G). For carbonized RHs, regardless their carbonization temperature, the D peak appears at lower wavenumber than MWCNT. Furthermore, FWHM of HTRH is also lower than MWCNT, which can be associated with sharper G peak of HTRH. This further indicates that the HTRH are in the high carbon crystalline quality form.

Table 1. Raman information for LTRH and HTRH.

Sample	D peak (cm^{-1})	FWHM for D (cm^{-1})	G peak (cm^{-1})	FWHM for G (cm^{-1})	I_D/I_G
LTRH	1339.7	-	1595.8	-	-
HTRH	1339.8	51.2	1569.9	23.4	0.23
MWCNT	1344.1	52.6	1577.1	52.9	0.96

4. Conclusions

Two types of carbon materials derived from RHs were prepared, namely RHs heat treated at 800°C (LTRH) and RHs heat treated at 2500°C (HTRH). Their structure, morphology, elemental composition and quality evaluations were carried out through TEM observation, XPS evaluation, and Raman measurement. Few and multilayer graphene can be obtained from high temperature treated RHs. Furthermore, high purity and crystallinity carbon material also can be obtained from high temperature treated RHs. From Raman results, RHs carbonized at 2500°C also can be considered as high carbon crystalline with minimal defects. The features of RHs carbonized at high temperature imply that they could be a prominent material to be manipulated in wide variety of applications.

References

[1] Wang Z, Ogata H, Morimoto S, Ortiz-Medina J, Fujishige M, Takeuchi K, Muramatsu H, Hayashi T, Terrones M, Hashimoto Y and Endo M 2015 *Carbon* **94** 479-84
[2] Alfarisa S, Bakar S A, Mohamed A, Hashim N, Kamari A, Isa I M, Mamat M H, Mohamed A R and Mahmood M R 2015 *Adv. Mat. Res.* **1109** 25-9
[3] Lee H K, Park Y G, Jeong T and Song Y S 2015 *J. Appl. Polym. Sci.* **132** 42043
[4] Yun Y S, Park M H, Hong S J, Lee M E, Park Y W and Jin H-J 2015 *ACS Appl. Mater. Interfaces* **7** 3684-90
[5] Wei L and Yushin G 2012 *Nano Energy* **1** 552-65
[6] Barroso-Bogeat A, Alexandre-Franco M, Fernandez-Gonzalez C, Macias-Garcia A and Gomez-Serrano V 2014 *Phys. Chem. Chem. Phys.* **16** 25161-75
[7] Yang T and Lua A C 2003 *Micropor. Mesopor. Mat.* **63** 113-24
[8] Guo Y, Yu K, Wang Z and Xu H 2000 *Carbon* **41** 1645-87
[9] Chen H, Wang H, Yang L, Xiao Y, Zheng M, Liu Y and Fu H 2012 *Int. J. Electrochem. Sci.* **7** 4889-97
[10] Wang L, Guo Y, Zhu Y, Li Y, Qu Y, Rong C, Ma X and Wang Z 2010 *Bioresource Technol.* **101** 9807-10
[11] Saito Y and Arima T 2007 *Carbon* **45** 248-55
[12] Wang Z, Ogata H, Morimoto S, Fujishige M, Takeuchi K, Hashimoto Y and Endo M 2015 *Carbon* **90** 154-9
[13] Liu N, Huo K, McDowell M T, Zhao J and Cui Y 2013 *Sci. Rep.* **3** 1919
[14] Liu Y, Guo Y, Gao W, Wang Z, Ma Y and Wang Z 2012 *J. Clean. Prod.* **32**, 204-9
[15] Bakar R A, Yahya R and Gan S N 2016 *Procedia Chem.* **19** 189-95
[16] Ferrari A C 2007 *Solid State Commun.* **143** 47-57
[17] Guerrero M, Ruiz M P, Millera A, Alzueta M U and Bilbao R 2008 *Energy Fuels* **22** 1275-84
[18] Li Y, Wang F, Liang J, Hu X and Yu K 2016 *New J. Chem.* **40** 325-9

Acknowledgements

This work was supported by grant (SLB0120-SG-2016) from Universiti Malaysia Sabah (UMS), Malaysia. The authors also would like to acknowledge members of Institute of Carbon Science and Technology, Japan for their technical supports.

A Comparative Study of Different Acids used for Sandstone Acid Stimulation: A Literature Review

Leong Van Hong[1,*], Hisham Ben Mahmud[1]
[1]Petroleum Engineering Department, Curtin University, Malaysia

E-mail: leongvanhong@postgrad.curtin.edu.my

Abstract. Matrix acidizing is an effective well stimulation technique, in which acids are injected at a pressure below the formation fracture pressure. The application of sandstone matrix acidizing has been widely used in the oil and gas industry for many decades. The application of mud acid, which is a combination of Hydrofluoric acid and Hydrochloric acid (HF:HCl) in well stimulation, has gained its popularity in improving the porosity and permeability of reservoir formation. In fact, this is driven by the effectiveness of HF in dissolving minerals in sandstone and HCl in controlling precipitation. Nonetheless, high temperature matrix acidizing approach is in growing need since many wells nowadays are producing from much deeper and hotter reservoir, with a temperature higher than 200°F. In such conditions, mud acid causes rapid reaction rates, hence becoming less efficient as the acids are consumed too early. Furthermore, mud acid is hazardous and very corrosive. On the contrary, previous studies had shown that Fluoroboric Acid (HBF_4) and Phosphoric acid (H_3PO_4) offered numerous advantages in comparison to the conventional mud acid. HBF_4 can hydrolyze to form HF whereas H_3PO_4 acts as a buffer acid; which is able to penetrate deeper into the formation before spending. Likewise, both acids cause more increase in the permeability, less change in the strength of core samples and significantly less corrosive. This paper had critically reviewed the experimental works which had been done on different types of acids. The advantages and disadvantages of these acids are evaluated. Therefore, a new acid combination (HBF_4:H_3PO_4) is developed and the future work which can be done on it is proposed.

Keywords: Well stimulation; Sandstone; Matrix acidizing; Mud acid; Fluoroboric acid; High temperature

1. Introduction

In this new era, the energy requirement of the world has increased. In 2020, prediction shows that 40% more energy will be demanded than in the current time [1]. High-temperature reservoir acidizing becomes the main target for new oil and gas reserve exploration [2]. This is because many reservoir condition are at high temperature of 200°F, whereas some even exist at ultra-high temperature of 500°F. Therefore, current technology is less suitable for reservoirs at such conditions. hence, all aspects of acidizing, from corrosion rates to treatment-fluid stability must be improved.

Mud acid (HF-HCl) is the most conventionally used acid in sandstone acid stimulation. However at high temperature of 200°F, mud acid can cause rapid reactions rates. Consequently, this would result in inefficiency of acidizing treatment because the acids are consumed too fast [2]. Mud acid is also very corrosive. In order to overcome the problems of the conventional mud acid at high temperature, new acid combination was proposed in this paper.

The aim of this paper is to review the state-of-art of the sandstone acid stimulation. The paper evaluated the advantages and disadvantages of different acid being used. Based on the literature review, fluoroboric acid, HBF_4 and phosphoric acid, H_3PO_4 were found to offer more advantages. For instance, HBF_4 can hydrolyse in aqueous form to form HF slowly whereas H_3PO_4 acts as a buffer acid; which allow deeper penetration into the formation. These acids increased the permeability of sandstone matrix and were less corrosive in comparison to mud acid. In this paper, the recommendation of the future work that can be done on the new acid was also suggested.

2. Background

2.1. Introduction to Sandstone Matrix Acidizing

Sandstone (also known as arenite) is a clastic sedimentary rock. It is mainly composed of Silica and Silicate minerals, which include quartz, feldspar, various forms of clay, and in rare cases zeolite [3].

Matrix acidizing is effective in increasing the production rate and reducing the skin of a sandstone reservoir. For many years, acids have been applied to stimulate reservoir rocks by changing their properties. The acid is injected into the formation at injection pressure below the formation fracturing pressure. By injecting the acid, the minerals present in the soluble reservoir formation will be dissolved, thus increasing the rate of fluid flow from the reservoir to the wellbore. The common practice is a combination of Hydrochloric Acid (HCl) - Hydrofluoric Acid (HF) with a composition of 3% HF – 12% HCl [4].

2.2. The Mechanism of Chemical Reaction of Sandstone Matrix Acidizing

Mud acid (HF-HCl) is able to dissolve clay minerals [5]. Acid used to stimulate sandstone formations contains Fluoride Ion (F^-), which is a very reactive ion and is the only ion that reacts with the sand and clay significantly. The reactivity of HF acid with silica makes it unique in sandstone acidizing applications. Other acids such as hydrochloric, nitric and sulphuric acid are unreactive with silica [6]. As HF enters sandstone core, almost all the minerals begin to dissolve, but at different rates depending on the intrinsic rates of heterogeneous reactions and the exposed surface areas. The reacting minerals can be divided into two distinct categories: slow and fast reacting. Quartz tends to act at a slower rate whereas feldspar and clay tend to react at a faster rate.

During sandstone acidizing, the following precipitation reactions take place and may lead to formation damage and reduction in the porosity and permeability [7]:

(i) Precipitation of hydrated silica:

$$4HF + SiO_2 \rightarrow SiF_4 + 2H_2O \quad (1)$$
$$6HF + SiO_2 \rightarrow H_2SiF_6 + 2H_2O \quad (2)$$
$$H_2SiF_6 + 4H_2O \rightarrow Si(OH)_4 + 6HF \quad (3)$$

(ii) Precipitation of sodium, Na silicates, potassium, K silicates, and calcium, Ca fluoride:

$$H_2SiF_6 + 2Na^+ \rightarrow Na_2SiF_6 + H_2 \quad (4)$$
$$H_2SiF_6 + 2K^+ \rightarrow K_2SiF_6 + H_2 \quad (5)$$
$$2HF + Ca^+ \rightarrow CaF_2 + H_2 \quad (6)$$

Sandstone acidizing is a challenging task due to multiple stages of fluids and reaction of these fluids with the minerals in porous media [8]. At high temperatures, these reactions cause precipitation reactions, which are dangerous as the formation can be damaged. To avoid reactions (1), (2) and (3), HCl or organic acid is used in the main acid stage. To avoid reactions (4), (5) and (6), ammonium chloride or hydrochloric acid is used as a pre-flush ahead of the main acid, which usually contains HF acid.

Typically, sandstone matrix acidizing consists of three stages [9], which are:

1. Pre-flush to dissolve Na, K and Ca ions that will react with the silica to form insoluble silicates.
2. Main-flush to dissolve the silicates, feldspar and clay.
3. After-flush to keep the wettability in its original state and clean the formation.

3. Literature Review of Experimental Studies

In the past, researchers had developed organic acids (Acetic, Formic), and powdered (Sulfamic, Chloroacetic) [10]. The choices of acid used and additives added depend on the reservoir formation characteristics and the intention of the stimulation. There are significant numbers of experimental studies that had been conducted to investigate the performance and effectiveness of different acid combinations in sandstone matrix acidizing. All these studies have been critically reviewed. Hence, the research outcomes have formed a basic framework for the experimental setup in the future work.

3.1. Mud Acid (HF:HCl)

Gomaa et al. in 2013 [11] had carried out core flood test to investigate the effect of different mud acid, HF-HCl ratio on the permeability enhancement of the sandstone core sample at 180°F. Four mud acid combinations (1.9% HF:15% HCl, 2.3% HF:10% HCl, 2.6% HF:5% HCl and 2.8% HF:3% HCl) were tested on the core sample. Based on the result, all four acid combinations can successfully enhance the permeability of the core sample. However, better permeability enhancement result was observed with increasing HF:HCl ratio. At the same time, the injected acid volume was also reduced.

Al-Harthy et al. in 2009 [2] stated that although mud acid was proven to be efficient and had been widely used, nevertheless it still caused rapid reaction rates at high temperature of 200°F. This resulted in inefficiency of acidizing process as the acids were consumed too early. It was undeniable that mud acid is also hazardous and corrosive to wellbore equipment. Therefore, all of these disadvantages should be taken into considerations in the experimental setup in the future as to overcome the research vulnerability. In view of the shortcoming of the mud acid, new acid combination (HBF_4:H_3PO_4) is foreseen to be a better choice on the other hand. It is inferred to not only be having better permeability improvement, but also less corrosive.

3.2. Chelating Agents

Ali et al. in 2008 [12] had carried out experimental investigation on the use of low pH solution, Sodium Hydroxyethylethylenediaminetriacetic Acid (Na_3HEDTA) in stimulating a high temperature well in West Africa. This chelating fluid showed a high efficiency in stimulating a high temperature formation by increasing its permeability. In addition, Parkinson et al. in 2010 [13] also performed a different approach in stimulating the production zone of Pinda formation, located in West Africa. This formation was having the bottomhole static temperature (BHST) of 300°F and was multilayered with carbonates. A low pH 4 HEDTA chelant was used as the main stimulation fluid for six producing wells from the formation. The result showed a high effectiveness of the stimulation acid at a high temperature. The production of all six wells had been doubled after the stimulation.

LePage et al. in 2009 [14] studied the reaction of Glutamic Acid N, N-Diacetic Acid (GLDA) with calcites in carbonate formation. Comparisons between GLDA and other chelants like Ethyllenediaminetetraacetic Acid (EDTA), Hydroxyethylethylenediaminetriacetic Acid (HEDTA), Nitrilotriacetic Acid (NTA) and Ethanoldiglycine Acid (EDG) are discussed. In short, GLDA is less corrosive than HCl and is effective as HEDTA. Apart from that, Mahmoud et al. in 2011 [15] also used GLDA to investigate its effect on sandstone cores stimulation in which various parameter such as injection rate, volume of GLDA, temperature, and GLDA initial pH value were investigated. The results revealed that GLDA had a strong ability in chelating calcium, iron, magnesium, and it chelated small amounts of aluminum ions from the sandstone cores. Moreover, it showed that the concentration of GLDA before and after core flood experiment was almost the same and showing a high thermal stability up to 300°F. Nasr-El-Din in 2013 [16] further proved that GLDA has high thermal stability and low corrosion potential. Based on the result, GLDA causes 21% and 84% increase in permeability at 200°F and 300°F respectively in comparison to HCl, which causes 42% reduction in permeability due to iron hydroxide precipitation.

Whereas Feng et al. in 2011 [17] used a high temperature deep penetrating (HTDP) acid in its investigation. A new corrosion inhibitor has been developed, which is a mixture of complex organo-phosphate hydrolyzed fluoride salts to generate HF and make HTDP acid suitable for formations with high temperatures. The HTDP acid also has much higher quartz solubility in comparison to the mud acid and will reduce precipitation. Rignol et al. in 2015 [18] tested on another chelant-based fluid system (low pH chelant + fluroboric acid, HBF_4) in sandstone stimulation under ultra-high temperature of 375°F. The core flow test and chemical analysis were performed on the core sample. According to the sequential dissolution results, the chelant-based fluid did not result in silica precipitation due to the absent of HCl. It is also highly effective in increasing the permeability of the core. Garcia et al. in 2016 [19] had discussed the advantages of Aminopolycarboxylic Acid (APCA) fluid, which contains 1 – 1.5% of HF acid. It is better than the conventional HF acid that showed ineffectiveness in acidizing sandstone at a high temperature of higher than 300°F due to the presence of sodium and potassium iron precipitation. This APCA/HF* blend fluid had been used in offshore reservoirs and successfully result in production enhancement of 30 to 50% BOPD for more than 12 months.

From the literature, many researchers had performed experimental studies using different chelants, which can be used in high temperature well stimulation. It is highly commendable that many researches had focused on the representativeness of those acids in high temperature wells. These acids have been proven to be very suitable and reliable to be applied and used in practice. However, chelating agents are generally more suitable for heterogeneous carbonates and clay-rich sandstones but less suitable for clean homogeneous sandstones due to the precipitation of silica. It is also very costly in comparison to mud acid, retarded and organic acids. On the contrary, the new acid combination (HBF_4:H_3PO_4) can eliminate this problem as it is less corrosive, stable and allow deep penetration due to slow hydrolytic reaction.

3.3. Retarded and Organic Acids

Al-harbi et al. in 2012 [20] performed an experimental work using different organic-HF acid mixtures to stimulate sandstone formations. The precipitations that take place during the reactions and the factors affecting these precipitations are investigated. These acids include Acetic (CH_3COOH)-HF, Formic (HCOOH)-HF and Citric ($C_6H_8O_7$)-HF combinations. The type and amount of precipitate were found to be mainly dependent on solution pH, organic-HF type, and initial free fluoride concentration. Also, the main parameter that is responsible for the aluminum-fluoride precipitation, is F/Al ratio. At a certain point over critical ratio, the aluminum fluoride precipitation took place.

Another experimental work was conducted by Yang et al. in 2012 [21] with a combination of HF-organic acids instead of mud acid to minimize the problems that are related to mud acid. The reaction kinetics and products are analyzed and the study outcome revealed the fact that the reactions depend on the type of minerals present in the core samples. Zhou and Nasr-El-Din in 2013 [22] investigated the performance of single stage sandstone acid system (HF – phosphonic acid) during sandstone acidizing in high temperature formation of 300°F. The effectiveness of low pH 3.8 GLDA, HEDTA and Formic acid, HCOOH in removing carbonate minerals from a sandstone core sample were also investigated. The results showed that the acid system increases the permeability of Berea sandstone core. In comparison to GLDA and HEDTA, HCOOH is more effective in removing carbonate minerals from Bandera sandstone cores.

Similar experimental study was presented by the same research team, Shafiq et al. in 2015 [23] and Shafiq and Ben Mahmud in 2016 [24] with different acid combinations, replacing conventional mud acid. These include the combinations of (1) Orthophosphoric (H_3PO_4) acid-HF, (2) Fluoboric (HBF_4) acid-HCOOH, and (3) HCOOH-HF. The sandstone core samples are saturated in these acid combinations and analyzed in term of the porosity, permeability, mineralogy and strength. Based on their findings, although all these acid combinations can be used as the main acid in sandstone acidizing, yet the best acid combination is 3% HF: 9% H_3PO_4. By showing permeability increase of 135.32%, this combination is even better than standard mud acid (3% HF: 12% HCl), which shows permeability increase of 101.76%. However, the experimental condition was only at ambient temperature, which is not representative at real field condition. Hence, it shows a research gap which can be bridged in the future.

Likewise, Zhou and Nasr-El-Din in 2016 [25] had also studied the performance of an alternative to mud acid, which is the phosphonic-based HF acid system. The interaction between the new acid systems with the clay minerals was investigated in term of acid concentration, reaction time and temperature. In comparison to mud acid, phosphonic-based HF acid system shows significantly better performance of permeability enhancement of 177.86% at 300°F.

Therefore, it is clear that there are many different retarded and organic acid combinations, which can be applied in sandstone stimulation apart from the conventional mud acid, which have been widely used. Out of all the acid combinations used and tested experimentally, (HBF_4:H_3PO_4) can be further tested from the research conducted by Shafiq and Ben Mahmud in 2016 [24] at the higher temperatures of 80°C and 100°C. Then, these results should be compared with the performance of mud acid at the same temperatures. In conclusion, the experimental procedures and outcomes from these literatures have provided some constructive approaches for the experimental setup and design in the future.

3.4. Comparison of Different Acid Combinations

Table 1. Comparison of Different Acid Combinations used in Matrix Acidizing

Acid Formulae	Authors and Years	Advantages	Disadvantages
Conventional Mud Acid HF:HCl	Al-Harthy et al. (2009) [2], Gomaa et al. (2013) [11]	- HF dissolves minerals in sandstone. - HCl controls precipitation. - Different HF:HCl concentration can successfully improve the permeability of the core.	- Cause rapid reaction rates at high temperature of 200°F. - Inefficiency of acidizing process as the acids are consumed too early. - Hazardous, difficult for safety control and corrosive to wellbore equipment.
Chelating Agents GLDA EDTA HEDTA NTA	Ali et al. (2008) [12], LePage et al. (2009) [14], Parkinson et al. (2010) [13], Mahmoud et al.	- Have high thermal stability up to 300°F. - Less corrosive than HCl. - Less corrosive and not hazardous. - Highly effective for heterogeneous carbonate and	- Less suitable for clean sandstone due to the precipitation of silica. - High cost in comparison to other acids.

EGD/HEIDA HTDP APCA	(2011) [15], Feng et al. (2011) [17], Nasr-El-Din (2013) [16], Rignol et al. (2015) [18], Garcia et al. (2016) [19]	clay-rich sandstone. - Successful application in offshore wells, with improved production of 30% - 50% BOPD.	
Retarded and Organic Acids HCOOH CH_3COOH HBF_4 H_3PO_4 $C_6H_8O_7$	Al-harbi et al. (2012) [20], Yang et al. (2012) [21], Zhou and Nasr-El-Din (2013) [22], Shafiq et al. (2015) [23], Shafiq and Ben Mahmud (2016) [24], Zhou and Nasr-El-Din (2016) [25]	- Act as a buffer and able to penetrate into deeper formation before spending. - Stable and not sensitive to undissolved fines. - Less corrosion and slow hydrolytic reaction.	- Reaction Mechanism with sandstone is not investigated up to date. - Less investigation shown at elevated temperatures. - Less practical results in field.

4. Conclusion

In the past, mud acid (HF:HCl) had been well proven for its effectiveness in improving the porosity and permeability of sandstone reservoir formation. This is due to the uniqueness of HF in dissolving silica minerals in sandstone and HCl in controlling precipitation. However, as the well temperature increases to more than 200°F, mud acid faces significant problem like rapid reaction rates, which results in early acid consumption and decreased acidizing efficiency. Mud acid is also hazardous and very corrosive. Therefore, Fluoroboric Acid (HBF_4) and Phosphoric acid (H_3PO_4) has been proposed in the future work as they offered numerous advantages in comparison to the conventional mud acid. Both acids result in more increase in the permeability and significantly less corrosive.

5. Future Work Recommendation

In the future, it is recommended that these two acids can be developed as a new acid combination. The newly acid combination is proposed to be experimentally tested using core flooding equipment at higher temperature at 80°C and 100°C. Different analytical tests such as Poro-Perm, FESEM, CT Scan, ICP and FNMR can be done to investigate the effectiveness of the new acid combination.

References

[1] Aboud R S, Smith K L, Forero Pachon L and Kalfayan L J 2007 *SPE Annual Tech. Conf. and Exh.* (Anaheim California USA)
[2] Al-Harthy S, Bustos O A, Fuller M J, Hamzah M E, Ismail M I and Parapat A 2009 *Oilfield Review* **4.20** 52-53
[3] Muecke T W 1982 Principles of Acid Stimulation. *Int. Pet. Exh. and Tech. Symp* (Beijing China)
[4] Smith C F and Hendrickson A R 1965 *J. Pet. Tech.* **17** 215-222

[5] Flood D T 1933 *Org. Synth.* **2** 295
[6] Ponce da Motta E, Plavnik B and Schechter R S 1992 *SPE Reservoir Engineering* **7** 149-153
[7] Ayotte P, Hebert M and Marchand P 2005 *J. Chem. Phys.* **123** 184501
[8] McLeod H O 1984 *J. Pet. Tech.* **36** 2055-2069
[9] Zeit B 2005 *SPE Technology Transfer Workshop (TTW)* A Presentation
[10] Farley J T, Miller B M and Schoettle V 1970 *J. Pet. Tech.* **22** 433-440
[11] Gomaa A M, Cutler J, Qu Q, Boles J and Wang X 2013 *SPE Euro. Formation Damage Conf. and Exh.* (Noordwijk The Netherlands)
[12] Ali S A, Ermel E, Clarke J, Fuller M J, Xiao Z and Malone B 2008 *SPE Euro. Formation Damage Conf.* (Sheveningen The Netherlands)
[13] Parkinson M, Munk T K, Brookley J G, Caetano A D, Albuquerque M A, Cohen D and Reekie M R 2010 *SPE Int. Symp. and Exh. on Formation Damage Control* (Louisiana USA)
[14] LePage J N, De Wolf C, Bemelaar J and Nasr-El-Din H A 2009 *SPE Int. Symp. on Oilfield Chemistry* (The Woodlands Texas USA)
[15] Mahmoud M A, Nasr-El-Din H A, De Wolf, C and Alex A 2011 *SPE Int. Symp. on Oilfield Chemistry* (The Woodlands Texas USA)
[16] Nasr-El-Din H A 2013 *J. Pet. Tech.* **65** 28-31
[17] Feng P, Wang D, Liu G, Wang H and Economides M J 2011 *SPE Western North American Region Meeting* (Anchorage Alaska USA)
[18] Rignol J, Ounsakul T, Kharrat W, Fu D, Teng L K, Lomovskaya I and Boonjai P 2015 *SPE Int. Symp. on Oilfield Chemistry* (The Woodlands Texas USA)
[19] Garcia E A R, LaBlanc A, Beuterbaugh A and Calabrese T 2016 *SPE Int. Symp. and Exh. on Formation Damage Control* (Louisiana USA)
[20] Al-harbi B G, Al Dahlan M N and Khaldi M H 2012 *SPE Int. Symp. and Exh. on Formation Damage Control* (Louisiana USA)
[21] Yang F, Nasr-El-Din H A and Harbi B A 2012 *SPE Production and Operations Symp.* (Doha Qatar)
[22] Zhou L and Nasr-El-Din H A 2013 *SPE Euro. Formation Damage Conf. and Exh.* (Noordwijk The Netherlands)
[23] Shafiq M U, Ben Mahmud H K and Hamid M A 2015 *IOP Conf. Ser.: Mater. Sci. Eng.* **78** 012008
[24] Shafiq M U and Ben Mahmud H K 2016 *IOP Conf. Ser.: Mater. Sci. Eng.* **121** 012002
[25] Zhou L and Nasr-El-Din H A 2016 *SPE Production and Operations Symp.* (Oklahoma USA)

Acknowledgements
This study was supported by the Fundamental Research Grant Scheme (FRGS) and Faculty of Engineering and Science (FOES) in Curtin University Malaysia.

Assessment of C-Type Darrieus Wind Turbine Under Low Wind Speed Condition

M S Misaran[1], Md. M Rahman, W K Muzammil and M A Ismail

Mechanical Engineering Program, Faculty of Engineering, Universiti Malaysia Sabah, Kota Kinabalu Sabah Malaysia

E-mail: [1]suffian@ums.edu.my

Abstract. Harvesting wind energy in in a low wind speed region is deem un-economical if not daunting task. Study shows that a minimum cut in speed of 3.5 m/s is required to extract a meaningful wind energy for electricity while a mean speed of 6 m/s is preferred. However, in Malaysia the mean speed is at 2 m/s with certain potential areas having 3 m/s mean speed. Thus, this work aims to develop a wind turbine that able to operate at lower cut-in speed and produce meaningful power for electricity generation. A C-type Darrieus blade is selected as it shows good potential to operate in arbitrary wind speed condition. The wind turbine is designed and fabricated in UMS labs while the performance of the wind turbine is evaluated in a simulated wind condition. Test result shows that the wind turbine started to rotate at 1 m/s compared to a NACA 0012 Darrieus turbine that started to rotate at 3 m/s. The performance of the turbine shows that it have good potential to be used in an intermittent arbitrary wind speed condition as well as low mean wind speed condition.

Keywords: Wind turbine, C-type Darrieus, Tip speed ratio.

1. Introduction

Wind turbine is analogues to renewable energy and have been widely accepted as one of the common way to produce electricity, cost effectively. This technology has supplied electricity amounting to 3% of total global electric consumption [1]. Essentially, wind turbine converts kinetic energy from rotary motion of the blade/shaft assembly to electrical energy and store them in a battery or other energy storage means [2].

Generally, there are two types of wind turbine; horizontal axis wind turbine and vertical axis wind turbine [3]. Horizontal axis wind turbine is more common as it can produce electricity at higher output than vertical axis wind turbine in the lower mean speed range. But this type of wind turbine is heavy and can't operate efficiently in turbulent wind condition. Vertical axis wind turbine can operate at arbitrary condition and it can harvest wind power from any direction due to the versatility of the blade construction [4].

Malaysia mean wind speed is 2 m/s with high wind speed recorded up to 6 m/s at certain locations. However, because of monsoon season the mean speed for a location varies at different month of the year. This is further emphasized by geographical setting of Malaysia which is naturally shielded in west Malaysia by the Indonesian island cluster. On the eastern seaboard, Malaysia is exposed to the South China Sea which caused the area to register higher wind speed at peak monsoon season. In sabah, one district that shows potential for wind turbine utilization as an alternative electricity source is the Kudat region. However, the mean wind speed registered are relatively low at 3 m/s and barely enough for an economically viable wind turbine electricity generation.

The vertical axis wind turbine is available in two design type; lift and drag type [3]. In a slow wind speed condition, vertical axis drag type wind turbine is better suit compared to both horizontal and vertical axis-lift type turbine. However, vertical axis lift-type turbine performs better at higher wind speed condition. The Darrieus wind turbine is an example of lift type vertical axis wind turbine that perform efficiently at high wind speed condition. Furthermore, the design requires smaller number of components; easily accessible for maintenance work; low noise emissions and ability to operate at near ground level [5]. However, it is well known that Darrieus wind turbine have a distinct disadvantage compared to both a horizontal wind turbine and drag type wind turbine; its inability to self-start in low wind speed condition. Thus, several solutions have been proposed to overcome this drawback; variable pitch mechanism [6], hybrid Savonius-Darrieus configuration [7] or custom blade profile that improves the self-start capabilities of Darrieus turbine [8].

The blades play an important role in determining the overall efficiency of the wind turbine, thus wind turbine manufacturer concentrated their effort to design and developing new blade design to increase the performance of their wind turbine [9]. The development of a blade with self-start capability is a challenging engineering task as effort is concentrated on improving the blade profile rather than the use of extra components. This is to keep the Darrieus blade design simple as possible as additional complex mechanism affect reliability and cost [5].

Designing of wind turbine blades is a true multi objective engineering task. The aerodynamic effectiveness of the turbine need to be balanced with the system loads caused by the rotor [10]. Recent study shows blade improvement such as turbine blade with feasible laminated layout and thin construction of blade improved the efficiency and durability of the turbine [11]. Light weight is one of the factor affecting wind turbine efficiency [12], thus material selection and blade/turbine design are an important phase to ensure that the turbine have lower cut-in speed and faster response time.

Most the wind turbine design available in the market today are design to have wind cut-in speed at 3.5 m/s. As a result, these turbines will not constantly run due to the low wind speed with intermittent high speed wind condition recorded in Malaysia. Thus, a Darrius wind turbine that able to self-start at low wind speed while operate efficiently at high wind speed is highly desirable. Hence, this paper objective is to develop a Darrius wind turbine blade capable to start at low wind speed while running efficiently at high wind speed. A C-type wind turbine is proposed in this paper as a solution to the current demand. The C-type blade is a hybrid design; a combination of Darrieus section with NACA 0012 profile which provide lift type characteristic while a hollow section provides drag type characteristic.

2. Methodology

The C-type Darrieus blade profiles are generated with the NACA 0012 profile as the reference. The design of this unique blade structure is that it is hollow and exposed at one side, while having the NACA 0012 profile as the cross section as shown in figure 1. The generation of the NACA 0012 profile were done using online tools (airfoiltools, 2014) and the major dimension are shown in figure 2.

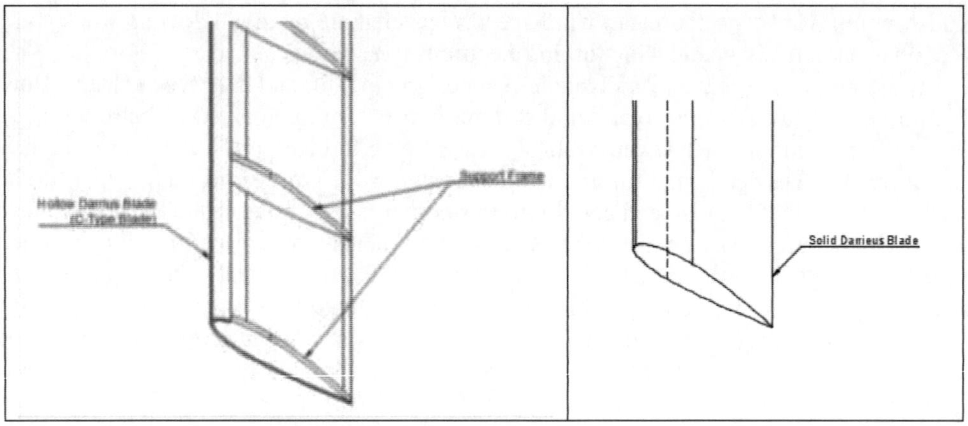

(a) (b)

Figure 1. Blade design for (a) C-Type (b) Solid Type.

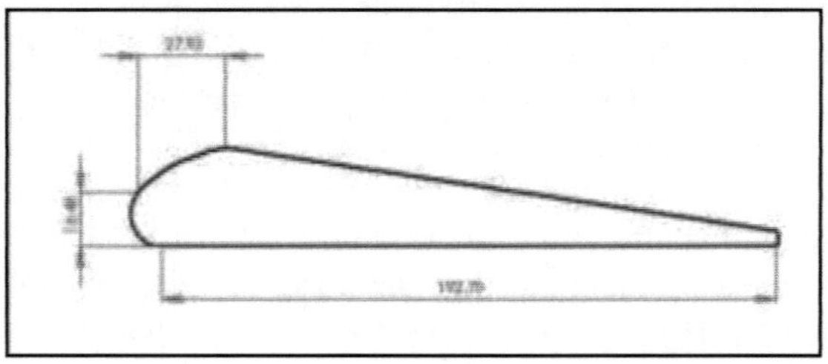

Figure 2. Dimension in mm for C-type with NACA 0012 profile.

To demonstrate the effect of the C-type design, the rotation of the turbine is divided into two phase as shown in figure 3. In front facing phase, the profile gave less drag to the overall assembly. However, it is noted that at the beginning of front facing phase the drag force is higher due to the C-profile exposed to wind stream. Thus, a three blade system is used to counter this issue. While in the rear facing phase, the C-type profile significantly increases the driving force of the blade thus giving higher torque.

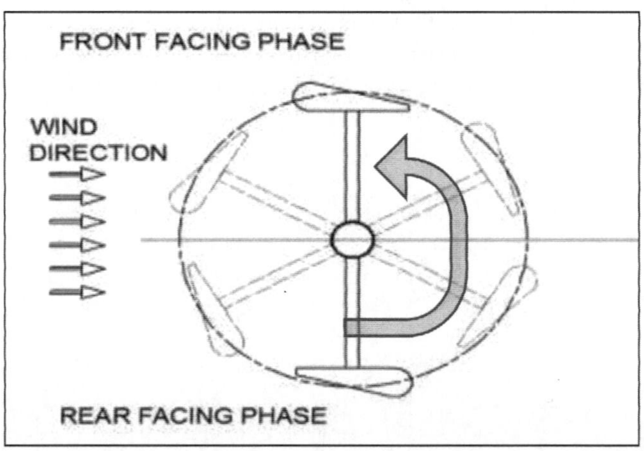

Figure 3. Rotation phase for C-type Darrieus turbine.

The turbine blade was designed using CAD software and fabricated in the faculty in-house manufacturing facility. It was designed in such a way that different angle of attack can be used, thus enable us to study the effect of angle of attack on the wind turbine performance. The material used for the blades prototype was mainly aluminum sheet as it has adequate yield strength to withstand the low wind force and easy to be shaped. Balsa wood were used as the supporting frame for the blade while the centre rotor hub was made from aluminum shaft. The final design of the wind turbine is as shown in figure 4.

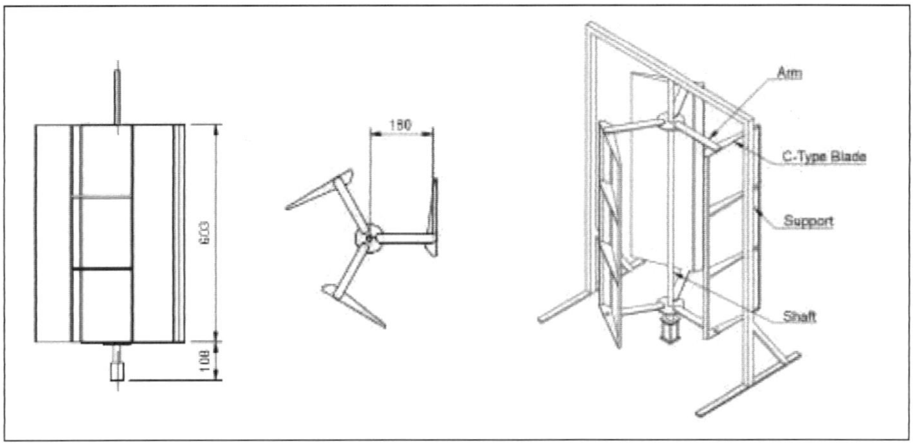

Figure 4. Schematic figure of C-type Darrieus wind turbine.

3. Experimental Work

To assess the performance of the C-type Darrieus turbine, the wind turbine undergoes simulated wind condition with rotation speed recorded at 1 m/s wind speed increment up to 5 m/s. Six different rotor pitches at -3°, -6°, 0, 3° and 6° representing different angle of attacks were tested. NACA 0012 Solid Darrieus blade was also tested at similar wind condition and angle of attack for comparison purpose. The rotor speed was measured using a digital tachometer and the wind speed is determined using a digital anemometer. The purpose of this study is to investigate the performance of the C-type Darrius blade turbine and compare it with the solid blade NACA 0012 turbine at different wind speed. Thus, the performance of the machine can be expressed in term of tip speed ratio at different wind speed. Tip speed ratio is defined as the ratio of blade tangential speed to actual wind speed and is written in equation (1);

$$TSR = \frac{\omega R}{v} \quad (1)$$

4. Discussion

A total of 6 experiments were conducted for each turbine; C-type Darrieus blade and NACA 0012 blade. For each experiment sets, the minimum wind speed is 1 m/s and increased up to 5 m/s. Figure 5 shows the tip speed ratio for C-type Darrieus turbine at various wind speed and angle of attack. Due to the C-type design, the turbine started to rotate as low as 1 m/s an incrementally increased its rotation speed as the wind speed increases. Notably, C-type Darrieus turbine with angle of attack 3° performs extremely well compared to other angle of attacks with maximum TSR at 1.8. As for NACA 0012 Darrieus blade turbine, the Tip Speed Ratio (TSR) data are shown in Figure 6. As expected, the

IOP Conf. Series: Materials Science and Engineering **217** (2017) 012019 doi:10.1088/1757-899X/217/1/012019

turbine only started to rotate at wind speed 3 m/s which is typical of this kind of blade. The TSR incrementally increased as the wind speed increases, however different angle of gave clear TSR deviation from each other. Such behavior were not observed in the C-type Darrieus turbine where the value of TSR does not differ much at all wind speed at the exception of angle of attack 3° and 0°. The best performance recorded for NACA 0012 Darrieus turbine is the 0° angle of attack configuration with maximum TSR value recorded at 1.15.

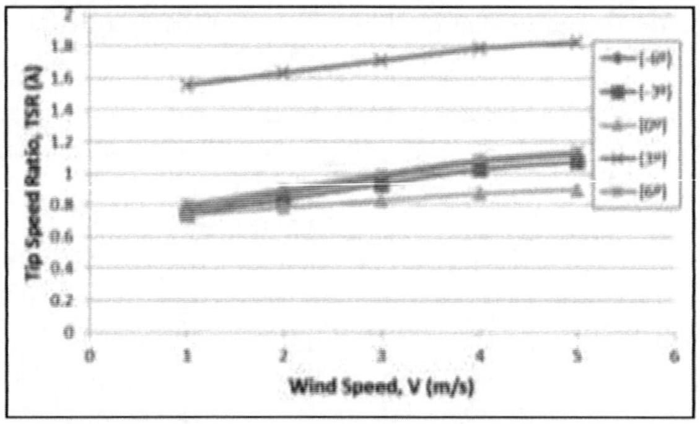

Figure 5. Tip speed ratio (TSR) for C-type Darrieus wind turbine.

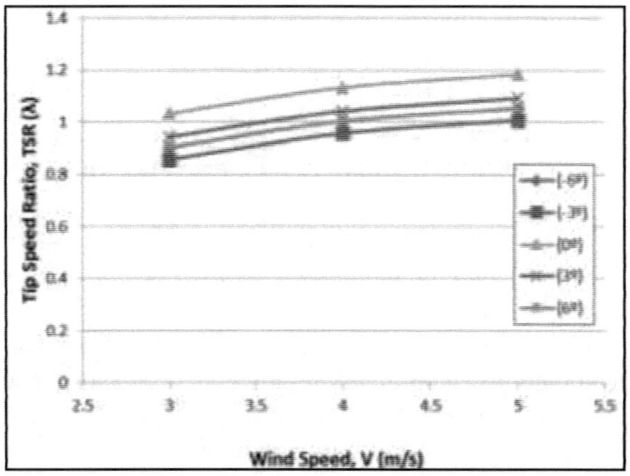

Figure 6. Tip speed ratio (TSR) for NACA 0012 Darrieus wind turbine.

The experimental result shows that the hybrid design of the C-type turbine have successfully aided the turbine to self-start at low wind speed. The hollow side of the C-type blade provide sufficient drag when is exposed to down-wind force during the rear-facing-phase while the NACA 0012 profile kept a low drag force in front-facing phase. Comparing this to a solid Darrieus blade design, the turbine only started to rotate at higher wind speed albeit at lower tip speed ratio.

5. Conclusion

In this work, a C-typed Darrieus wind turbine was designed, build and tested against a NACA 0012 Darrieus wind turbine at various wind speed and pitch angle. Test result shows that;

a) C-type Darrieus wind turbine started to rotate at lower wind speed of 1 m/s compared to NACA 0012 Darrieus turbine which only start to rotate at 3 m/s.

b) Generally, C-type Darrieus turbine TSR value have small difference in value with NACA 0012 Darrieus turbine at similar wind speed. However, for angle of attack 3° the TSR is significantly higher.

As a conclusion, the C-typed Darrieus wind turbine shows good potential as an alternative design specifically for arbitrary wind speed condition as observed in Malaysia. However, the flimsy blade construction might pose a challenge to design a reliable wind turbine especially in larger size. Future study need to focus on studying the effect of different NACA C-type profile on the wind turbine performance as the current study only focus on a single NACA 0012 profile.

References

[1] Igba J, Alemzadeh K, Anyanwu-ebo I, Gibbons P, and Friis J 2013 A Systems Approach towards Reliability-Centred Maintenance (RCM) of Wind Turbines. *Procedia Computer Science*. **16** pp 814-823

[2] Sathiyamoorthy S, and Bharathi N 2012 Hybrid energy harvesting using piezoelectric materials, automatic rotational solar panel,vertical axis wind turbine. *Procedia Engineering*. **38** pp 843-852

[3] Pramod J 2011 *Wind Energy Engineering* Mc Graw (Hill New York)

[4] Castillo J 2011 *Small- scale Vertical Axis Wind Turbine Design*. Bachelor Thesis. (Tampere University)

[5] Batista N C, Melicio R, Matias J and Catalao J P S 2011 New blade profile for Darrieus wind turbines capable to self-start. *Conf. Renewable Power Generation*

[6] Paraschivoiu I, Trifu O, and Saed F 2009 H-Darrieus wind turbine with blade pitch control. *Int. J. of Rotating Machinery*, **2009**, pp 1-7

[7] Gupta R, Biswas A, and Sharma K K 2008 Comparative study of three bucket savonius rotor with a combined three bucket savonius-three bladed darieus rotor. *Energy*. **33**, pp 1974-1981

[8] Kirke B K 1998 Evaluation of self-starting vertical axis wind turbines for stand alone applications, *PhD dissertation*, Griffith University, Australia

[9] Kim B, Lee Y, Kim W, Lee S, and Bae S 2013 Developement and verification of a performance based optimal design software for wind turbine blades *Renewable Energy* **54** pp 166-172

[10] Savill M, Fischer G R, and Kipouros T 2014 Multi-objective optimisation of horizontal axis wind turbine structure and energy production using aerofoil and blade properties as design variables *Renewable Energy* **62** pp 506-515

[11] Ameku K, Nagai B M, and Roy J N 2008 Design of a 3 kW wind turbine generator with thin airfoil blades *Exp. Thermal and Fluid Science* **32** pp 1723-30

[12] Tong W 2010 *Wind power generation and wind turbine design* (Southampton; Boston: WIT Press)

Acknowledgements
The authors wish to convey a sincere gratitude to Universiti Malaysia Sabah (UMS) for supporting the current research work thru the SBK research grant, project no. SBK0086-TK-2013.

IOP Conf. Series: Materials Science and Engineering **217** (2017) 012020 doi:10.1088/1757-899X/217/1/012020

Performance Investigation of A Mix Wind Turbine Using A Clutch Mechanism At Low Wind Speed Condition

M J Jamanun[1], M S Misaran[2], M Rahman, W K Muzammil
Mechanical Engineering Programme, Faculty of Engineering, Universiti Malaysia Sabah, 88450 Kota Kinabalu, Sabah, Malaysia.

E-mail: [1]jamrye@ymail.com, [2]suffian@ums.edu.my

Abstract. Wind energy is one of the methods that generates energy from sustainable resources. This technology has gained prominence in this era because it produces no harmful product to the society. There is two fundamental type of wind turbine are generally used this day which is Horizontal axis wind turbine (HAWT) and Vertical axis wind turbine (VAWT). The VAWT technology is more preferable compare to HAWT because it gives better efficiency and cost effectiveness as a whole. However, VAWT is known to have distinct disadvantage compared to HAWT; self-start ability and efficiency at low wind speed condition. Different solution has been proposed to solve these issues which includes custom design blades, variable angle of attack mechanism and mix wind turbine. A new type of clutch device was successfully developed in UMS to be used in a mix Savonius-Darrieus wind turbine configuration. The clutch system which barely audible when in operation compared to a ratchet clutch system interconnects the Savonius and Darrieus rotor; allowing the turbine to self-start at low wind speed condition as opposed to a standalone Darrieus turbine. The Savonius height were varied at three different size in order to understand the effect of the Savonius rotor to the mix wind turbine performance. The experimental result shows that the fabricated Savonius rotor show that the height of the Savonius rotor affecting the RPM for the turbine. The swept area (SA), aspect ratio (AR) and tip speed ratio (TSR) also calculated in this paper. The highest RPM recorded in this study is 90 RPM for Savonius rotor 0.22-meter height at 2.75 m/s. The Savonius rotor 0.22-meter also give the highest TSR for each range of speed from 0.75 m/s, 1.75 m/s and 2.75 m/s where it gives 1.03 TSR, 0.76 TSR, and 0.55 TSR.

Keywords: Wind speed performance, Clutch system, Savonius-Darrieus turbine, low wind speed.

1. Introduction

Wind energy is one of energy resources that is available in abundance and can be utilized to produce mechanical work. A wind turbine is used to convert the kinetic energy from air movement to mechanical work, thus a lot of research have been done to increase the efficiency of a wind turbine at different wind condition. The technology has evolved and continually developed throughout the years and now it becomes one of the economically viable and sustainable sources of energy [1].

Wind turbine technology is commonly used in a suitable location that has strong consistent wind speed. In the Malaysian context, studies showed that there is a limited application for wind turbine technology as an electric generator. Generally, a site which has a mean wind speed exceeds 6m/s can be considered as a suitable location for a wind farm. Several locations that have minimum mean wind speed have been identified in Malaysia. Per Jha, 2011 [2], the locations that haves a mean speed of 6 m/s in Malaysia includes; Kelantan, Terengganu, Pahang, Sarawak, and Kedah. In the context of Sabah and its immediate shores, Kudat and Labuan is a suitable place for a small-scale wind turbine because it located at the northern site of Sabah that has an average wind speed of 6m/s and meets the requirement for wind farm development [3].

Wind turbines are categorized by two type which is Horizontal Axis Wind Turbine (HAWTs) and Vertical Axis Wind Turbine (VAWTs) [4]. Each of these turbine types has distinct properties; HAWT is much more suitable at low wind speed while VAWT is better suited for a higher wind speed condition [5]. The general different between these two type is the orientation of their axis of rotation. For HAWT, the horizontal axis always pointed in the horizontal direction into the wind or away from the wind to get the maximum efficiency in order to produce power while for VAWT rotational axis is perpendicular to the wind direction [6]. This study focuses on the VAWT which generally have better tolerance to wind direction change compared to the HAWT. A vertical axis wind turbine (VAWT) do not have to be orientated into the wind and can operate in any wind direction [7]. Subsequently, two types of VAWT are available; Savonius and Darrieus turbine [4]. A Savonius turbine is a drag type wind turbine while a Darrieus turbine is a lift type turbine.

The savonius wind turbine has an ability to self-start [8] and able to generate torque due to the drag force of its own geometry, but the efficiency is (20%) lower than darrieus rotor [9]. The ability to self-start is one of the critical factors for wind turbine considering the place that has low wind speed and varied from time to time.

Darreius turbine has the highest efficiency among the other VAWTs but do not have a self-start ability and has a low starting torque [9]. According to Daut 2012 [4], the darrieus wind turbine normally requires self-starting assist mechanism. To overcame these problem Darrieus- Savonius mix wind turbine were introduced and the starting torque and performance show dramatic improvement [10]. Almotairi 2011 [11] also state that Darrieus- Savonius mix turbine can generate more power and able to self-start in low wind condition. Combining these two type of blade on a single rotor can improve and compliment the disadvantage of each type of blade [12].

2. Methodology

The noise-free clutch system as shown in Figure 1 functioned as a coupling to connect Savonius and Darrieus blade in a single rotor. This clutch allows the Savonius blade to rotate at low wind speed and helps to drive the Darrieus turbine. At higher wind speed, the Darrieus turbine realizes higher rotation speed and the clutch will mechanically disconnect the Savonius blade from the Darrieus rotor. At this stage, the Darrieus turbine drives the turbine while the Savonius blade rotates freely. To investigate the effectiveness of the design Savonius rotor the performance of the turbine will be compared with stand-alone Darrieus turbine.

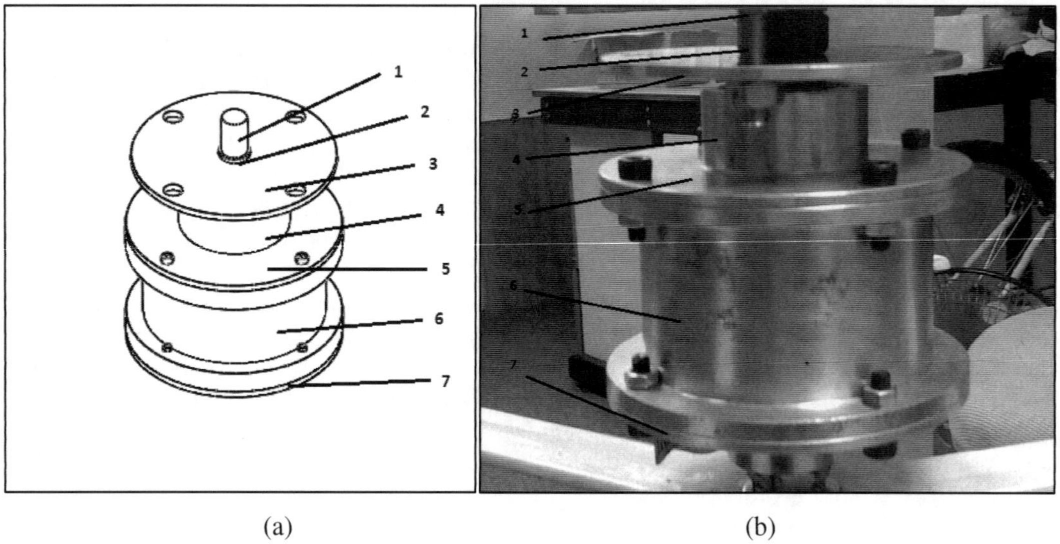

Figure 1. (a) Schematic of the design clutch, (b) photograph of fabricate noise-free clutch, considered in this work it consists of; 1 -shaft, 2 -c-clip, 3 -top plate, 4 - top cover, 5 -top base, 6 - base, 7 -bottom base.

Figure 2 shows the Savonius blade design that is used in this study. Researchers [8,14] have reported that the Savonius design is more stable when using two blade rotor. The power coefficient for the Savonius rotor increases when the aspect ratio increases and two stage rotor is more efficient compare to single stage rotor. The addition of end plates on a Savonius rotor also increases the overall efficiency and is included in this design.

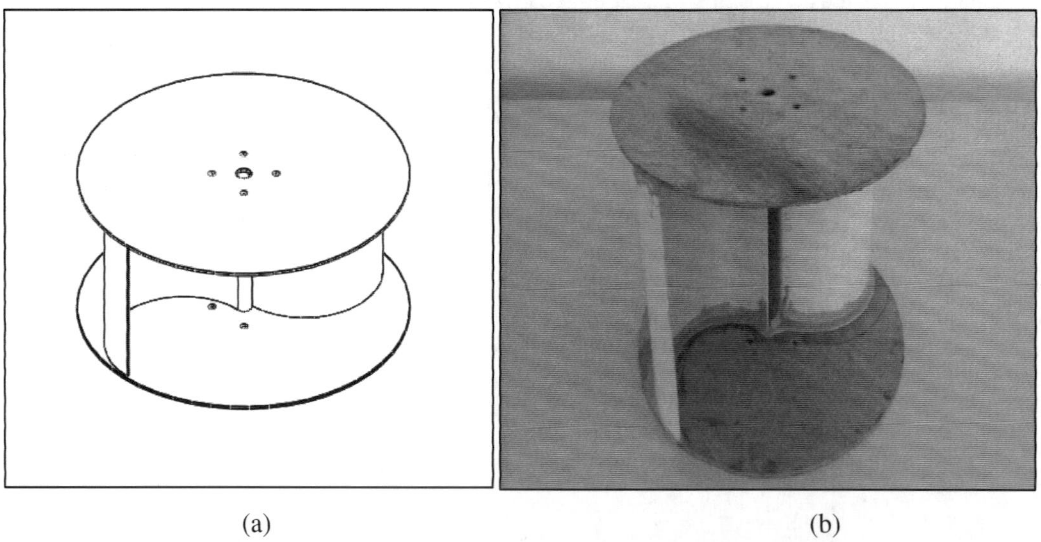

Figure 2. (a) Savonius rotor, (b) photograph of Savonius rotor.

The Darrieus blade used in this project are as shown in Figure 3. A Darrieus wind turbine powered by the lift force from the vertical axis blade; thus, the shape of the blade is identical to an airfoil [15]. The lift force created from the shape of the blade which is an asymmetric shape. A pressure difference is created from the blade when it cut the air with an angle of attack driven by wind flow creating a lift force. Also, Mohammed 2014 [16] reported that 0° (zero degree) blade pitch angle gave the best efficiency compared to other pitch angle.

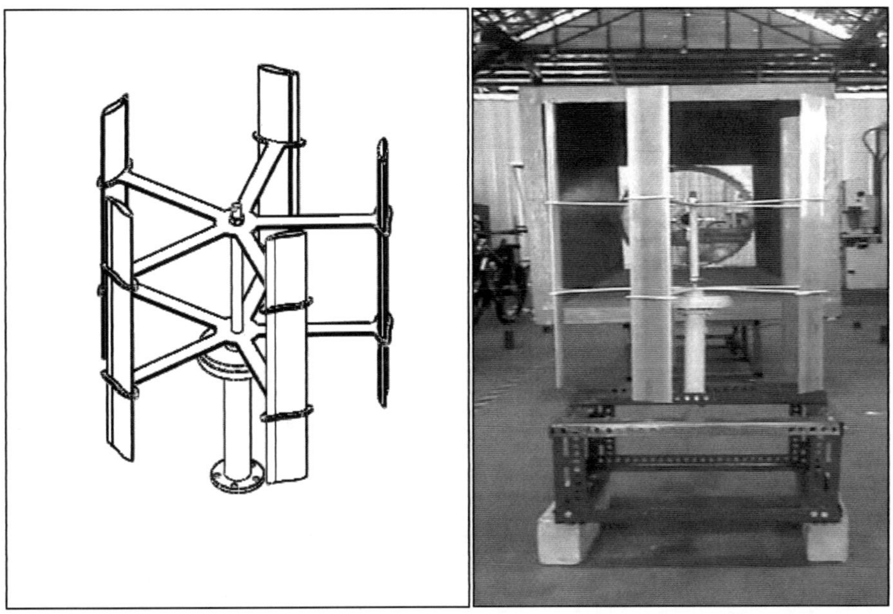

(a) (b)

Figure 3. (a) Darrieus turbine, (b) photograph of Darrieus turbine.

A mix wind turbine was designed and fabricated using the three main components; Savonius rotor, Darrieus blade and a clutch. Different Savonius heights were experimented to study the effect of Savonius size to the wind turbine performance. The material used for the clutch must have high yield strength so that it can withstand the weight from Savonius rotor and the wind force from the blower while able to resist weather element. Thus, an Aluminium 5000 series was used as the material as it corrosion resistant, machine-able and attainable in the market. Figure 4 shows the complete CAD model and fabricated model of the mix wind turbine with the clutch system. Table 1 and table 2 below shows the specification of the Darrieus and Savonius rotor used in this study. The design Savonius rotor has a fix diameter for each design which is 0.16 meter.

IOP Conf. Series: Materials Science and Engineering **217** (2017) 012020 doi:10.1088/1757-899X/217/1/012020

Table 1. Darrieus Turbine.

Model	SAV-75W Wind Turbine Generator
Weight	10.5 kg
Height	0.745 meter
Rated Power	50 Watt
Rotor Diameter	0.56 meter
Rated RPM	260 RPM

Table 2. Savonius rotor.

Savonius rotor	Savonius 0.16-meter	Savonius 0.19-meter	Savonius 0.22-meter
Height (meter)	0.16	0.19	0.22
Aspect ratio (H/D)	1.00	1.19	1.39
Swept Area (m^2)	0.026	0.030	0.035

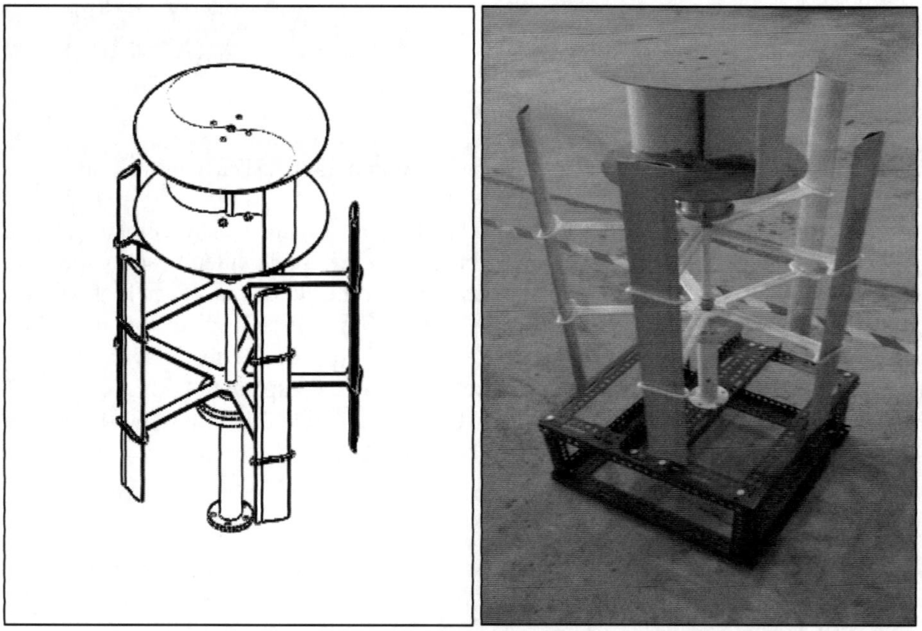

(a) (b)

Figure 4. (a) Complete CAD model, (b) photograph of complete model.

3. Experimental Work

To assess the performance of the design blade related to RPM performance using design noise-free clutch for MVAWT, three configurations and three different speed were planned and tested. Each set of the experiment were repeated three times to get an average reading. The rotor rotation speed (RPM) was recorded and swept area (SA), aspect ratio (AR) and tip speed ratio (TSR) was calculated the formula as shown as bellow.

$$TSR = \frac{\omega R}{v} \tag{1}$$

$$\omega = \frac{2 \times \pi \times n}{60} \tag{2}$$

$$SA = H \times D \tag{3}$$

$$AR = \frac{H}{D} \tag{4}$$

Where, TSR (Tip Speed Ratio), ω ω is angular velocity, R is the radius of the rotor, n is revolution per minutes, H is the height of the rotor and D is the diameter of the rotor.

The performance of the mix wind turbine is measured at three different wind speed; 0.75 m/s, 1.75 m/s and 2.75 m/s. These speeds are chosen because the experiment is designed to investigate the wind turbine performance at low wind speed condition. The Savonius rotor were varied in heights; 0.16 meter, 0.19 meter and 0.22 meter. The low wind speed is generated by a custom-made wind blower as shown in Figure 5. The wind speed is measured using smart sensor AR83 Anemometer Plus while the rotation speed of the wind turbine is measured using a MASTECH MS6208B non-contact Type Digital Tachometer. Three readings were made for each experiment and the average value is used to calculate the TSR.

Figure 5. Wind Blower

4. Results

The test shows that the highest RPM recorded was at 2.75 m/s which is 90 RPM for Savonius 0.22-meter. Savonius 0.22-meter, has the largest value of swept area which is 0.035 m^2 compare to the other design; Savonius 0.16-meter and Savonius 0.19-meter which have 0.026 m^2 and 0.030 m^2 swept area. As the swept area of the rotor increase, the amount of air which strikes the blades of Savonius rotor also will increases. Thus, the rotation per minutes (RPM) also increases. As shown in Figure 6, Stand-alone Darrieus turbine started to rotate at 1.5 m/s as the RPM recorded for this speed is only 6 RPM, compare to the other configuration shown in Figure 7 that used the clutch system as connecting device for Savonius and Darrieus turbine the turbine started to rotate at 0.75 m/s for each design of Savonius rotor. As shown in Figure 7 as the wind speed of the blower increases from 0.75 m/s to 1.75 m/s and to 2.75 m/s the RPM for each design increases for Savonius 0.22-meter the RPM increases from 46 RPM to 76 RPM to 90 RPM while for Savonius 0.19-meter and Savonius 0.16-meter the RPM increases from 44 RPM to 70 RPM to 86 RPM and 40 RPM to 68 RPM to 82 RPM. It can be concluded that using coupling as the connecting device for Savonius-Darrieus turbine increase its performance and able to self-start at low wind speed. Figure 8 show the highest value for tip speed ratio (TSR) recorded in this experiment which is 1.03 TSR at 0.75 m/s for Savonius 0.22-meter at wind speed 1.75 m/s and 2.75 m/s Savonius 0.22-meter also have the highest value for TSR compare to the other design which is 0.76 TSR and 0.55 TSR. When the RPM of the turbine increase, the TSR also will increases. It can be concluded that the Savonius 0.22-meter has shown good performance in term of RPM and TSR at low wind speed condition. Figure 6, Figure7, and Figure 8 shows more details about the comparison between Savonius 0.16-meter, Savonius 0.19-meter and Savonius 0.22-meter in term of RPM and TSR.

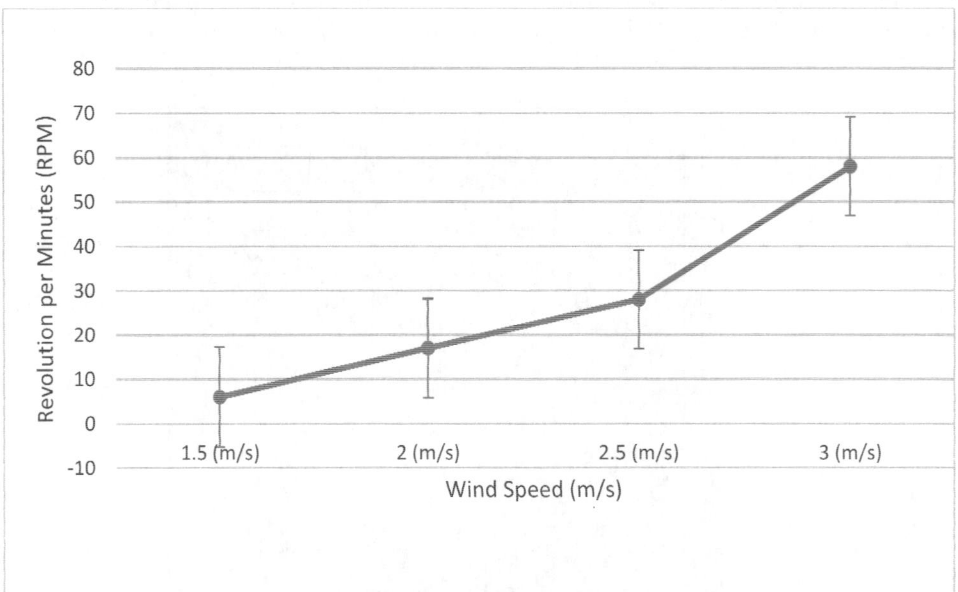

Figure 6. Data of Revolution per Minutes (RPM) vs Wind Speed (m/s).

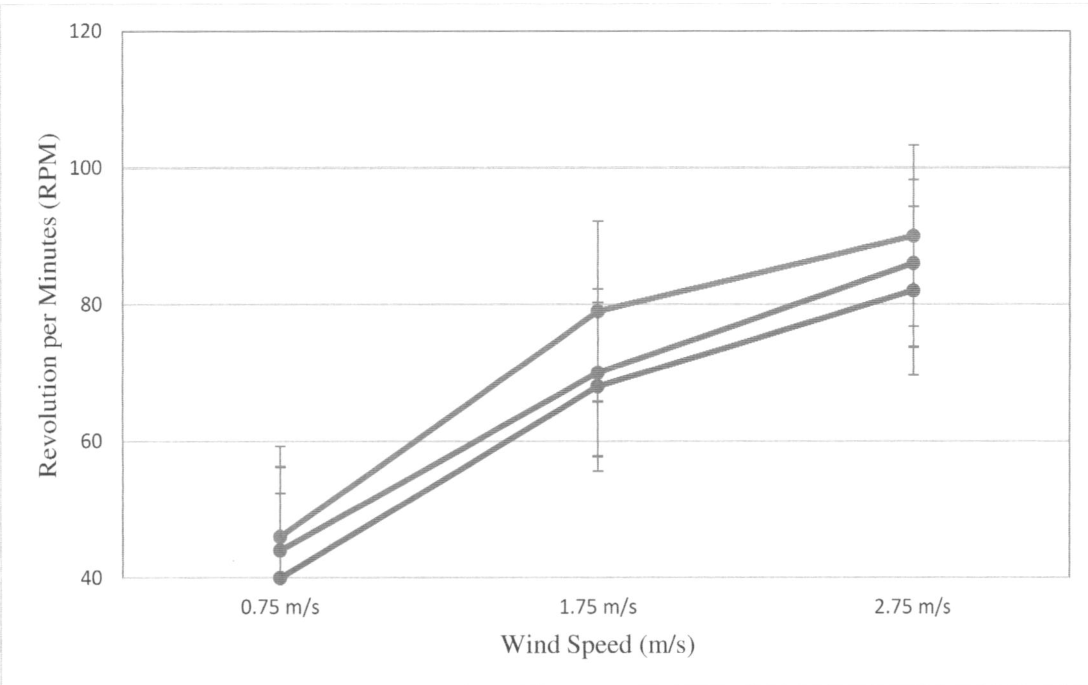

Figure 7. Data of Revolution per Minutes (RPM) vs Wind Speed (m/s).

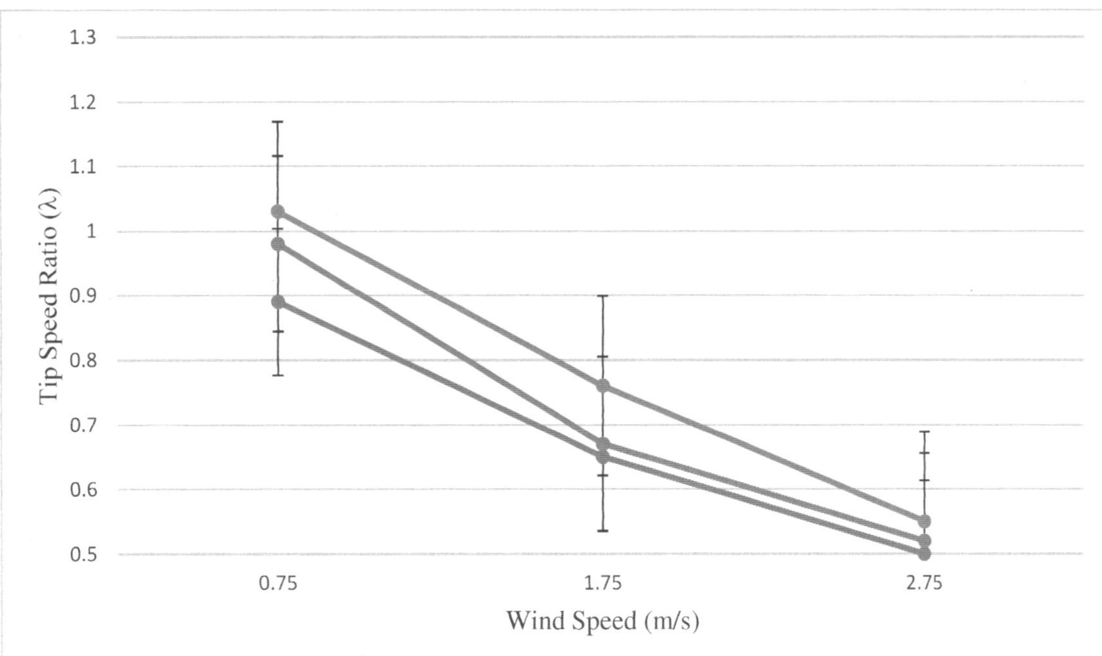

Figure 8. Data of tip speed ratio (λ) vs wind speed (m/s).

IOP Conf. Series: Materials Science and Engineering **217** (2017) 012020 doi:10.1088/1757-899X/217/1/012020

5. Conclusion

In this study, a mix wind turbine with a new coupling system and at various Savonius rotor height were investigated. The results show that when the Savonius rotor has a bigger swept area the TSR of the turbine will increase. As the swept area increase the air which strikes the blades of Savonius rotor also will increase and affecting the performance of the turbine in term of RPM and TSR. Figure 8 show the TSR of each design decrease as the wind speed increases, this due to the weight of the Savonius rotor that affecting the RPM of the turbine. The experiments show that;

- Highest RPM recorded is 90 RPM for Savonius 0.22-meter.
- Highest TSR recorded for this experiment is 1.03 which for Savonius 0.22-meter at 0.75 m/s.
- The addition of a clutch system in a mix wind turbine design have been proven to give the turbine a self-start capability.
- The Savonius rotor main function is to assist the Darrieus blade self-start capability as well as to maintain rotation at low wind speed condition. Thus, Savonius 0.22-meter have shown good rotation speed and TSR at low wind speed compared to another configuration.

Moreover, the design should be able to function normally at high wind speed condition as the clutch system is expected to disengage and allow the Darrieus turbine to rotate faster than the Savonius rotor. Further study needs to focus on better design and a full characterization of the noise-free clutch so it can increase the performance of the turbine. To improve the feasibility of wind turbine as an alternative power, further study like this project need to be conducted so that energy from a wind turbine can be generated at the region where the wind speed is not suitable for wind turbine technology.

Reference

[1] Jha A 2011*Wind Turbine Technology* (Taylor & Francis Group)

[2] Asmida and Nizam k 2016 Wave power potential in Malaysia territorial waters. *Earth and Environmental Science.* p 37

[3] Saidur R, Islam M and Rahim N 2011 Assessment of wind energy potentiality at Kudat and Labuan, Malaysia using Weibull distribution function. *ELSEVIER.* pp 985-992

[4] Daut I, Razliana A, Irwan A and Farhana Z 2012 A Study on the Wind as Renewable Energy in Perlis, Northern Malaysia. *ELSEVIER.* pp 1428-33

[5] Pramod J 2011 *Wind Energy Engineering* Mc Graw Hill (New York)

[6] Magedi M, Saad M and Asmuin N 2014 Comparison of Horizontal Axis Wind Turbine and Vertical Axis Wind Turbine. *IOSR Journal of Engineering.* pp 27-30

[7] Al-Bahadly and Ibrahim 2009 Building a wind turbine for rural home. *ELSEVIE.* pp 159-165

[8] Kadam A and Patil S A Review Study on Savonius Wind Rotors for Accessing the Power Performance. *IOSR Journal.* pp. 18-24

[9] Bhutta A, Nasir H, Farooq A and Ali Z 2012 Vertical axis wind turbine- A review of Various configurations and design techniques. *ELSEVIER.* pp. 1926-39

[10] KYOZUKA and Yusaku 2008 An Experimental Study on the Darrieus-Savonius Turbine for the Tidal Current Power Generation. *Journal of Fluid Science and Technology.* pp vol **3** no 3

IOP Conf. Series: Materials Science and Engineering **217** (2017) 012020 doi:10.1088/1757-899X/217/1/012020

[11] Almotairi A, Mustapha F, Ariffin M and Zahari R 2016 Synergy of Savonius and Darrieus types for vertical axis wind turbine. *IASE.* pp 25-30

[12] Kumar A and Nikhade A 2014 Hybrid Kinetic Turbine Rotors: A Review. *IJESAT.* pp 453-463

[13] Abid M, Karimov K, Wajid H, Farooq F, Ahmed H and Khan O 2014 Design, Development and Testing of a Combined Savonius and Darrieus Axis Wind Turbine. *Iranica Journal of Energy & Environment.* pp 1-4

[14] Mahmoud N, El-Haroun A,Wahba E and Nasef M 2012 An experimental study on improvement of Savonius rotor performance. *Alexandria Engineering Journal.* pp 19-25

[15] Tapan H, Hitesh J and Mayur K 2015 A Review on Straight Bladed VErtical Axis H-Type Darrieus Wind Turbine. *International Journal of Innovative Research in Science.* pp 1873-76

[16] Mohamed M, Ali A and Hafiz A 2014 CFD analysis for H-rotor Darrieus turbine as a low speed wind energy converter. *ELSEVIER.* pp 1-13

Acknowledgments
We would like to express our sincere gratitude to the Faculty of Engineering, Universiti Malaysia Sabah for their financial and facility supports provided in completing this project thru project no. SBK0086-TK-2013.

IOP Conf. Series: Materials Science and Engineering **217** (2017) 012021 doi:10.1088/1757-899X/217/1/012021

Working parameters affecting earth-air heat exchanger (EAHE) system performance for passive cooling: A review.

D Darius[1], M S Misaran[2], Md. M Rahman, M A Ismail, A Amaludin

Mechanical Engineering Programme, Faculty of Engineering, Universiti Malaysia Sabah, 88450, Kota Kinabalu, Sabah, Malaysia.

E-mail: [1]darryldarius1993@gmail.com, [2]suffian@ums.edu.my

Abstract. The study on the effect of the working parameters such as pipe material, pipe length, pipe diameter, depth of burial of the pipe, air flow rate and different types of soils on the thermal performance of earth-air heat exchanger (EAHE) systems is very crucial to ensure that thermal comfort can be achieved. In the past decade, researchers have performed studies to develop numerical models for analysis of EAHE systems. Until recently, two-dimensional models replaced the numerical models in the 1990s and in recent times, more advanced analysis using three-dimensional models, specifically the Computational Fluid Dynamics (CFD) simulation in the analysis of EAHE system. This paper reviews previous models used to analyse the EAHE system and working parameters that affects the earth-air heat exchanger (EAHE) thermal performance as of February 2017. Recent findings on the parameters affecting EAHE performance are also presented and discussed. As a conclusion, with the advent of CFD methods, investigational work have geared up to modelling and simulation work as it saves time and cost. Comprehension of the EAHE working parameters and its effect on system performance is largely established. However, the study on type of soil and its characteristics on the performance of EAHEs systems are surprisingly barren. Therefore, future studies should focus on the effect of soil characteristics such as moisture content, density of soil, and type of soil on the thermal performance of EAHEs system.

Keyword: Earth-air heat exchanger, soil characteristics, thermal performance.

1 Introduction

The utilization of geothermal energy to meet the heating and cooling needs in buildings has received increasing attention during recent years, as this technology will help the global community reduce its dependence on fossil fuel. Geothermal energy is widely used for the heating and cooling of commercial and residential buildings in North America and Europe [1]. Even so, this sustainable geothermal technology has not been widely implemented in other countries, including Malaysia. Chin [2] stated that the growth for electricity use in Malaysia has risen from 3.1% in 2011 and to 3.7% in the following year. Moreover, the total number of residents with air-conditioning in Malaysia has increased exponentially, up to 16.2% in 2000 from a mere usage in 1970s as mentioned by Kubota et al., [3]. Therefore, Tang [4] proposed that there is a need to study alternative ways to provide thermal comfort in buildings, in the effort to save energy and to minimize harmful effects on the environment. One of the ways to overcome this is by introducing the earth air pipe heat exchanger (EAPHE) as a passive cooling system in Malaysia.

An earth air pipe heat exchanger (EAPHE) is a long metal or plastic pipe, which are buried a few meters deep that utilizes the ground as a heat sink for cooling or heating purposes. EAPHE as a passive cooling system works when hot air flows from a room into the EAPHE system through length of the pipe, where heat is transferred from the surrounding soil into the air, thus causing the air to be cooled and enters the room again as cooler air during the cooling period. [5] [6]

As stated above, the utilization of the geothermal energy in Malaysia has yet to be explored. In recent years, there are several studies conducted on the potential of shallow geothermal technology in Peninsular Malaysia. Noor Aziah et al., [7] have conducted a study concerning the effect of pipe materials on the EAPHE system as a passive ground cooling technology, specifically for the hot-humid climate in Malaysia. Sanusi et al., [8]have conducted a study to investigate the characteristics of Malaysian soil temperature to demonstrate the potential of applying the EAHE technology in Malaysia. Consequently, Yusof et al. [9] have recently reviewed the potential and benefit of ground heat exchanger (GHE) implementation in the Malaysian climate for cooling applications to reduce the energy used in buildings and greenhouse gas emission. Amaludin et al., [10] have conducted a study to determine the effect of moisture content on thermal conductivity, thermal resistivity and volumetric specific heat values of various cohesive soils to assess the viability of shallow geothermal energy pile system installed in the Malaysian environment.

2 Type of earth-air-heat exchanger EAHE

Most EAHE systems are either classified as an open loop or closed loop EAHE system [11]. In an open loop EAHE system, fresh ambient air is drawn through buried pipes, which is moderated to the ground temperature and supplied to the building in order to meet the building's heating/cooling requirement. Meanwhile, for closed loop EAHE systems, warm air flows from the building through the buried pipes, and cooler air is recirculated to the building. The closed loop EAHE system is not preferred over open loop EAHE system because it does not meet the building's fresh air requirement as stated by [6] [9]. The subsections below will explain the details of both types of the EAHE system.

2.1 Open-loop system

This type of system uses wells or surface body water as the heat exchange fluid which circulates directly through the ground heat pump (GHP) system. Once it has circulated through the system, the water returns to the ground through the well, a recharge well, or surface discharge. This option is practical only where there is an adequate supply of relatively clean water, and all local codes and regulations regarding groundwater discharge are met. This system is usually used for larger installations. Figure 1 below shows an open-loop system.

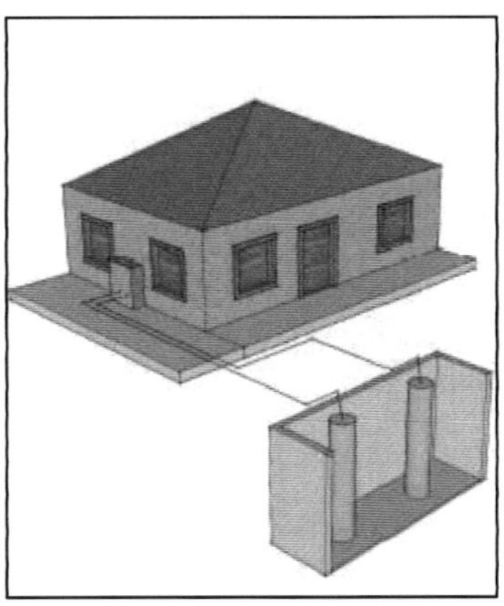

Figure 1. Open-loop system in EAHE

2.2 Closed-loop system
This type of system uses heat exchangers that are located underground, either in horizontal or vertical position and a heat carrier medium is circulated inside the heat exchanger within a closed loop. It transfers heat from the ground to a heat pump or vice versa. For the horizontal type ground heat exchanger, there are three types of horizontal ground which differs with the ways of which the pipes are connected, namely the series, parallel or spiral connections. Figure 2 below shows the closed loop system in an EAHE system.

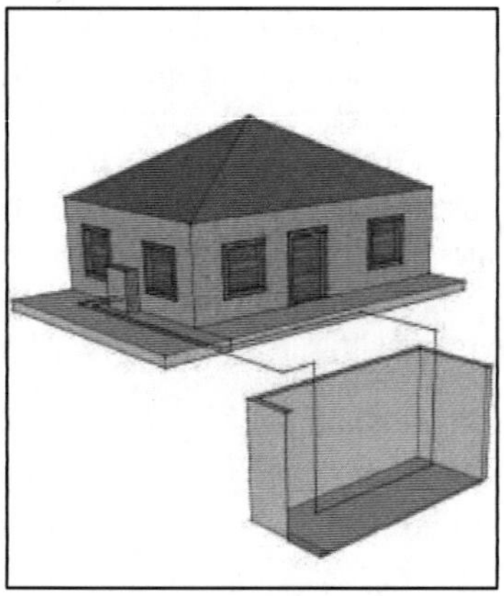

Figure 2. Closed-loop system in EAHE

Vertical EAHE are widely used when there is a need to install sufficient heat exchange capacity under a confined surface area. Examples include when the earth is rocky and close to the surface, or where minimum disruption of the landscaping is desired. This is possible because the temperature below a depth of 15 m to 20 m remains constant over the year. In a standard borehole, which in typical applications is 50 to 150 meters deep, plastic pipes (polyethylene or polypropylene) are installed, and the space between the pipe and the hole is filled with an appropriate material to ensure good contact between the pipe and the undisturbed ground and reduce the thermal resistance.

3 Recent studies conducted on modelling of the EAHE systems
In recent decades, a lot of research has been carried out to develop analytical and numerical models to analyse EAHE systems. The performance analysis of EAHE involves the calculation of conductive heat transfer from the pipe to the groundmass, or the calculation of convective heat transfer from the circulating air to the pipe and changes in the air temperature and humidity. In order to model the EAHE systems, there are a few types of computer modelling tools that are readily available to be used. For example, EnergyPlus and TRNSYS have EAHE modules that work well; but it is not suitable for design purposes and is more suitable to be used as an analysis tool for the EAHE system. Nowadays, a lot of researchers have used computational fluid dynamics (CFD) to model and study the performance analysis of the EAHE systems. This is because CFD employs a very simple rule of discretization of the whole system in small grids and governing equations applied on these discrete elements, which are done to obtain numerical solutions concerning flow parameters, pressure distribution, and temperature gradients in less time and at reasonable cost because of reduced required experimental work [12] [13]. A lot of research papers on different design methods of EAHE systems have been published. Previous research conducted on the calculation models of EAHE systems are presented in the subsequent three subsections; namely, one-dimensional models, two-dimensional models, and three-dimensional models of EAHE systems.

3.1 One-dimensional Models of EAHE systems.

Kabashnikov et al., [14] have developed a mathematical model to calculate the soil and air temperature in a soil heat exchanger for ventilation systems. The model was based on the representation of temperature in the form of the Fourier integral. The study was carried out to analyse the effect of air flow rate, variation in length, diameter of tubes, depth of the buried pipe, and spacing between tubes on the thermal performance of the EAHE systems. Then the results of the calculation was validated against the experimental data. The developed mathematical model does not involve difficult calculations and can therefore be used for design considerations.

De Paepe and Janssens, [15] have presented a one dimensional analytical method to analyse the effect of design parameters of the heat exchanger on thermo-hydraulic performance. Parametric study for tube length, tube diameter, and number of parallel tubes of the heat exchanger on thermo-hydraulic performance have been carried out. Based on the results, it was found that both thermal performance of the heat exchanger and pressure drop increases with the increase of tube length. However, in the case of tube diameters, it was found that smaller tube diameters give better thermal performance, but also larger pressure drop. Other than that, it also found that an increase in the number of tubes in parallel gives both better thermal performance, albeit with a larger pressure drop. Based on the specific pressure drop, the correlation between thermal effectiveness and the pressure drop of the air inside the tube was made. This correlation allows a design method to be formulated, in order to find the characteristic dimensions of the EAHE system that permits optimal thermal effectiveness to be reached with an acceptable pressure loss. Therefore, the choice of characteristic dimensions are independent of the soil type and climatological conditions, which allows designers to select the EAHE configuration with the best performance.

Hollmuller, [16] had demonstrated the complete analytical solution for the heat diffusion of a cylindrical air/soil heat exchanger with adiabatic or isothermal boundary condition, submitted to a constant airflow with harmonic temperature signal at input. The analytical results were verified against the finite-difference numerical simulation model with an experimental setup.

Lee and Strand, [5] developed a new module and performed it in the EnergyPlus program for the simulation of earth tubes. A parametric analysis was conducted using the new module to study the effect of pipe radius, pipe length, pipe depth and air flow rate on the overall performance of the earth tube under various conditions during cold weather. The model was validated against the data from other theoretical and experimental studies and showed good agreement with both theoretical and experimental data. It reveals that pipe length and pipe depth appeared to affect the overall cooling rate of the earth tube, whereas pipe radius and air flow rate mostly affect earth tube inlet temperature. It was deduced that with a proper design, an earth tube can save more than 50% of the total cooling load of the earth tube, depending on the weather and soil conditions. Even so, the earth tube alone cannot replace conventional air-conditioning system in these case studies, but it can greatly help reduce the cooling load in buildings.

Cucumo et al., [17] have proposed a transient one-dimensional analytical model which can be used for the correct installation and the calculation of the performances of EAHEs buried at different depths. With acceptable theory, the model is able to calculate the length and humidity ratio at the outlet section of a buried pipe at different depths of installation and to evaluate the performance of a buried pipe of assigned length. The model was obtained by the solution of heat and mass balances for air through the buried pipe, considering a suitable temperature profile in the ground. This was achieved by two methods: the first is based on Green's functions and the second, simplified, was based on the principle of superposition. The numerical and the experimental data showed a very satisfactory agreement with each other.

A one-dimensional numerical model has been proposed by Sehli et al., [18], to check the performance of EAHEs installed at different depths. Parametric study on the effects of Reynolds number, installation depth, and form factor on the performance of an earth-to air heat exchanger (ETHE) system were analyzed. Ratio of pipe length to pipe diameter of the EAHE systems was defined as a form factor. From the result, it is found that as the installation depth increases, the form

IOP Conf. Series: Materials Science and Engineering **217** (2017) 012021 doi:10.1088/1757-899X/217/1/012021

factor of the outlet air temperature decreases, while the increase in Reynolds number also causes the outlet air temperature to increase. The study finds that EAHE systems alone are not sufficient to create thermal comfort, but it can be used to reduce the energy demand in buildings in south Algeria, if used in combination with conventional air-conditioning systems.

De Jesus Freire et al., [19] presented a study examining the use of a heat exchanger with a multiple layer configuration and comparing it with a single layer of pipes and reporting the major performance differences. Other than that, a parametric analysis of the effect of the main input parameters on the heat exchanger was also performed and it was concluded that the heat exchanger power increases with the layers depth until 3m and the distance between layers should be kept at 1.5m to make sure the heat exchanger is more efficient. It was found that a one-dimensional discrete model can respond faster to a performance analysis of compact buried pipes systems compared to the two dimensional model and still producing results within a good accuracy.

In a one-dimensional model, the description of the pipe to derive a relation between its inlet and outlet temperature is used and therefore it is very simple to solve to obtain the design parameters. It can be concluded that 1D models are simplest to solve within a short time period, but the analysis is not able to analyze the EAHE system completely.

3.2 Two-dimensional Models of EAHE systems

Zhao et al., [20] performed a research to evaluate the thermal performance of saturated soil around coaxial ground-coupled heat exchanger (GCHE). The heat-transfer experiments were conducted based on some artificial glass microballs as porous medium. A theoretical model with Darcy's natural convection was developed and numerical solutions were obtained by using Keller's shooting method. There was a better correlation between experimental and numerical results, which further verified the validation of theoretical model. Based on the results, the heat transfer usually happens near the outer wall of coaxial GCHE and inclines to stabilization at far-field. It was found that, the major factors affecting heat transfer were inlet temperature, initial temperature of porous medium, and the flow rate. It was reported that there was a linear relationship between the dimensionless temperature gradient along the outer wall of GCHE and the dimensionless height.

Tittelein et al., [21] proposes a new numerical model of earth-to-air heat exchangers. The discretized model was solved using the response factors method in order to reduce computational time. Each response factor was calculated using a finite elements program which solves two-dimensional (2D) conduction problems. Based on the result, there are a few advantages of the new model, firstly the calculation time was reduced by the use of response factor. Moreover, it was precise for a short solicitation period (1-day) and a long solicitation period (1-year). Secondly, every type of soil characteristic (inhomogeneous, anisotropic etc.) and of geometry can be considered due to the response factor calculation in a 2D finite elements program (not as analytical models). Lastly, multiple pipes exchangers could be considered with their interaction by calculating more response factors.

To conclude, 2D models allow calculation of soil temperature at the surface and at different depths. The finite element programs are used in 2D models to solve two-dimensional conduction problems of EAHE systems by using finite element programs.

3.3 Three-dimensional Models of EAHE systems

To conduct the performance analysis of EAHE systems using 3D heat transfer and energy balance equations, the computational fluid dynamic (CFD) method will be used. In recent years, CFD has become a popular and a powerful method to study heat and mass transfer. The complex fluid flow and heat transfer processes involved in any heat exchanger can be solved by using CFD software that uses partial differential equations governing airflow and heat transfer.

Ramırez-Davila et al., [22] have carried out a numerical study to predict the thermal behavior of an earth-to-air heat exchanger (EAHE) for three cities in Mexico. Based on the finite volume method, a computational fluid dynamics code has been developed in order to model the EAHE systems. Simulations have been conducted for sand in the city of Ciudad Juarez, silt in the Mexico city, and clay soil in the city of Merida. It was found in simulation results that the thermal performance of the EAHE is better in summer than in winter. For both Ciudad Juarez and Mexico City, the air temperature decreases at an average of 6.6 and 3.2 °C for summer and increases by 2.1 and 2.7 °C for winter, respectively. On the contrary, the EAHE thermal performance in winter is the highest, where the air temperature increases by 3.8 °C for Merida. Therefore, it was concluded that the use of EAHEs is acceptable for either heating or cooling of buildings in lands of extreme and medium temperatures, where the thermal inertia effect in soil is higher.

Flaga-Maryanczyka et al., [23] have demonstrated the experimental measurements and numerical simulation of a ground source heat exchanger operating at a cold weather for a passive house ventilation system. The study carried out for a passive house was located in the South of Poland. The house and its components were fitted with a data acquisition system that was running from 2011 and records 139 points at an interval of 1 minute. The calculations were performed using the CFD ANSYS FLUENT software package. Based on the experimental data on the CFD simulations done for the ground source heat exchanger that was operating for the passive house ventilation system on February, it showed a good correlation with the measured values. Moreover, the difference between measured and calculated values at the level of 1.7°C on average leads. Therefore, it was concluded that CFD tool is acceptable to be used for simulations of the ground source heat exchanger working in cold climates for the passive house ventilation system.

In conclusion, 3D models are dynamic and technically more advanced, which can provide room for all types of grid geometry to produce detailed thermal analysis of EAHE systems. However, their applicability for design is limited to the people who are able to use the calculation codes. In 3D models, finite volume method are used to solve 3D conduction equations. It is concluded that CFD is an effective tool for predicting the behavior and performance of a wide variety of heat exchangers.

4 Factors affecting the thermal performance of the EAHE systems

The energy performance of an earth air pipe (EAP) system can be mainly influenced by a few parameters such as; pipe materials, geometric dimension and design of the EAPHE and the air velocity (air flow), pipe burial depth, air temperature, soil temperature and type of soil [6]. To design an efficient system of EAHE, a study should be conducted for the parameters affecting thermal performance of EAHE. Therefore, in the last two decades, investigational work have been carried out intensively on the parameters affecting the thermal performance of the EAHE system. In the subsections below, previous studies conducted on each of the parameters were stated and discussed.

4.1 Pipe Materials

As mentioned before, one of the parameters affecting the performance of the EAPHE system is the pipe material. Recent studies have not clearly explained the major reason behind the selection of pipe materials for their respective studies. However, it can be concluded that the selections were made based on the availability of materials and cost. The materials of higher thermal conductivity have higher heat transfer rate, and therefore can reduce the buried pipe outlet temperature more efficiently. The impact of different pipe materials were analysed through a number of studies as summarized below.

Bansal et al., [24] studied the EAPHE system performance during winter in Ajmeer (Western India). The pipes were buried at a depth of 2.7 m in a flat land with dry soil. They experimented on steel and PVC pipe materials with the air velocity ranging from 2-5 m/s which flowed through 23.42 m length of 150 mm diameter pipe. The result showed that the system gives heating in the range of 4.4 - 4.8 °C. In another study, Bansal et al., [25] further analyzed the EAPHE system performance using CFD simulation during summer with similar parameters and showed that the system gives a cooling range of 8 - 12.7 °C in reduction compared to the ambient temperature. He concluded that the performance of EAPHE system is not affected by the material, therefore the cheaper material may be used to construct the pipe.

Ascionne et al., [26] evaluated the energy performances achievable by using an earth-to-air heat exchanger for an air-conditioned building for both winter and summer for the Italian climate. Three different Italian climates (cities of Naples, Rome, Milan) and an air-conditioned building have been analyzed for both winter and summer. The energy requirements of these systems have been evaluated as a function of the main boundary conditions (such as the typology of soil, tube material, tube length and depth, velocity of the air crossing the tube, ventilation airflow rates, and control modes). The obtained results shows that concrete, plastic or metallic materials have very similar energy performances. In fact, due to the small thickness of the tubes (5 mm in the case of PVC, 7 mm for the metallic material, 7 cm for the concrete), the different thermal conductivity values are almost negligible and does not significantly influence the heat exchange, if the right depths and lengths are used as the earth tube model.

Sanusi, et al., [27] have studied the potential of this system in the hot-humid climate in Malaysia. The study uses 3 polyethylene (PE) pipes with 76 mm in diameter and 30 m long. All three pipes were buried separately at depths of 0.5 m, 1.0 m and 1.5 m. Fan blowers were installed at each inlet of the pipes which provided air velocity at 5.6 m/s. The result showed the best reduction of temperature of up to 6.4°C during the wet season, and 6.9°C during the hot and dry season, where the pipes were buried at 1 m depth underground. Meanwhile, 30m long PVC pipes were also buried at 0.5m, 1.0m and 1.5m depth underground at the same time. However, due to the increasing weight of the soil as it goes deeper underground, the PVC pipes below 0.6m were found broken just before measurements were taken. It concluded that the PVC pipes have low durability as compared to PE pipes. Furthermore, a parametric study was carried out in EnergyPlus in comparing other pipe materials, namely PVC, PE, Brick and Clay [8].The results had shown that Clay pipes provide the lowest pipe outlet air temperature among the four pipes. This research extends the type of buried pipes for its parametric analysis on buried pipe materials.

Hatraf et al., [28] have investigated the parameters influencing the performance of earth to air heat exchanger through modeling and experimentation. They have done the calculation using a simple model the distribution of the air temperature and varied several parameters such as nature of the ground and the ground depth diameter of the duct. One of their finding was that the pipe material has no effect on the performance of the heat exchanger, which is in agreement with the findings of Bansal et. al (2010).

Aziah et al., [7] have studied the materials for the EAPHE system using computer simulation to find the best pipe material for the EAPHE system for the hot and humid climate in Malaysia. The study utilizes the EnergyPlus environmental simulation program to investigate the performances of three pipe materials system; which includes single pipe material, hybrid pipes (combination of pipe materials) and insulated hybrid pipes system (combination of pipe materials with insulation). It is found that for single pipe material polyethylene (PE), it has the maximum temperature reduction followed by steel (St), copper (Cu) and polyvinyl chloride (PVC). Through an exhaustive enumeration process, the study found that the insulated hybrid pipes system gave the best temperature reduction, which indicates a promising cooling and energy savings potentials. It also concluded that the choice of the pipe material does not have any significant consequence on the temperature reduction by the EAPHE system in hot and humid climate.

Jahkar et al., [29] have designed and simulated an earth water heat exchanger (EWHE) in transient analysis tool TRNSYS (v17.0) in Pilana, Rajashtan (India). Parametric study was conducted to analyse the effect of mass flow rate, pipe length, diameter of buried pipe and pipe material on the performance analysis of the EWHE. To conduct the performance analysis, the pipe length, pipe diameter and mass flow rate was fixed to 90m, 25mm and 0.02kg/s, respectively. Three types of material have been used to conduct the analysis such as high density polyethylene (HDPE), Galvanized Iron (GI) and steel pipe (St). Based on the result, GI and steel is at 0.2 °C and for HDPE is around 1.6 °C for the temperature difference at the outlet of the EWHE system with the same inlet conditions. This slight difference occurred because the HDPE pipes coefficient of friction is lower than the other pipes. Even though the thermal conductivity for both of GI and steel is higher than HDPE, the coefficient of friction for

HDPE is higher compared to both materials. Hence, a lower outlet temperature is observed for both, GI and steel pipes compared to HDPE pipe.

Ahmed et al., [30] developed a thermal model for the horizontal earth pipe cooling (HPEC) system using the simulation model, FLUENT 15.0 in a hot humid subtropical climate at Rockhampton, Australia. The studied mainly focused on the parameters affecting the thermal performance of the earth pipe cooling system, which are the air velocity, pipe length, pipe diameter, pipe depth and also pipe material. Materials such as PVC, polyethylene, concrete and clay have been chosen to study the effect of the pipe material on the thermal performance of HPEC. It is found that the clay PVC pipe produces higher temperature at the pipe outlet that varies from 19.5°C to 24.9°C, compared to the other material and while the clay pipe provides the lowest average pipe outlet temperature as 20.5°C.

Serageldin et al., [31] studied the thermal performance of an earth air heat exchanger (EAHE) used for heating and cooling under the Egyptian weather conditions. They have developed a mathematical model based on unsteady, one-dimensional, quasi-state for energy conservation equation, and also a three dimensional, steady and double precision Computational Fluid Dynamics (CFD) ANSYS Fluent simulation model that is established to predict the air and soil temperature. A parametric study was done to investigate the impact of different parameters such as pipe diameter, pipe material, pipe space, pipe length and flowing fluid velocity. Three different types of pipe materials were used, namely PVC, steel and copper. The outlet air temperature was 19.7 °C in PVC pipe, and 19.8 °C for both steel and copper respectively. Therefore, it is concluded that the change in outlet air temperature for various pipe material is too small and hence it is negligible.

Based on previous studies, the pipe material does not significantly affect the thermal performance of the EAHE system. However, pipe materials with higher thermal conductivity can increase the thermal performance of EAHE, but it is not reliable to use it when compared to the price of the materials which have higher thermal conductivity. This is because materials such as copper and steel are more expensive and the availibilty in market is quite low compared to PVC pipes, even though it has lower thermal conductivity as reported by previous researchers [24-31]. Therefore, it can be concluded that pipe material selection does not significantly affect the thermal performance of EAHE, and is therefore negligible. Types of pipe materials used in the past studies are listed in Table 1.

Table 1. List of pipe materials have been used in the past studies

Authors	Type of Pipe Material used	References
Bansal et al.,	PVC	[24]
Ascionne et al.,	PVC, Metallic, Concrete	[26]
Sanusi et al.,	Polyethylene (PE), PVC, Brick, Clay	[27]
Aziah et al.,	Polyethylene (PE), Copper, PVC	[7]
Jahkar et al.,	High Density Polyethylene (HDPE), Galvanized Iron, Steel	[29]
Ahmed et al.,	PVC, PE, Concrete, Clay	[30]
Serageldin et al.,	PVC, Steel, Copper	[31]

4.2 Geometric dimension and design of the EAHE pipe system

Geometric dimension and design of the EAHE pipe system also plays a big role in determining the thermal performance of the system. A number of studies have been conducted by researchers on different geometric design of the EAHE system which includes; length of pipe, burial depth of the pipe, diameter of the pipe and pipe arrangements.

Ghosal and Tiwari, [32] have developed a thermal model to analyse the potential of using thermal energy of the ground for greenhouse heating and cooling, where the earth to air heat exchanger (EAHE) system was integrated with the greenhouse located in the premises of IIT, Delhi, India. Experiments were conducted extensively throughout 2003, and the developed model was validated against typical clear and sunny day experiments. Parametric studies were carried out to study the effects of buried pipe length, pipe diameter, mass flow rate of air, depth of ground and types of soil on the greenhouse air temperatures. Based on the results, the effect of the length of buried EAHE pipes, the increase of the length of buried pipes from 30m to 50m directly impacted the temperature changes of the air inside the greenhouse; where an increase in the winter period and a decrease during the summer period were observed. This is mainly because the longer pipe length allows a longer time for the thermal heat exchange between the air and the ground to take place. Meanwhile, for the effects of diameter of buried pipe of the EAHE, an increase in diameter results in lower and higher greenhouse air temperatures in the winter and summer, respectively. This is caused by the decrease in the heat transfer of heat from earth or lower convective heat transfer coefficient due to increase of pipe surface and slower air flow.

Maerefat and Haghighi, [33] have proposed and studied a passive solar system comprising of solar chimneys and earth to air heat exchangers (EAHEs). In this study, it was shown that the optimum diameter for cooling pipes is 0.5m (with the lowest number of solar chimney) and the EAHE was needed to achieve the cooling demand required. It also reveals that the pipe length of more than 20m should be used in the EAHE system to provide a better thermal comfort condition.

Pohstiri et al., [34] have developed a mathematical model based on energy conservation equations and solved by iterative method to study the capability of the solar chimney and earth air heat exchanger system to meet the thermal needs of individuals and also the dependence of the system performance on environmental and geometrical issues. Based on the results, EAHE pipe lengths of less than 35m should be used to provide a better thermal comfort condition. Other than that, the result of the effect of EAHE pipe diameter on system performance reveals that the increase of the EAHE diameter up to 0.5m does not increase the room air temperature.

Misra et al., [35] developed a transient and implicit numerical model based on coupled simultaneous heat transfer and turbulent flow, to conduct the study on the effect on time duration of continuous operation, thermal conductivities of soil pipe diameter and flow velocity on thermal performance of earth air tunnel heat exchanger (EATHE) system. Three different pipe diameters were chosen; 0.10m, 0.15m, and 0.20m and the air flow velocity was kept constant to investigate the effect on transient performance of EATHE system to work in long continuous operation. Based on the result, thermal performance of EATHE system with larger pipe diameter had dropped faster after 24h of continuous operation especially the pipe length which are 20m or less from the inlet. Other than that, pipes with smaller diameter and pipe length are beyond 30m experiencing less deterioration in thermal performance.

Jahkar et al., [29] have designed and simulated an earth water heat exchanger (EWHE) in transient analysis tool TRNSYS (v17.0) in Pilana, Rajashtan (India). Parametric study was carried out to analyse the effect of mass flow rate, pipe length, diameter of buried pipe and pipe material on the performance analysis of the EWHE. To study the effect of variation of length, the mass flow rate of 0.02 kg/s with diameter of 25mm for HDPE pipe was set to be constant. By varying the length of the pipe from 50 m to 90 m, the result was obtained. It was found that the increase in length caused an increase in the temperature drop. When the pipe length was at 90 m and the fluid temperature inlet at 90°C, the outlet temperature recorded was 31.9 °C. The effect of variation in pipe diameter ranging from 25 mm to 50 mm of HDPE pipe at the same flow rate of 0.02 kg/s and 90 m length was recorded. It reveals that the increase of the pipe diameter caused the outlet temperature to decrease gradually over a period of time. After 10 hours of operation, the outlet temperature differences were 31.80 °C for 50 mm and 36.16 °C for 25 mm. In another study, Jahkar et al., [36] presented a model of earth air pipe heat exchanger that was generated using TRYNSS 17 simulation tool and was validated using experimental investigation in an experimental set up in Ajmer, India. The experiment was done during winter season, where the system was examined for the effect of different inlet flow velocities, pipe length and depth of burial on the outlet air temperatures of the EAHE systems. The effect of pipe length with respect to changes in flow velocities has been conducted. It was found that about 82–85%

of the total increase in the temperature of air along the EAHE pipe is achieved at a length of 34m from the inlet.

Ahmed et al., [37] have developed a thermal model to compare the earth pipe cooling performance between two different piping systems for a hot and humid subtropical climatic zone in Queensland, Australia using ANSYS Fluent. Vertical earth piping cooling VEPC system and horizontal earth piping cooling HEPC systems were laid in the ground in order to compare the cooling performance as shown in Figure 3. The developed model was validated at the pipe inlet and different points of the room with the measured data. The inlet air velocity and air temperature was set at the pipe inlet to predict their effect on the room temperature. The simulation result shows that the minimum temperature reductions of approximately 1.82 °C and 1.02 °C in the modelled room of vertical and horizontal earth pipe cooling system, respectively. Based on the results, the vertical earth pipe cooling system is better than the horizontal earth pipe cooling system even though there was no large temperature reductions found for both systems.

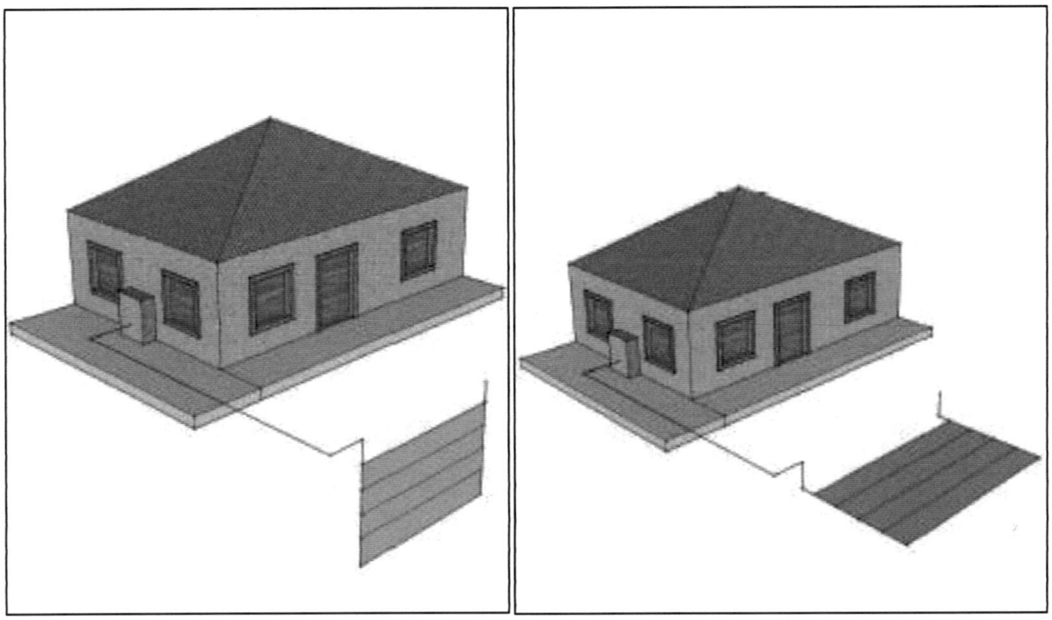

(a) (b)

Figure 3. (a) -Vertical earth piping cooling VEPC system, (b) - Horizontal earth piping cooling HEPC systems

Gan [38] has developed models of coupled heat and moisture transfer in soil and simulation of the thermal performance of an earth air heat exchanger for building ventilation, considering the dynamics variations of climatic, load and soil conditions. The dynamic interactions between soil, ambient environments and heat exchanger have been analysed. Other than that, the effect of the heat exchanger length on thermal performance are also investigated. It has been found that the heat transfer rate decreases along the heat exchanger and the rate of decrease is non-linear. Thus, the heat transfer rate and temperature rise of supply air per unit length decrease with increasing length of the heat exchanger for preheating.

According to the recent studies conducted, the increasing depth of burial of the EAHE pipes making the outlet air temperatures of EAHE systems to increase in winter and decrease in summer. Whereas for the pipe diameter, it shows that an increase in diameter resulted in lower thermal performance in the EAHE systems. The result is caused by the decrease in the heat transfer of heat from earth or lower convective heat transfer coefficient due to increase of pipe surface and slower air flow. For the effect of pipe length on the thermal performance of EAHE system, each of the recent study has different optimum lengths. However, the increase of pipe length does increase the thermal performance of EAHE system until certain length of the pipe, which causes the thermal performance of EAHE system to decrease.

4.3 Air flow

Air flow or volume flow rate of air is one of the parameter which affects the thermal performance of EAHE system. In this subsection, previous studies focussed on the air flow rate is reviewed.

Ghosal and Tiwari, [32] have developed a thermal model to analyse the potential of using thermal energy of the ground for greenhouse heating and cooling, where the earth to air heat exchanger (EAHE) system was integrated with the greenhouse located in the premises of IIT, Delhi, India. Parametric studies were carried out to study the effects of buried pipe length, pipe diameter, mass flow rate of air, depth of ground and types of soil on the greenhouse air temperatures. It is found that for the effects of mass flow rate of flowing air in EAHE, the increase in mass flow rate leads to the decrease and increase of the greenhouses air temperature in the winter and summer, respectively. This may be caused by the air that has a shorter time in contact with the soil, causing lower thermal exchange rate between soil and air.

Misra et al., [35] developed a transient and implicit numerical model based on coupled simultaneous heat transfer and turbulent flow to conduct the study on the effect on time duration of continuous operation, thermal conductivities of soil pipe diameter and flow velocity on thermal performance of earth air tunnel heat exchanger EATHE system. Three different velocities were chosen; 2.0m/s, 5.0m/s, and 8.0m/s and the pipe diameter was kept at 0.1m for all different flow velocities to investigate the effect on transient performance of EATHE system to work in long continuous operation. From the result, the increase in flow velocity cause the thermal performance of EATHE system to deteriorate. This is because when the flow velocity increases, the amount of heat is transferred to the soil from air per unit time also increases. Later, this causes more heat to be accumulated in the soil layers in the immediate vicinity of pipe surface for soil with less thermal conductivity. It is concluded that increasing the flow velocity has a detrimental effect on the thermal performance of EATHE system.

Jahkar et al., [29] have designed and simulated an earth water heat exchanger EWHE in transient analysis tool TRNSYS (v17.0) in Pilana, Rajashtan (India). Parametric studies were done to analyse the effect of mass flow rate, pipe length, diameter of buried pipe and pipe material on the performance analysis of the EWHE. To study the effect of mass flow rate on the performance of EHWE, the length of pipe and pipe diameter have been fixed to 90m and 25mm, respectively. Five values of mass flow rate have been used in the analysis; 0.008 kg/s, 0.02kg/s, 0.03kg/s, 0.04kg/s and 0.05kg/s. It is found that with an increase in mass flow rate, the outlet temperature of EHWE increases. It is noticed that the highest temperature drop occurred when the mass flow rate is set at 0.008 kg/s. In another study, Jahkar et al., [36] have presented a model of earth air pipe heat exchanger that was generated using TRNSYS 17 simulation tool, and was validated using experimental investigation on an experimental set up in Ajmer, India. The experiment was done during winter season, where the system was examined for the effect of different inlet flow velocities, pipe length and depth of burial on the outlet air temperatures of the EAHE systems. There are three different air velocity used; 2.5m/s, 3.5m/s and 5.0m/s to study the effect of air flow velocity on the EAHE outlet temperature and it reveals that, with the increase in air velocity, EAHE outlet temperature decreases. This is mainly because the convective heat transfer coefficient have increased by 2.3 times. Moreover, the increase of air velocity reduces the time to which the air remains in contact with the ground by a factor of 2.5. It is concluded that, the velocity at 5m/s has the best increase in EAHE outlet temperature compared to velocity at 2.5m/s and 3.5m/s.

Based on the recent studies conducted on the effect of air flow velocity in EAHE systems, it is found that the increase of air velocity leads to the decreasing thermal performance of the EAHE system. This is caused by the shortened time in contact between air and the ground, therefore the heat from the air has not enough time to achieve thermal equilibrium with the ground. Therefore, the later effect was dominant and causing the thermal performance of EAHE to decrease at a higher flow rate compared to lower flow rate.

4.4 Site characteristic

Research work on EAHE system comes from various location around the world. Each site offers different climate condition which affects the EAHE performance and suitable applications. A list of research papers published in different locations are summarized in Table 2.

Table 2. Summary of Site Characteristics for various EAHE System Installations

Authors	Weather types	Location	Type of ground heat exchanger	Application	References
Bansal et al.	Winter	Ajmer, India	Earth air pipe heat exchanger (EAPHE)	Passive heating	[24]
Jahkar et al.	Winter	Ajmer, India	Earth water heat exchanger (EHWE)	Passive heating	[36]
Jahkar et al.	Winter and summer	Pilana, Rahjastan, India	Earth water heat exchanger (EHWE)	Passive cooling and heating	[39]
Ghosal and Tiwari.		Premises ITT, Delhi, India	EAHE	Passive cooling and heating	[32]
Ahmed et al.	hot and humid subtropical	Queensland, Australia	Horizontal earth pipe cooling (HEPC) and Vertical earth pipe cooling (VEPC)	Passive cooling system	[37]
Ahmed et al.	hot and humid subtropical	Rockhampton, Australia	Horizontal earth pipe cooling (HEPC)	Passive cooling	[30]
Aziah et al.	Hot and humid	Malaysia	EAHE	Passive cooling system	[7]
Sanusi et al.	Hot and humid	Malaysia	EAHE		[27]
Ascionne et al.	Winter and summer	cities of Naples, Rome, Milan (Italy)	EAHE	Passive cooling and heating	[26]

Ramırez-Davila et al.	Winter and summer	Ciudad Juraez, Mexico City, Merida, (Mexico)	EAHE	Passive cooling and heating	[22]
Flaga-Maryanczyka et al.	Cold	South of Poland	Ground Source heat exchanger	Passive heating	[23]

Based on Table 2, it can be concluded that researchers in India have investigated the EAHE system based on the weather/climate of its location compared to any other country. It also can be deduced that the EAHE system can be implemented to provide thermal comfort either in hot or cold weather using the EAHE passive cooling system. Even though there are a limited number studies conducted on different locations with different weather conditions, there is still no studies conducted in Malaysia, particularly in Sabah. Therefore, there is a need to assess the performance of EAHE system in the Sabah weather condition to determine its viability as a thermal comfort mechanism.

4.5 Soil characteristic

Ahmed et al., [30] conducted an experimental study of thermal and moisture behaviours of dry and wet soils heated by buried capillary plaits. A prototype, which is similar to an agricultural tunnel greenhouse was used to carry out the experiment. There are three different operational conditions of the capillary plaits: heating at 70°C, heating at 40°C and without heating in summer, in order to understand the greenhouse climate effect on soil behaviour. It is revealed that, in unsaturated moist soils the transport of heat is complicated by the fact that heat and mass transfer is a coupled process. It was found that the surface temperature amplitude was higher in wet soil compared to dry soil during the daily soil temperature variation. Even though thermal diffusivity, calculated according to the phase delay method between 0 and 20 cm, in wet soil was higher than in dry soil, the wet soil surface temperature was higher during daytime. The formation of a superficial thin dry soil layer will thermally insulate the soil, causing both the surface heat flux to fall and the surface temperature to rise.

Ghosal and Tiwari, [32] have developed a thermal model to analyse the potential of using thermal energy of the ground for greenhouse heating and cooling, where the earth to air heat exchanger (EAHE) system was integrated with the greenhouse located in the premises of IIT, Delhi, India. Parametric studies were carried out to study the effects of buried pipe length, pipe diameter, mass flow rate of air, depth of ground and types of soil on the greenhouse air temperatures. With the same operational parameters of the EAHE experiment (length of buried pipe 39 m, diameter 0.06 m, mass flow rate 0.027 kg/s and depth 1 m), the experiments were conducted to analyze the effect of the types of soils on the greenhouse air temperatures. Based on the result of the effect of different soil types on the greenhouse air temperatures, it was seen that both of the greenhouse air temperatures are highest either in winter or summer day especially in sandy soil followed by sandy loam, gravelly sand and silt loam. The higher thermal conductivity of sandy soil followed by sandy loam, gravelly sand and silt loam soil are responsible for the higher temperatures of greenhouse air temperature.

Bansal et al., [40] have conducted study on the effect of thermal conductivity of soil on thermal performance of earth air tunnel heat exchanger under transient operating conditions in predominantly hot and dry climate of Ajmer (India) using experimental and computational fluid dynamics modeling with FLUENT software. From the result, maximum air temperature drop obtained under steady state operation of EATHE for pipe length 100 m is 18.4 °C, 18.7 °C and 18.4 °C for soil thermal conductivity of 0.52, 2.0 and 4.0$Wm^{-1}K^{-1}$ respectively. However, the maximum air temperature drop under transient conditions for 24 h of operation varies between 18.3°C and 14.0 °C, 18.3 °C and 17.2 °C and 18.6 °C and 18.0 °C for soil thermal conductivity of 0.52, 2.0 and 4.0Wm^{-1} K^{-1} respectively. This is due to higher soil thermal conductivity results into better thermal performance of EATHE system under transient conditions, even after longer period of operation. It is deduced that for better thermal performance of EATHE, the soil placed in the immediate vicinity of the EATHE pipe should have higher thermal conductivity. On the other hand, while optimizing the thermal performance of EATHE, the thermal conductivity of soil, duration of continuous operation of EATHE and length of pipe should be taken into account so as to confirm that the EATHE would be able to give a steady thermal performance for longer duration of operation.

Gan [38] has developed models of coupled heat and moisture transfer in soil and simulation of the thermal performance of an earth air heat exchanger for building ventilation, considering the dynamics variations of climatic, load and soil conditions. The dynamic interactions between soil, ambient environments and heat exchanger have been analyzed. It has been found that the direct thermal and moisture interaction between heat exchanger, soil and atmosphere have a significant impact on the heat transfer through the heat exchanger.

Studies done by the above researchers shows that thermal conductivity of soil generally affect the outlet air temperature of EAHE system. As reported by Bansal et al., [40], the thermal conductivity of soil increases the thermal performance of EAHE system increases. Moreover, Ghosal and Tiwari, [32] also reported the higher thermal conductivity of sandy soil followed by sandy loam, gravelly sand and silt loam soil are the reason for the higher temperatures of the greenhouses air temperature. More study on soil properties should be conducted since soil characteristic is very unique. Below is the summary of soil properties that have been used in past studies conducted by the researchers, as shown in Table 3.

Table 3. List of soil properties that have been used in the past studies

Authors	Soil Properties	References
Ahmed et al.,	Wet and Dry Soil	[30]
Ghosal and Tiwari	Sandy loam, Gravelly sand, Silt loam	[32]
Bansal et al.,	Varies the soil thermal conductivity of 0.52, 2.0, 4.0 $Wm^{-1}k^{-1}$	[39]

5 Conclusion

This paper has reviewed the recent discoveries related to EAHE and factors affecting it performance. The method to assess the performance of EAHE is first discussed. Research trends show that with the advancement of computational technology, investigational work have evolved from lengthy experimental investigation to computer powered simulation work. This is expected as computational method matures and produces more reliable data, thus reducing cost and time.

Reviews of published research work on parameters affecting EAHE performance have shown that the performance are governed by four parameters, namely the pipe dimension, pipe material, depth, air velocity and soil properties. Interestingly, pipe material does not have a significant impact on the thermal performance of the EAHE system. Whereas for the effect of the depth of pipe installation, it can be deduced that, as the depth of the EAHEs pipe installation increases, the thermal performance of EAHE system also increases. However, an increase in diameter of the EAHEs pipes have resulted in a lower performance in the EAHE systems. For the effect of pipe length on the thermal performance of EAHE system, each of the recent studies has different optimum lengths with varying results. Nevertheless, the researchers have found that the increase of pipe length does increase the thermal performance of EAHE system until an optimum length of the pipe is achieved and subsequently causing the thermal performance of EAHE system to decrease. For the effect of air velocity on the thermal performance, it can be deduced that the increase of air velocity leads to the decreasing thermal performance of the EAHE system.

Several EAHE performance assessments were done in different parts of the world. This trend shows that the EAHE system can be used to provide thermal comfort in various seasons. According to recent studies, the effect of soil characteristics on the thermal performance shows that as the thermal conductivity of soil increases; the thermal performance of EAHE system also increases. However, thus far there are no investigations conducted in East Malaysia. Therefore, a study on EAHE performance in East Malaysia region is highly sought as different types of soil have unique properties in different parts of the world. To implement an efficient EAHE system in Malaysia especially in East Malaysia (Sabah), both of these opportunities should be pursued to gain a profound understanding of EAHE system performance and its potential use in the Island of Borneo.

References

[1] Barry-Macaulay D, Bouazza D, and Singh R, 2011 Study of a Balastic Clay. *Geo-Frontiers.* pp 480-487

[2] Chin F.K 2012 The Future of Energy in Malaysia in *Minister of Energy. Green Technology and Water at the Malaysian International Chamber of Commerce & Industry* (Micci)

[3] Kubota T, Jeong S, Toe, and Ossen D 2011 Energy Consumption and Air Conditioning Usage in Resedential Building in Malaysia. *Journals of International Development and Cooperation.* pp 61-69

[4] Tang C, 2008 A re-examination of the relationship between electricity consumption and economic grwoth in Malaysia. *Energy Policy* pp 3077-3085

[5] Lee K H and Strand R K 2008 The cooling and heating potential of an earth tube system in buildings. *Energy and Buildings.* pp 486-494

[6] Bisoniya T S, Kumar A and Baredar P, 2013 Experimental and analytical studies of earth-air heat exchanger (EAHA) systems in India: A review. *Renewable and Sustainable Energy Reviews.* pp 238-246

[7] Mohd, Noor A, Aliyah N Z, Sanusi and Aminuddin M, 2014 Materials for earth air pipe heat exchanger (EAPHE) system as a passive ground cooling technology for hot-humid climate. *International Conference on Emerging Trends in Scientific Research*, Pearl International Hotel (Kuala Lumpur)

[8] Sanusi and Aliyah N Z 2012 The potential of low energy earth pipe cooling in hot and humid Malaysia. *International Conference on Green in the built environment*

[9] Yusof T, Anuar S and Ibrahim H, 2015 A review of ground heat exchanger for cooling application in the Malaysian climate. *Journal of Mechanical Engineering and Sciences.* pp 1426-1439

[10] Amaludina A, Martoc A, Hatta M, Hassanel A and Salinah D, 2016 Thermal Properties of Malaysian Cohesive Soils. *Jurnal Teknologi (Sciences & Engineering)* pp 54-58

[11] Mands and S B Shallow geothermal energy.

[12] Kanaris A G, Mouza A A and Paras S V, 2006 Flow and heat transfer predicition in a corrugated plate heat exchanger using a CFD code. *Chemical Engineering & Technology.* pp 923-930

[13] Wang Y, Dong Q and Liu M, 2007 Characteristic of fluid and heat transfer in shell side of heat exchangers with longitudinal flow of shell side fluid with different

supporting structures. *Proceedings of International Conference on Power Engineering.* (Hangzhou, China)

[14] Kabashnikov V P, Danilevskii L N, Nekrasov V P and Vityaz I P, 2002 Analytical and numerical investigation of the characteristics of soil heat exchanger for ventilation systems. *International Journal of Heat and Mass Transfer.* pp 2407-2418

[15] Paepe M D and Janssens A, 2003 Thermo-hydraulic design of earth-air heat exchangers. *Energy and Buildings.* pp 389-397

[16] Hollmuller P 2003 Analytical characterization of amplitude dampening and phase-shifting in air/soil heat-exchangers. *International Journal of Heat and Mass Transfer.* pp 4303-4317

[17] Cucumo M, Cucumo S, Montoro L and Vulcano, 2008 A dimension transient analytical model for earth-to-air-heat exchangers, taking into account condensation phenomena and thermal perturbation from the upper free surface as well as around the buried pipes. *International of Heat and Mass Transfer.* pp 506-916

[18] Sehli A, Hasni A and Tamali M, 2012 The potential of earth-air heat excahngers for low energy cooling of buildings in South Algeria. *Enegy Procedia.* pp 496-506

[19] Freire A D J, Alexandre J L C, Silva V B, Couto N D and Rouboa A, 2013 Compact buried pipes system analysis for indoor air conditioning. *Applied Thermal Engiineering.* pp 1124-1134

[20] Zhou J, Wang H, Li X and Dai C, 2008 Experimental investigation and theoretical model of heat transfer of saturated soil around coaxial ground coupled heat exchanger. *Applied Thermal Enginnering.* pp 116-125

[21] Tittelein P, Achard G and Wurtz E, 2009 Modelling earth-to-air heat exchanger behaviour with the convolutive response factors method. *Applied Energy.* pp 1638-1691

[22] Ramırez L, Xaman J A J, Alvarez G and Hernandez P 2014 Numerical study of earth-to-air heat exchanger for three different climates. *Energy and Buildings.* pp 238-248

[23] Flaga M, Schnotale J, Radon J and Was K, 2014 Experimental measurements and CFD simulation of a ground source heat exchanger operating at a cold climate for a passive house ventilation system. *Energy and Buildings.* pp 562-570

[24] Bansal V, Misra R, Agrawal G D and Mathur J, 2009 Performance analysis of earth-air-pipe-air heat exhanger for winter heating. *Energy and Buildings.* pp 1151-1154

[25] Bansal V, Misra R, Agrawal G D and Mathur J, 2010 Performance analysis of earth-pipe-air heat exchanger for summer cooling. *Energy and Buildings.* pp 645-648

[26] Ascione F, Bellia L and Minichiello F, 2011 Earth-to-air heat exchangers for Italian climates. *Rnewable Energy.* pp 2177-2188

[27] Sanuzi, Aliyah N Z, and Ibrahim M N, 2013 Passive ground cooling system for low energy buildings in Malaysia (hot and humid climates). *Renewablle Energy.* pp. 193-196

[28] Hatraf N and Foued C, 2014 Parametric Study to Design an earth-to-air heat exchanger with experimental validation. *Engineering Journal.* pp 42-52

[29] Jakhar S, Soni M S and Gakkhar N, 2016 Performance analysis of earth water heat exchanger for concentrating photovoltaic cooling. *Energy Procedia.* pp 145-153

[30] Ahmed S, Amanullah M, Khan M, Rasul M and Hassan N, 2016 Parametric study on thermal performance of horizontal earth pipe cooling system in summer. *Energy Conversion and Management.* pp 324-337

[31] Serageldin A, Abdelrahman A K and Ookawara S, 2016 Earth-Air Heat Exchanger thermal performance in Egyptian conditions: Experimantal results, mathematical model, and Computational Fluid Dynamics simulation. *Energy Conversion and Management.* pp 25-38

[32] Ghosal M K, Towari G, Das D and Pandey K P, 2005 Modeling and comparative thermal performance of ground air collector and earth air heat exchanger for heating of greenhouse. *Energy and Buildings.* pp 613-621

[33] Maerefat M and Haghighi A P, 2010 Passive cooling of buildings by using integrated earth to air heat exchanger and solar chimney. *Renewable Energy.* pp 2316-2324

[34] Poshitri A H, Gilani N and Zamiri F, 2011 Faeasibility study on using solar chimney and earth-to-air heat exchanger for natural heating of buildings. *World Renewable Energy Congress* (Linkoping, Sweden)

[35] Misra R, Bansal V, Agrawal G D, Mathur J and Aseri T K, 2013 CFD analysis based parametric study of derating factor for Earth Air Tunnel Heat Exchanger. *Applied Energy.* pp 266-277

[36] Jakhar S, Misra R, Soni M and Gakkhar N, 2016 Parametric simulation and experimental analysis of earth air heat exchanger with solar air heating duct. *Engineering Science and Technology.* pp 1059-1066

[37] Ahmed S, Khan M, Rasul M,Amanullah M and Hassan N, 2014 Comparison of earth pipe cooling system performance between two different piping system. *Energy Procedia.* pp 1897-1901

[38] Gan G, 2015 Simulation of dynamic interactions of the earth–air heat exchanger with soil and atmosphere for preheating of ventilation air. *Applied Energy.* pp 118-132

[39] Jahkar S, Soni M and Gakkhar N, 2016 Performance analysis of earth water heat exhanger for concentrating photovoltaic cooling. *Energy Procedia.* pp 145-153

[40] Bansal V, Misra R, Agarwal G D and Mathur J, 2013 Transient effect of soil thermal conductivity and duration of operation on performance of Earth Air Tunnel Heat Exchanger. *Applied Energy.* pp 1-11

Acknowledgement

The authors wish to convey a sincere gratitude to Universiti Malaysia Sabah (UMS) and Ministry of Science and Technology (MOSTI) for supporting the current research work thru the FRGS research grant, project no. FRGS/2/2014/TK06/UMS/03/2.

Investigate a Gas Well Performance Using Nodal Analysis

Hisham Ben Mahmud[1,*], Aman Abdullah[2]

[1]Department of Petroleum Engineering, Curtin University, Malaysia
[2]Chemical Engineering Department, Universiti Malaysia, Malaysia

Email: hisham@Curtin.edu.my

Abstract. Gas condensate well has unique reservoir characteristics and ups and downs in well behaviour affect the production rate significantly. A proper optimization can reduce the operating cost, maximize the hydrocarbon recovery and increase the net present value. Well level optimization can be achieved through optimizing well parameters, such as wellhead, tubing size, and skin factor. All of these factors have been investigated using a real field of Thrace Basin and PROSPER simulation program. The history matching data are validated to identify the future performance prediction for the same reservoir deliverability following the period changes. Therefore, predicted results are compared and validated with measured field data to provide the best production practices. Moreover, the results show that the skin factor has a large influence on the production rate by 45% reduction. The reduction in the reservoir pressure declines the production rate dramatically resulted in 70% decline. While manipulating the wellhead pressure shows minor decline compare to tubing size that does not show any significant change to production rate.

Keywords: Gas condensate; Well optimization; Production rate; PROSPER

1. Introduction

Inconsiderable fraction of the hydrocarbon can be produced by the natural drive of the reservoir. Practical knowledge has proven that when the reservoir pressure is depleted, the recovery factor nearly reaches 20%. Some of heavy fluid reservoirs cannot be produced by all natural energy drivers [1]. Gas condensate reservoir acts differently than a typical oil field in which two-phase oil-gas flow in porous media is often taken place during oil production. Traditionally such flow is modelled by extending the Darcy's law to two-phase flow by employing the concept of saturation dependent on relative permeability [2]. While Retrograde gas fields always exist beyond the critical temperature (as shown in figure 1). In this type of reservoir, the pressure is the main factor to yield any liquid and this would influence the production rate at surface. According to Darcy's law of steady state radial flow of single phase, the production rate is directly proportional to the pressure difference or drawdown between the reservoir and the wellbore. Thus the productivity index is constant over the pressure, which is principally termed as ideal case. However, the ideal case does not exist in the real situation that is related to the reduction of pressure followed by viscosity and gas-oil ratio changes. The ideal case would not able to stabilize when two-phase (liquid and gas) flow presented in a reservoir [3].

To determine the production capacity of a well for a set of well conditions, it's important to determine the quantitative effect and the importance of each variable within the system performance. Therefore, one of the advantages of the system analysis approach is the ability to predict the result due to changes in design variables [4]. Currently nodal analysis is used to accomplish this study as nodal analysis involves calculating the pressure drop in individual components within the production system, thus that the pressure value at a given node in the production system (e.g., bottom hole pressure) can be calculated from both ends (separator and reservoir) (See figure 2). The rate at which pressure is calculated at the node from both ends must be the same. This is the rate at which the well produces. Once the rate under existing conditions is obtained, by adjusting individual components, the sensitivity of individual components on the overall production can be investigated; hence an optimum selection of components can be obtained at a given time [5,7,8].

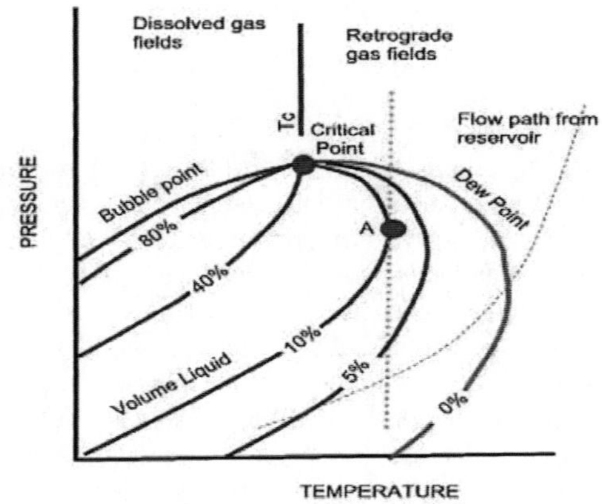

Figure 1. A typical gas condensate reservoir phase curve [6]

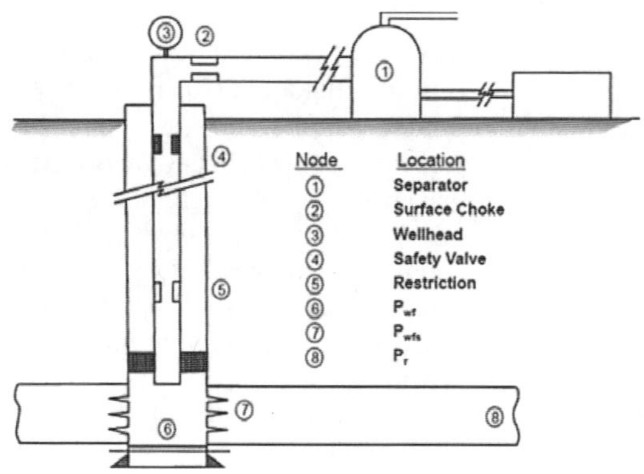

Figure 2. Location of various nodes in the production system

2. Dynamic Nodal Analysis

The estimation of the relationship between the bottomhole pressure and the flowrate is very important as it assists to analyze and to predict the individual well performance. This type of flow process is known as the inflow performance relationship (IPR). The first presentation of IPR concept was made by Gilbert. However, it is only a concept of gas flow for 24 hours versus the flowing well head pressure to obtain graphical results, which are not well detailed [9].

In 1954, Gilbert presented the same concept of inflow performance relationship (IPR) for the purpose of enhancing oil field production rate and flowing bottomhole pressures. The concept focused on the depletion pressure and flow rates only, which are simplistic and not consider many other factors such as skin factor, reservoir condition, restrictions due to production line components and sand screens. Therefore, the actual inflow performance shows linear relationship between the well head pressure and production rates [10]. Nonetheless, the nodal system analysis is developed and begins to using systematic numerical simulations tool in order to obtain best results for optimizing productions and reservoir sustainability. Nodal system analysis then refers to the systems approach for the optimization of the production operations of oil and gas wells thorough evaluation of the complete well production system [11]. It involves engaging correlations to forecast multiphase flow behaviour through pipe lines, well completions components, restrictions due to skin factors and the reservoir in order to analyse the inflow and outflow performance in the complete production [12]. The results obtained were non-linear relationship curves between inflow and outflow performance in multi-phase flow behaviour.

However, Kosmidis et al. [13] highlighted that nodal analysis is restricted only to oil fields with a small number of wells due to its trial-and-error nature during forecasts. In contrast, such as gas lift optimization system, injection system and tubing size adjustments using nodal analysis have been invented in gas wells as well. Mustafa Al Lawati [14] has investigated gas lift nodal analysis model, which explains the mature well produces an extra 153 BOPD after optimization process. By analysing the VLP/IPR relationships, running some sensitivity tests and calculate the gas mandrel depth via nodal analysis results in positive changes to the well, which produces 170 BOPD tend to improve the production up to 323 BOPD.

Therefore, nodal analysis plays an essential part with integrated production models, recommends a cost effective means for optimize performance, therefore; offering more economic systems. For instance, to optimize the production in a mature Niger Delta field, PROSPER's nodal analysis tool has been used in order to enhance the choke performance, which obtained better production results. The study of sensitivity tests and choke performance analysis tend to address the issues relating production enhancements since choke size is an important indicator of productivity [15].

Furthermore, in Nigerian gas field, nodal analysis was applied to determine the pressure drop and optimum tubing size from erosion through simulations. The bottomhole pressure and well head pressure can be easily determined by having any one of the data to find out others or vice-versa. In a natural flowing gas well, the flow velocity is very high, thus erosion could take place inside the production tubing. The IPR and VLP relationship is applicable to examine the pressure drop in the system, pressure drop across perforation and respective tubing size effects [16]. Therefore, the nodal analysis system is applicable to gas or oil field to obtain well production performance for inflow and outflow performance. Nevertheless, there are many other elements have to be considered into the calculation method and approach for extract accurate results.

The aim of this study is to evaluate the best production practice of gas condensate reservoir in Thrace Basin in order to meet the economical aspect of the future field development where the challenging part is to maximize the accumulated recovery since the reservoir pressure is in a rapid depletion due to weak reservoir drive mechanism. An educational software known as PROSPER is used for this study to perform history matching and conduct appropriate sensitivity analysis.

3. Reservoir Description

3.1. Geological Characteristics of Thrace Basin

Thrace basin originated from tertiary age and triangular intermontane basin filled with middle Eocene to Pliocene. In early Oligocene the transgression reached maximum level. Deep basin troughs mostly filled with clastic termed as turbiditic clastic and overlain by thick marine clastic on the northern shelf. Some andesitic tuffs were in inter-bedded formation depicting active volcanism. Overall, the wide-spread Thrace basin contains mostly clastic sediments, marginal marine and terrestrial environments [17]. Figure 3 shows the overall view of Thrace Basin location and the nearby fields.

3.2 Field Data

The W-1 well located in Thrace Basin, which was discovered in year 2000. W-1 well production data available in Appendix A. The formation is mainly sandstone and small indications of shale and has an anticlinal structure. The production mechanism mainly functions due to volumetric depletion. Table 1 shows production W-1 well and reservoir data as well as gas well composition.

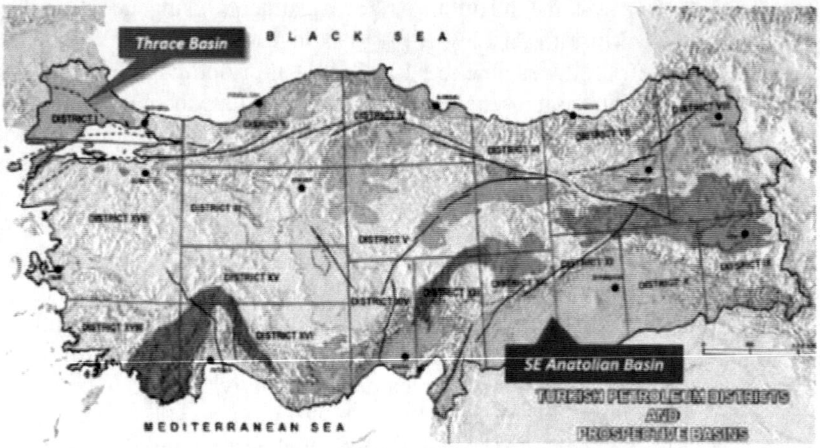

Figure 3. Thrace Basin map view [17]

Table 1. Production well and reservoir data [16].

Well ID	Value	W-1 Composition	%
Reservoir pressure	1,875 psi	C_1	93.25
Reservoir temperature	135oF	C_2	3.25
Datum depth	4,200 ft	C_3	1.45
Water-gas contact	3,788 ft	i-C_4	0.36
Gas specific gravity	0.611	n-C_4	0.45
Tubing size, OD	2.875 in	i-C_5	0.16
Tubing size, ID	2.44 in	n-C_5	0.13
Formation layer height	20 ft	C_6	0.16
Reservoir radius	2,107 ft	N_2	0.76
Oil density	48.56 lb/ft3		
Average porosity	15%		
Average permeability	13 md		
Well radius	0.583 ft		

4. Well model construction

The production system has been modelled using PROSPOR software. Actual field data were entered to the model. The input data including the rock fluid, reservoir properties as well as gas composition as listed in table 1. First of all, PVT data, downhole equipment and IPR were modelled, and then the nodal analysis was conducted on interesting nodes in order to gather enough data for any alteration purpose. Figure 4a and b describe the real and model in software, respectively.

5. Result analysis and discussion

5.1. PVT data analysis

PVT data predicts the formation of condensate when the fluid flows to surface conditions, which would be integrated into reservoir data to obtain the dew point, critical point and bubble point. Table 1 data was used to analyse the PVT data and gas composition using Lee et al correlation, which would comply the gas condensate behaviour. The Dew point pressure at the reservoir temperature was obtained by approximation and trial and error method, which is about 810 psig.

5.2. IPR construction

The IPR curve was constructed Initially without considering formation damage (skin = 0) since the early production starts without any restriction. The skin effect may take place near the wellbore due to condensate banking and PVT changes. At figure 5, IPR curve has an absolute open flow (AOF) about 17 MMscf/day. This indicates the maximum amount of gas can be produced if there is no pressure drop between production tubing components.

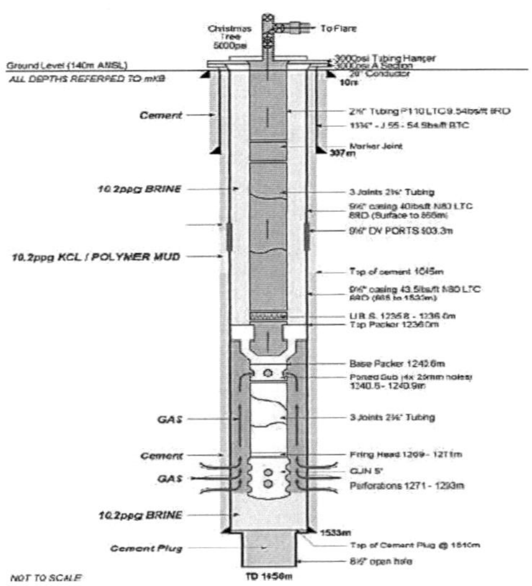

Figure 4a. W-1 Well diagram [17]

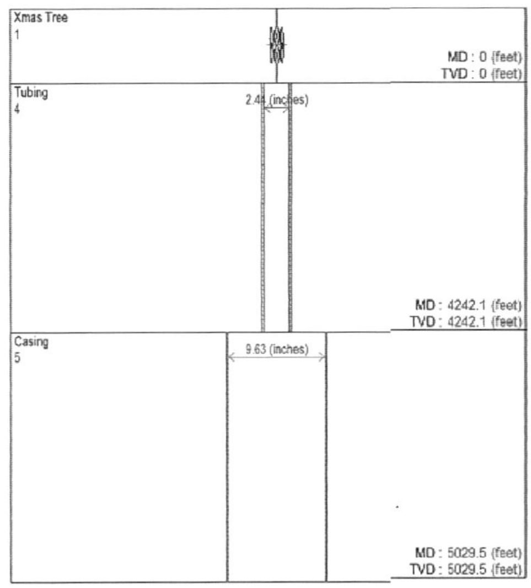

Figure 4b. Well completion from PROSPOR.

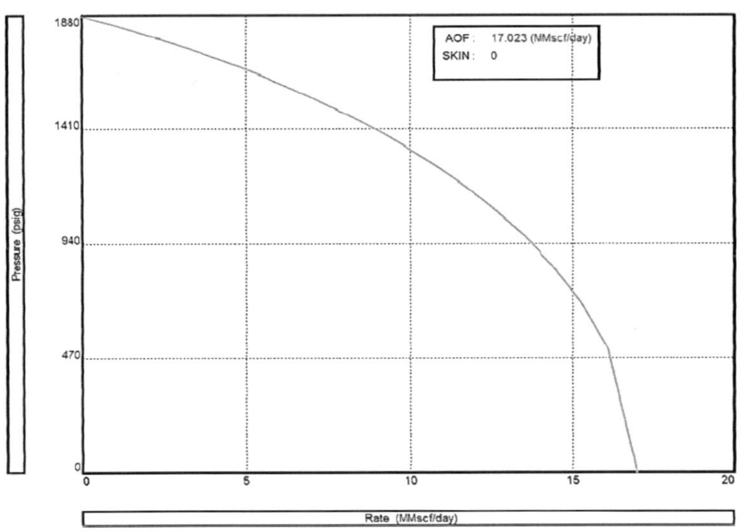

Figure 5. IPR Curve for initial reservoir properties.

However, the solution gas drive mechanism not having the same IPR curve over time because the reservoir pressure drops drastically than other reservoir drive mechanisms. For the gas condensate reservoir, the reservoir pressure drop has high probability of changing at a short period due to its behaviour. The result analysis of estimating reservoir pressure changes with respect to water-gas ratio (WGR) and condensate gas ratio (CGR) is shown in figure 6. It indicates that the variation of WGR (0.0002 to 0.0008 stb/Mscf) and CGR (0.007 to 0.01 stb/Mscf) does not have significant effect on IPR curve. Indeed, the drops in reservoir pressure (1,875 to 1,000 psig) changes considerably IPR and AOF reduced to approximately 6 MMscf/day when the reservoir pressure becomes 1000 psig. Therefore, IPR curve is too sensitive towards the reservoir pressure rather than CGR and WGR.

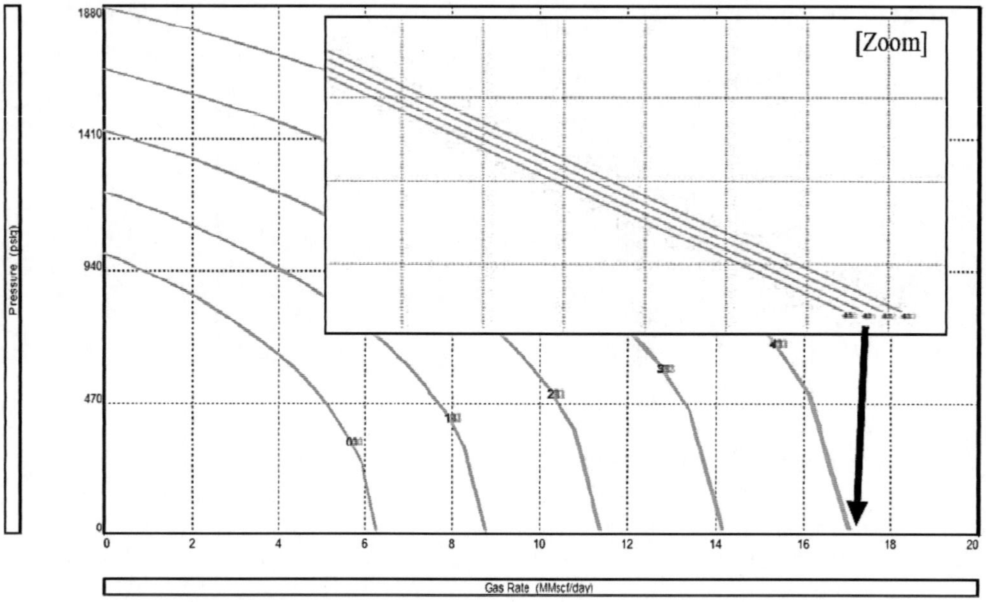

Figure 6. Effect IPR with reservoir pressure, WGR and CGR changes.

5.3. Operating point

According to production data, W-1 well has been produced since June 2002 to January 2005. In early production time, the reservoir pressure supported by a reservoir drive mechanism, thus maintaining VLP and IPR relationship. The reservoir pressure initially is 1,875 psig and well head pressure (WHP) is maintained at 1,590 psig. Implementing Jones reservoir model and Orkiszewski vertical gradient correlation, IPR versus VLP curves were constructed to obtain the bottomhole pressure and the production rate, which are 1,600 psig and 6.4 MMscf/day, respectively (Figure 7). While the field production rate is about 6.1 MMscf/day with less than 5% error. The model can be used for history matching and further analysis.

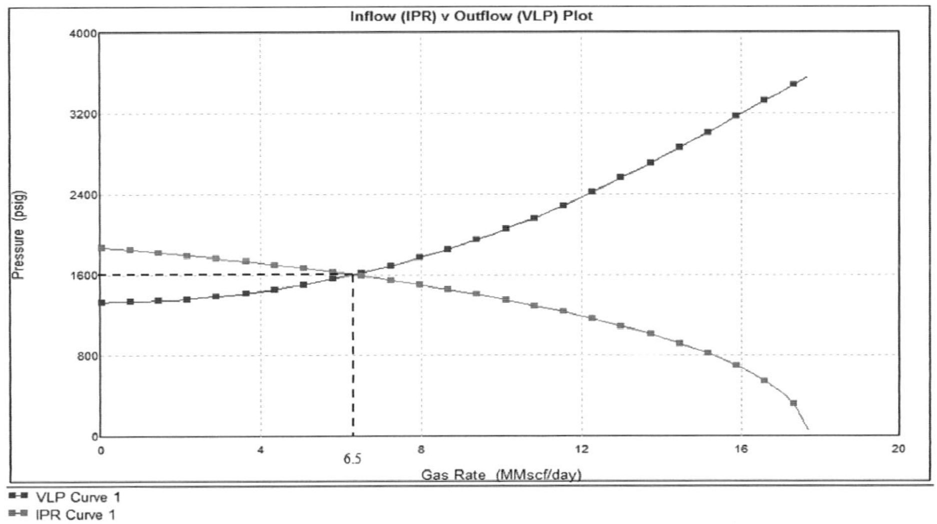

Figure 7. IPR versus VLP curve for initial reservoir condition of W-1.

5.4. History matching

In this case study, the production data depicts that W-1 well had been in operation since June 2002 until January 2005. The production data shows the inconsistency of gas and condensate production within this period. The production data changes due to many factors including changes of reservoir pressure, water cut increases, condensate banking, wellhead pressure adjustments and skin effects.

The condensate gas ratio (CGR) for overall production varies in negligible amount, which is only four percent (4%) changes from 0.0071 stb/Mscf to 0.0074 stb/Mscf. It indicates that CGR does not influence the overall production significantly. Therefore, the possibility of production data changes could be due to reservoir properties. The retrograde gas reservoir often affected by significant reservoir pressure drop due to its weak drive mechanism.

The history matching focuses on reservoir deliverability changes to find the recent gas and condensate production as function of well behaviour. Four particular test points were chosen for this purpose. The production data and WHP were taken from field production data while the reservoir pressure was assumed to be reducing gradually by 100 psig each year. Constructing the VLP versus IPR relationship for assumed reservoir data. The new solution rate is used to compare with field gas production data as presented in table 2. It indicates that the calculated reservoir pressure and field gas production data are in agreement with a minor error of less than 10%. Therefore, the estimated pressure can be used to predict the future performance for well behavior.

Table 2. Matched reservoir pressure for test points.

Test point	Date	WHP psig	Field production rate MMscf/d	Matched P_{re} psig	Calculated Q_g MMscf/d	Error %
1	1/06/2000	1590	6.438	2096.3	6.437	0.006
2	1/01/2003	1329	5.425	1772.5	5.456	0.571
3	1/01/2004	989	2.564	1259.8	2.552	0.468
4	1/12/2004	781	1.808	996.1	1.792	0.885

5.5. Well performance prediction

As the model was validated with field production data, the well behaviour can be analysed considering the wellhead pressure, tubing size, water cut, condensate banking, choke size, and skin factor.

5.5.1. Wellhead pressure

The initial setup of WHP is about 1,590 psig and decreased till 781 psig to sustain the production rate. By reducing the wellhead pressure leads to increase the pressure drawdown consequently it increases the production rate where the reservoir pressure has less influence to lift up the gas condensate to the surface facilities. The future prediction performance for W-1 well was analysed from pressure range of 765 to 500 psig as shown in figure 8.

The obtained results are tabulated in table 3, the reduction of WHP shows an increase in gas production rate. The predicted rate is significantly improved to 3.860 MMscf/day by reducing WHP to 500 psig compared to the production rate of 1.808 MMscf/day at 781 psig, which is slightly more than double of production rate. Thus, WHP variations play an important role in optimize and sustain the production.

5.5.2. Tubing size analysis

One of the important factor for optimizing the production over time is the installation of optimum tubing with appropriate internal diameter size. The tubing size allows the gas and condensate to lift up to the surface with a particular volume and pressure drop. The internal tubing size of W-1 well is a standard 2.441 inch. Therefore, the analysis is done by inserting different tubing sizes ranging from 2.0-4.92 inch to determine the optimum rate. Figure 9 shows the results of production gas rate for various tubing sizes. It is noted that gas production rate has not increased significantly, only 0.25 MMscf/d with respect to tubing size as listed in table 4. Therefore, the tubing variation should not be considered for future field development plan.

Table 3. Predicted result of wellhead pressure analysis.

No.	WHP (psig)	Predicted Q_g (MMscf/d)
6	765	2.14
5	712	2.74
4	659	3.06
3	606	3.18
2	553	3.28
1	500	3.86

Table 4. Predicted result of tubing size analysis.

No.	Tubing size (inch)	Predicted Q_g (MMscf/d)
1	2.00	1.99
2	2.44	2.14
3	2.99	2.21
4	3.96	2.24
5	4.92	2.24

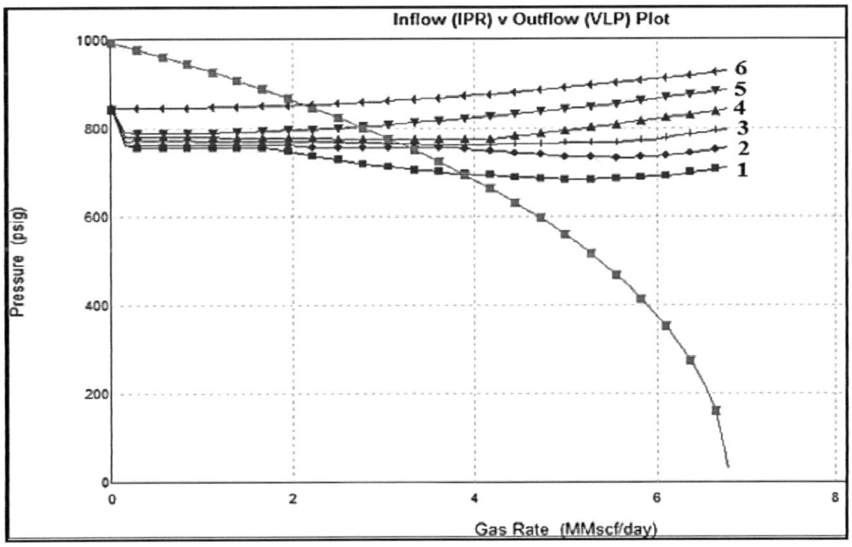

Figure 8. Sensitivity analysis of wellhead pressure.

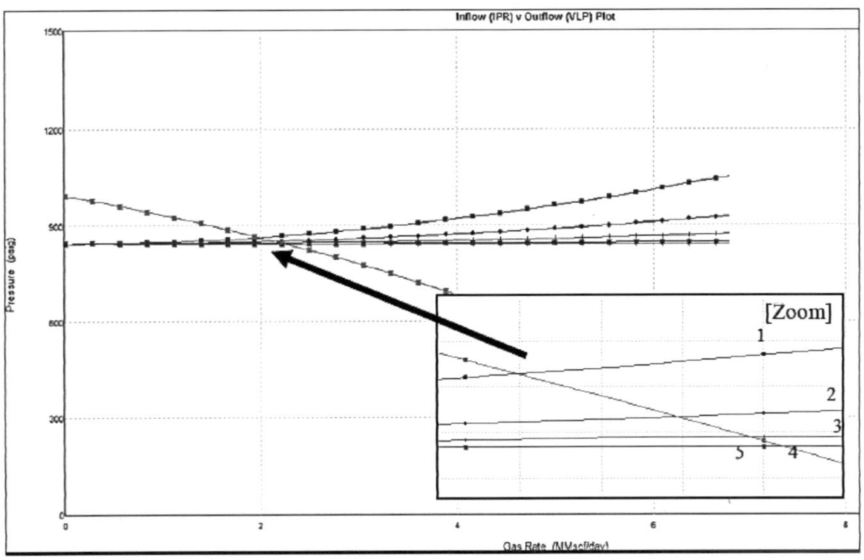

Figure 9. Results for different Tubing size.

5.5.3. Water cut impact

In this gas condensate reservoir, the solution drive mechanism acts as reservoir support where the water invasion is not in liquid form such as water drive but as vapour molecules mixed with hydrocarbon gas. As water gas ratio (WGR) increases theoretically it may have overall production effects due to increase the overall fluid density as water molecules have higher density than gas molecules. According to W-1 well history, WGR accumulated at the surface in a small quantity, which is only about 0.48 stb/MMscf with increment of only 0.03 stb/MMscf since June 2002. Thus, increasing WGR to 0.8 stb/MMscf, it shows less effect on the well production rate as shown in figure 10. Therefore, WGR is not an important factor in such reservoir characteristics.

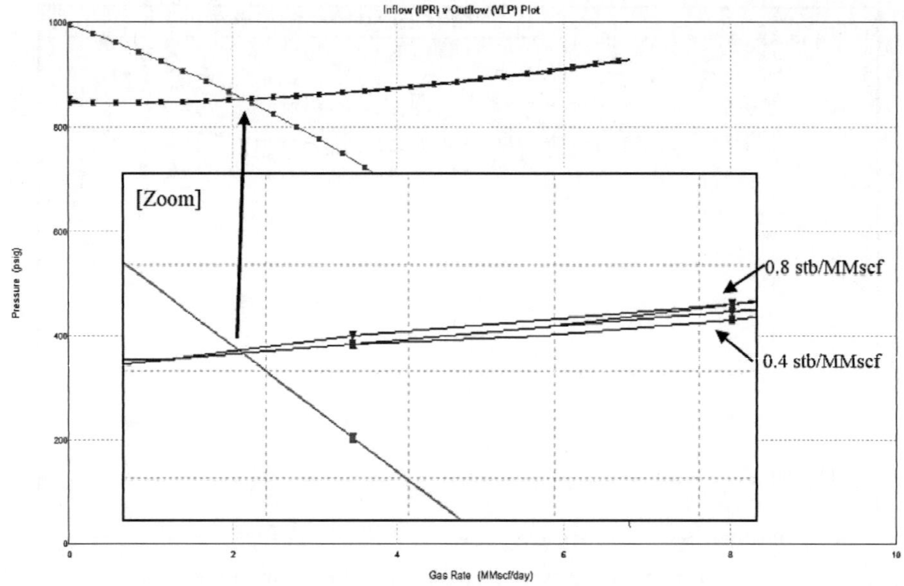

Figure 10. Results of Water Gas Ratio analysis.

5.5.4. Effect of gas condensate ratio

In the production of gas condensate wells there are two important components are profitably expected, which are hydrocarbon gas and condensate. Both produced fluids estimate the payback period in developing a field. In this case, the effect of gas condensate ratio (GCR) is considered for optimization purpose. Field production data shows GCR is 139.104 Mscf/stb in December 2004. Therefore, the considered range of GCR is between 120 and 160 Mscf/stb to predict the potential change of overall production.

Figure 11 displays the effect of gas condensate ratio on the well production. As can be seen that increasing GCR leads to increase the production rate but it is not significant at a small increment ratio. Thus, increasing GCR reduces the hydrostatic pressure gradient as well as the slip velocity. The overall results depicted that the gas rate production proportional to the GCR amount.

5.5.5. Skin effect

The skin effect is exaggerated by condensate banking and well deliverability nearby the wellbore. The condensate banking affects the gas flow within the tubing since the condensate fluid takes place near the perforated area. This leads to gas flow restriction, which increases the skin influence. Therefore, the effect of skin is analysed using positive values from +2 to +10, which represent the damage occurs nearby the wellbore area.

Figure 12 demonstrates the effect of skin on inflow performance relationship. As can be seen that the damage has a significant impact on the reservoir deliverability by achieving approximately 1 MMscf/d at a skin value of +10 in comparison to lower value of +2 with a production rate of 1.8 MMscf/d as listed in table 5. In gas condensate reservoir, the skin factor can be more than +20 in some cases if the well is not managed properly. Thus, the reservoirs deliverability is proportional to the skin factor that can affect the production rate significantly [18].

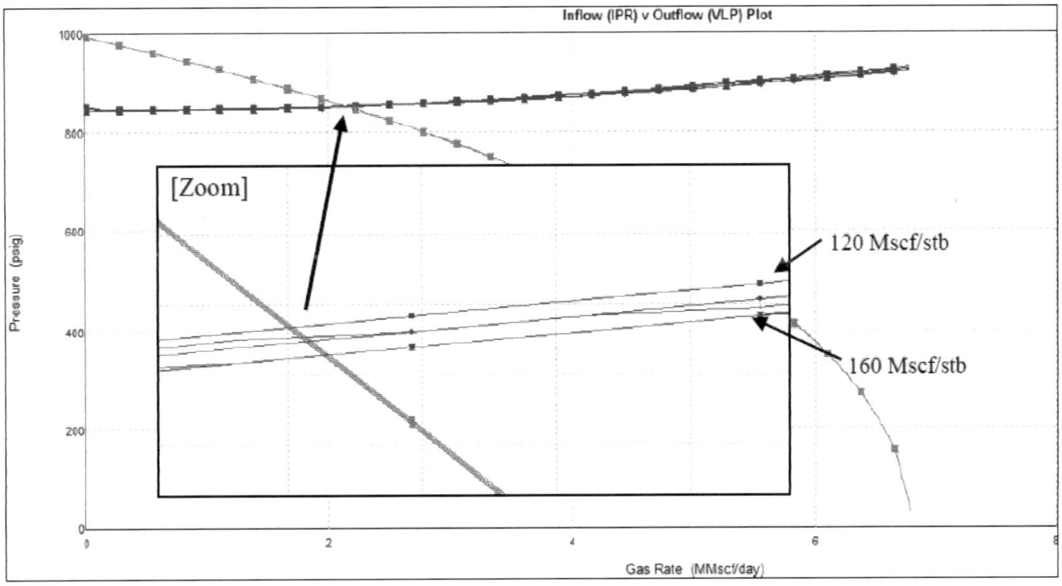

Figure 11. Results of IPR VLP relationship changes due to GCR.

Table 5. Predicted production from skin variations.

Skin factor	Predicted Q_g (MMscf/d)	Production decline %
+2	1.8	0
+4	1.5	16.7
+6	1.3	27.8
+8	1.1	38.9
+10	0.98	45.6

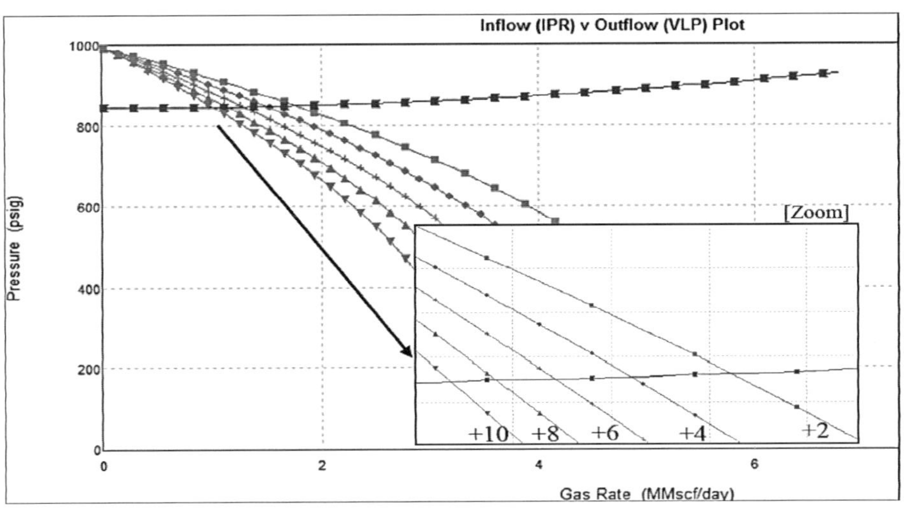

Figure 12. Results of overall skin effects on reservoir deliverability.

5.5.6. Surface choke size

The choke is mechanical equipment which can restrict the flow area by adjusting the bean size. It applies back pressure to the system to control the fluid flow nearby the wellbore and achieve stable flowing tubing head pressure (FTHP) on the surface. Therefore, the effect of bean size variations had been analysed in order to obtain the optimum rate of production.

Figure 13 shows the results of VLP and IPR relationship while changing the bean size from 0.5 (32/64) inch to 2.0 (128/64) inch. As the bean size increases the production also increases since it allows more gas flow easily to surface facilities by reducing the restrictions on pressure drop. Using choke size of 0.5 inch can yield about 1.78 MMscf/d, which is almost same as the average production. Moreover, as the bean size opened up to 2.0 inch, it enhanced the production to 2.14 MMscf/d, which is almost same as bean size of 1.0 inch that produces 2.1 MMscf/d. When the bean size increases reduce the back pressure towards to the reservoir, which leads to increase the drawdown pressure and fluid production.

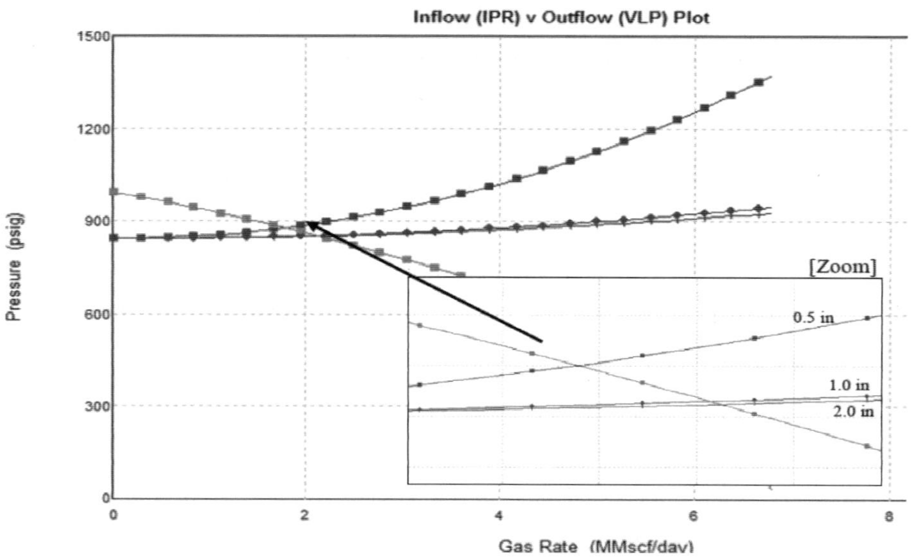

Figure 13. Choke size analysis.

6. Conclusion

Dynamic nodal analysis technique allows to perform sensitivity analysis of future performance for gas condensate wells after a satisfactory match of the previous production performance is obtained. The major contribution of this study is that it provides a comprehensive analysis the well performance changes as a function of time when the production parameters are altered.

The variation of tubing size can be optimized until 3.92 inch to improve the production rate from 1.8 MMscf/d to 2.24 MMscf/d, which is unprofitable due to production increment. However, changing tubing size is not economically in such well completion.

The well head pressure and choke size should be controlled to achieve and maintain the critical production flow rate. These two factors are able to predict how far each valve will open under current reservoir conditions to achieve optimum performance.

It is demonstrated that the skin factor has a large influence on the production performance and it is shown that the reduce of skin factor can improve the fluid flow without restrictions due to condensate banking.

Last not least, water gas ratio does not effectively affect the production performance of such reservoir characteristics. While increasing gas condensate ratio improved the well deliverability reduces the hydrostatic pressure gradient as well as the slip velocity.

References

[1] Ben Mahmud H Kh, Sheng I H and Shafiq M U 2016 *ARPN J. Eng. App. Sci.* **11** 24
[2] Brij B M and Bashir B 2010 *AIP Conf. Proc.* **1254** 103
[3] Evinger H and Muskat M 1942 AIME **146** 126-139
[4] Dmour H N 2013 *J. Pet. Sci. and Tech.* **31** 11 1109-1122
[5] Brown K E 1984 *Pennwell Publishing Company*, Tulsa, OK, volume **4**
[6] Perez G and Kelkar B G 1990 *Journal of Canadian Petroleum Technology* **29** 1
[7] Bitsindou A 1998 Gas well production optimization using dynamic nodal analysis, M.S. Thesis, University of Tulsa, Tulsa.
[8] Vogel J 1968 *J. Pet. Tech.*, 83-92.
[9] Rawlins E and Schellhardt M 1935 USBM 7
[10] Gilbert W 1954 *Drilling and Production Practice*, 126
[11] Brown K and Lea J 1985 *J. Pet. Tech.* **37** 11 1751-1763
[12] Brill J 1987 *J. Pet. Tech.*, **39** 1 15-21
[13] Kosmidis V, Perkins J and Pistikopoulos E 2004 *Industrial and Engineering Chemistry Research*, **43** 1520-1525
[14] Mustafa Al Lawati 2014 *SPE Artificial Lift Conf. & Exh-North America* (Houston, Texas, USA)
[15] Ebuka U G, Ogbunude B and Egelle E 2014 *Int. J. Sci. and Res.* **3** 8 1520-1525
[16] Dala J A, Agbaka O O, Olafuyi O A and Adewole E S 2015 *SPE Nig. Annu. Int. Conf. and Exh.* (Lagos Nigeria)
[17] Mindek, C 2005 Production optimization of gas condensate reservoir using black oil simulator and nodal system analysis: A case study. Middle East Technical University.
[18] Xiao J J and Al-Muraikhi A 2004 *SPE Annu. Tech. Conf. and Exh* (Houston, Texas)

IOP Conf. Series: Materials Science and Engineering **217** (2017) 012022 doi:10.1088/1757-899X/217/1/012022

APPENDIX A: Well production data

W1 well production data shown in Table B.1 below from June 2002 until January 2005.

Table A.1: Well 1 production data [17]

Date	Gas Production (MMscf)	Condensate Production (bbl)	WHP (psi)
Jun-02	0	0	1,590
Jul-02	94	691	1,595
Aug-02	164	1183	1,595
Sep-02	219	1587	1,595
Oct-02	342	2491	1,519
Nov-02	508	3698	1,428
Dec-02	664	4833	1,390
Jan-03	829	6021	1,329
Feb-03	995	7230	1,312
Mar-03	1,089	7,927	1,278
Apr-03	1,217	8,854	1,228
May-03	1,334	9,716	1,193
Jun-03	1,438	10,464	1,160
Jul-03	1,540	11,204	1,120
Aug-03	1,646	11,790	1,100
Sep-03	1,752	12,548	1,070
Oct-03	1,851	13,254	1,050
Nov-03	1,932	13,787	1,030
Dec-03	2,021	14,400	1,010
Jan-04	2,099	14,946	989
Feb-04	2,167	15,408	977
Mar-04	2,240	15,940	956
Apr-04	2,304	16,417	940
May-04	2,356	16,811	918
Jun-04	2,414	17,235	879
Jul-04	2,443	17,784	865
Aug-04	2,485	18,003	855
Sep-04	2,504	18,269	843
Oct-04	2,532	18,396	821
Nov-04	2,566	18,609	800
Dec-04	2,621	18,842	781
Jan-05	2,742	19,954	765

Investigation of Changing Pore Topology and Porosity During Matrix Acidizing using Different Chelating Agents

Mian Umer Shafiq[1], Hisham Khaled Ben Mahmud[1], Reza Rezaee[2] and Nadia Testamanti[2]

[1]Petroleum Engineering Department, Curtin University, Sarawak, Malaysia
[2]Department of Petroleum Engineering, Curtin University, Kent Street, Bentley, Perth, Australia

E-mail: mian.umer@curtin.edu.my

Abstract. Core flooding acidizing experiments on sandstone/carbonate formation are usually performed in the laboratory to observe different physical phenomena and to design acidizing stimulation jobs for the field. During the tests, some key parameters are analyzed such as pore volume required for breakthrough as well as pressure. Hydrochloric acid (HCl) is commonly used in the carbonate matrix acidizing while Mud acid (HF: HCl) is usually applied during the sandstone acidizing to remove damage around the well bore. However, many problems are associated with the application of these acids, such as fast reaction, corrosion and incompatibility of HCl with some minerals (illite). To overcome these problems, chelating agents (HEDTA, EDTA and GLDA) were used in this research. Colton tight sandstone and Guelph Dolomite core samples were used in this study. The experiments usually are defined in terms of porosity, permeability, dissolution and pore topology. Effluent samples were analyzed to determine dissolved iron, sodium, potassium, calcium and other positive ions using Inductively Coupled Plasma (ICP). Meanwhile Nuclear Magnetic Resonance (NMR) was employed to determine porosity and pore structure of the core sample. Core flood experiments on Berea sandstone cores and dolomite samples with dimensions of 1.5 in × 3 in were conducted at a flow rate of 1 cc/min under 150oF temperature. NMR and porosity analysis concluded that applied chemicals are effective in creating fresh pore spaces. ICP analysis concluded that HEDTA showed good ability to chelate calcium, sodium, magnesium, potassium and iron. It can be established from the analysis that HEDTA can increase more amount of permeability as compared to other chelates.

1. Introduction

Acids play a key role in boosting production or increasing injectivity in hydrocarbon fields. Aboud [1] predicted that 40% more energy resources will be required in 2020 than in the present because the world demand for energy is increasing. In order to fulfill this demand, acidizing of high temperature reservoirs gain importance because the new reserves explores are much deeper and hotter reservoirs [2]. In the process of acidizing, acids have been injected to react with sandstone and carbonate formations. The injection of different acids cause reaction of acids with the minerals present in the rock. Different minerals like quartz, carbonates and feldspar can be dissolved by the injection of acids, thus increasing the permeability of the reservoir around wellbore, which ultimately increases the flow rate of hydrocarbon fluid from to the wellbore [3] and [4, 5]. These reactions can cause dissolution of rock matrix and formation of precipitates, which might bring about changes in the permeability and porosity of a reservoir. Therefore, the main goal of sandstone acidizing is to remove the formation damage while for carbonate is to create new wormholes [6].

The mixture of hydrofluoric acid (HF) and hydrochloric acid (HCl), known as mud acid [7] is normally applied during the sandstone acidizing process to remove the damage around the wellbore while hydrochloric acid has been injected to create new wormholes in case of carbonate acidizing. Sandstone formation has been acidized using mud acid by [7, 8] and hydrochloric acid has been applied on carbonate formation by [9, 10] to dissolve carbonates and other positive ions. Stimulating carbonate rocks typically involves a reaction between an acid and the minerals like calcite ($CaCO_3$) or dolomite $CaMg(CO_3)2$. The primary reaction of carbonate with acid is presented in equation 1 [11].

Carbonate Reaction: $$HCl + CaCO_3 \longrightarrow CaCl_2 + H_2O + CO_2 \quad (1)$$

Sandstone reactions [12]: $$SiO_2 + 6HF \longrightarrow H_2SiF_6 + 2H_2O \quad (2)$$

$$H_2SiF_6 + 4H_2O \longrightarrow Si\,(OH)_4 + 6HF \quad (3)$$

However, using hydrochloric acid is not successful in some cases, even some field studies reported damages [11]. First is the presence HCl sensitive mineral of (illite), this may choke the pore throat and thus decreasing the permeability and porosity which is a serious issue. Secondly, corrosion rate is higher, also the reaction rate will be high at high-temperature conditions which may cause other problems like precipitation.

Because of mentioned problems faced using HCl acid, other approaches have been studied and used as an alternative to HCl acid in carbonate and sandstone acidizing. [13, 14] performed experimental study with different acid combinations, replacing conventional mud acid. HCl acid has been replaced with organic acid to acidize sandstone formation. Chunming [15] applied novel emulsified acid to acidize Chinese oil fields and discovered that the increase in permeability of oil saturated cores by 96.1% while by only 10.1% for water saturated cores. [12, 16] applied Organic-HF to overcome the problems faced during sandstone matrix acidizing. Different chelating agents have been used at high temperature conditions i.e., Hydroxyethylenediaminetetraacetic acid (HEDTA) and Ethylenediaminetetraacetic acid (EDTA) used at high temperature without the use of HF-containing fluids resulted in the increase of gas production. EDTA and HEDTA created wormholes at 400°F explained by [17] and discovered the significance of chelating agent Na_3HEDTA compared to mud acid. Various stimulating studies on sandstone formation using HEDTA chelating agent have been conducted at higher temperatures by [17, 18, 19] and showed its significance in increasing permeability compared to mud acid. [20] compared the use of different acids developed for sandstone acidizing. Slow reaction rate and less corrosivity have allowed the petroleum industry to apply formic and acetic acids for many years as alternatives to HCl in carbonate acidizing. The rapid HCl spending prevents deep penetration of acid into the formation [11].

The main aim of this study is to overcome the problems related to the presence hydrochloric acid. While the advantages of hydrochloric acid presented by McLeod [21] should be maintained. In this study different methods such porosity and NMR measurements and ICP analysis has been performed to monitor rock and fluid alterations at different stages of acidizing process. NMR analysis have been applied by [22] for the very first time on sandstone samples. This technique proved to be very useful in determining the pore space topology after acidizing.

2. Nuclear Magnetic Resonance

Nuclear magnetic resonance (NMR) has proved its significance in different fields like chemistry, physics, and biomedicine. The same principle used in these fields can also be applied for the imaging of reservoir rock. "NMR logging is a subcategory of electromagnetic logging [23], that can measure the induced magnet moment of hydrogen nuclei (protons) contained within the fluid-filled pore space of porous media (reservoir rocks)". Conventional logging measurements respond to both rock matrix and fluid properties and are mineralogy dependent while NMR-logging respond to the presence of hydrogen protons. Information related to the quantities of fluids present, the properties of these fluids, and the sizes of the pores containing these fluids can be obtained using NMR analysis. From this information, it is possible to infer or estimate.

- The volume (porosity) and distribution (permeability) of the rock pore space
- Rock composition
- Type and quantity of fluid hydrocarbons
- Hydrocarbon producibility

Cutoff-BVI (CBVI) model is used to determine the moveable fluids (large pores) and immobile fluid (capillary pores) based on a fixed-T2 value (T2cutoff) which divides the free-fluid index (FFI) and BVI. The T2 signal from the rock matrix is actually very rapid, even modern logging tools cannot detect it. Accordingly, the recorded T2 distribution only represents the porosity occupied by immovable BVI, movable (FFI) water and clay-bound-water (CBW) components. This value (T2cutoff) is used to divide the T2 distribution into movable (i.e., producible or free fluids and FFI) and immovable (i.e., bound-fluid, BVI, and CBVI) components.

3. Experimental studies

3.1 Core Flood Setup

Figures 1a and 1b shows the core-flooding experimental setup and schematic diagram respectively, which were used for the core flooding of Berea sandstone and Guelph dolomite core samples with chelates. This core flooding setup is capable of handling a pressure up to 10,000 psi and a temperature up to 200°F. The wetted parts of the apparatus are made of corrosion resistive material (e.g., hastelloy), which makes this system rust free even at higher concentrations of acids. The experimental setup consists of five basic parts which are explained below with detailed function.

- Core holder is used to confine the core sample at desired pressure and temperature. The material of this core holder is hastelloy. It has two inlets and one outlet. The confining pressure applied during the experiments is 1000 psi while the temperature is 150°F.
- The High Performance Liquid Chromatography pump as shown in Figure 1a and 1b, is used to inject the acid into the core holder at desired and controlled rate. It can inject liquid at a maximum rate of 6000 cc/min. The injection rate for this research is 1 cc/min. The less injection rate is used in order to avoid high back pressure during the experiment.
- The Temperature controller (Figure 1a) with a tape is used to heat the core holder at desired temperature value. The maximum temperature attained using this controller is 500°F.
- The syringe pump (Figure 1a) is used to create a high pressure inside the core holder. It uses fresh water to create a high pressure and can create up to 20,000 psi pressure.
- Collection beaker is used to collect the reacted acid sample at the collection point as seen in Figure 1a.

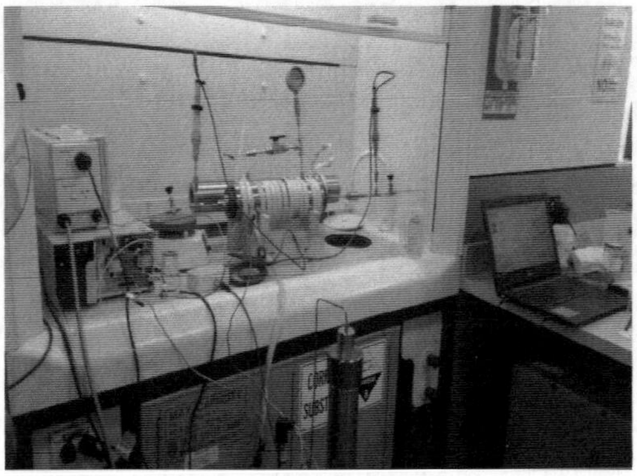

Figure 1a. Core Flood Setup

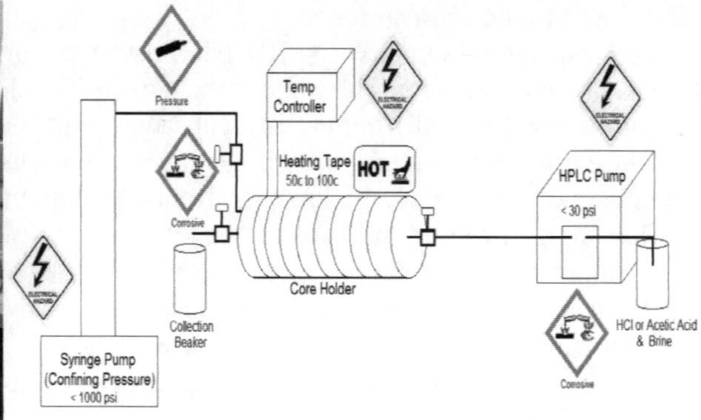

Figure 1b. Schematic Diagram

3.2 Core flood experiments

The core plug were weighed initially and dimensions were obtained using Vernier calliper. The core was put inside the core holder and core flood instrument was setup as shown in Figure 1a. The core was then confined under pressure of 1,000 psi using the syringe pump as shown in Figure 1a. The core holder was then heated to 150°F using the temperature controller and tape. The chelate was injected using HPLC pump at a rate of 1 cm^3/min where the pressure variation was monitored at pressure transducers. The constant pressure value at HPLC pump and transducer shows the completion of the reaction. Injection of acid is terminated and one hour residence time was allowed to complete the reaction of the chelate with the core sample. Distilled water is injected to remove the remains of the acid present in the core sample.

Table 1. Sample name and composition

Sample Name	Type of rock	Chelate
Sample A		EDTA
Sample B	Colton Sandstone	HEDTA
Sample C		GLDA
Sample A		EDTA
Sample B	Guelph Dolomite	HEDTA
Sample C		GLDA

3.3 Laboratory Measurements / Procedures

Apart from the core flooding experiments, there were variety of other measurements and preparation work carried out as well, which included:

- Measuring permeability, porosity of the core samples before and after core flooding experiments.
- Drying of core samples before and after each test at 80°C for 24 hours.
- Weight of all dry core samples was measured before and after core flooding experiments.
- Saturation of core samples using brine (5% NaCl) for NMR analysis using manual saturator.
- Running NMR tests on the brine saturated samples before and after each core flooding test.
- The reacted acid was collected from the core flood test and was analyzed for various ions using ICP analysis.

IOP Conf. Series: Materials Science and Engineering **217** (2017) 012023 doi:10.1088/1757-899X/217/1/012023

4. Result and Analysis

4.1. Porosity and Permeability Analysis

The porosity is defined as the measure of empty spaces in the rock and it provides information about the storage capacity of the reservoir rock. It is considered as one of the most critical parameter for the determination of oil and gas initial in place. It can be measured or determined in different ways i.e., manually and automatically. In this research Automatic Porosity Permeability Apparatus (Figure 2) is used to measure both porosity and permeability. Permeability and porosity values before and after acidizing are presented in table 2.

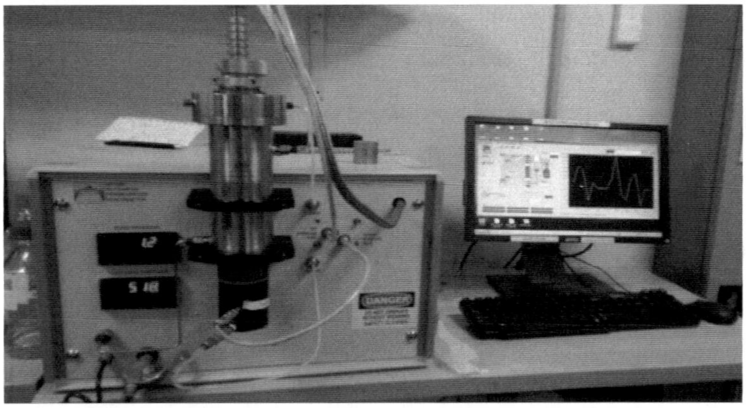

Figure 2. Automatic Porosity permeability apparatus

The main purpose of acidizing is to remove the damage around the wellbore and to create new channels in the reservoir rock. The main purpose of the application of HCl is to remove calcium, potassium, sodium, magnesium and iron minerals. Creating new pores by dissolving these minerals increase the porosity and permeability. Usually, sandstone samples contain very less amount of sodium, potassium and calcium ions, therefore; as a result, the change in porosity is also less. The removal of these ions is very necessary in order to get better acidizing results because these minerals can cause precipitation later during the main acid stage of sandstone acidizing.

Table 2. Porosity (ø) and Permeability (K) results

	Sample Name	Chelate used	Initial ø %	Final ø %	% change
	Sample A	EDTA	10.74	11.66	8.50
Tight sandstone	Sample B	HEDTA	11.49	12.65	10.10
	Sample C	GLDA	12.13	13.01	7.25
	Sample A	EDTA	16.80	16.82	0.11
Carbonate	Sample B	HEDTA	14.19	15.09	6.30
	Sample C	GLDA	21.20	21.63	2.03
	Sample Name		Initial K (md)	Final K (md)	% change
Tight sandstone	Sample A	EDTA	0.7118	0.6379	-10.30
	Sample B	HEDTA	0.4633	0.9542	105.67
	Sample C	GLDA	0.8069	0.7842	-2.80
Carbonate	Sample A	EDTA	104.27	93.76	-10.07
	Sample B	HEDTA	9.80	18.11	84.79
	Sample C	GLDA	405.55	399.41	-2.00

IOP Conf. Series: Materials Science and Engineering **217** (2017) 012023 doi:10.1088/1757-899X/217/1/012023

HEDTA proved to be very useful chelating agent as compared to EDTA and GLDA for both sandstone and carbonate samples. The increase in permeability is almost double when HEDTA was allowed to react with sandstone and carbonate samples. Although other chelates were also very effective in increasing porosity of tight sandstones but the permeability increment is not in favour. This maybe attributed towards the formation of precipitate and residing of these precipitates in the larger pores which choke the pore spaces. Also, these chelates are soluble in water at very high temperature ranges, decrease in temperature can cause formation of precipitates and settling of solid particles present in chelate.

4.2 NMR measurements

NMR measurement has been applied on all core samples before and after acidizing to investigate the change in pore size distribution and change in pore topology. Figure 3 represent the T2 spectrum for each dolomite core sample reacted with chelates while figure 4 represent the spectrum analysis of each sandstone core sample. Change in pore size distribution can be seen clearly in spectrum of sample C and B, it means that the new pores have been created while spectrum of sample A represents decrease in pore sizes. The reduction of pore space sizes can be attributed to dissolution of minerals and re-precipitation in the pores. These precipitates can be formed due to ineffective acidizing by chelate and can reduce the porosity. Sample B (reacted with HEDTA) spectrum showed maximum increase in number of large pore spaces, validating permeability and porosity results discussed in table 2.

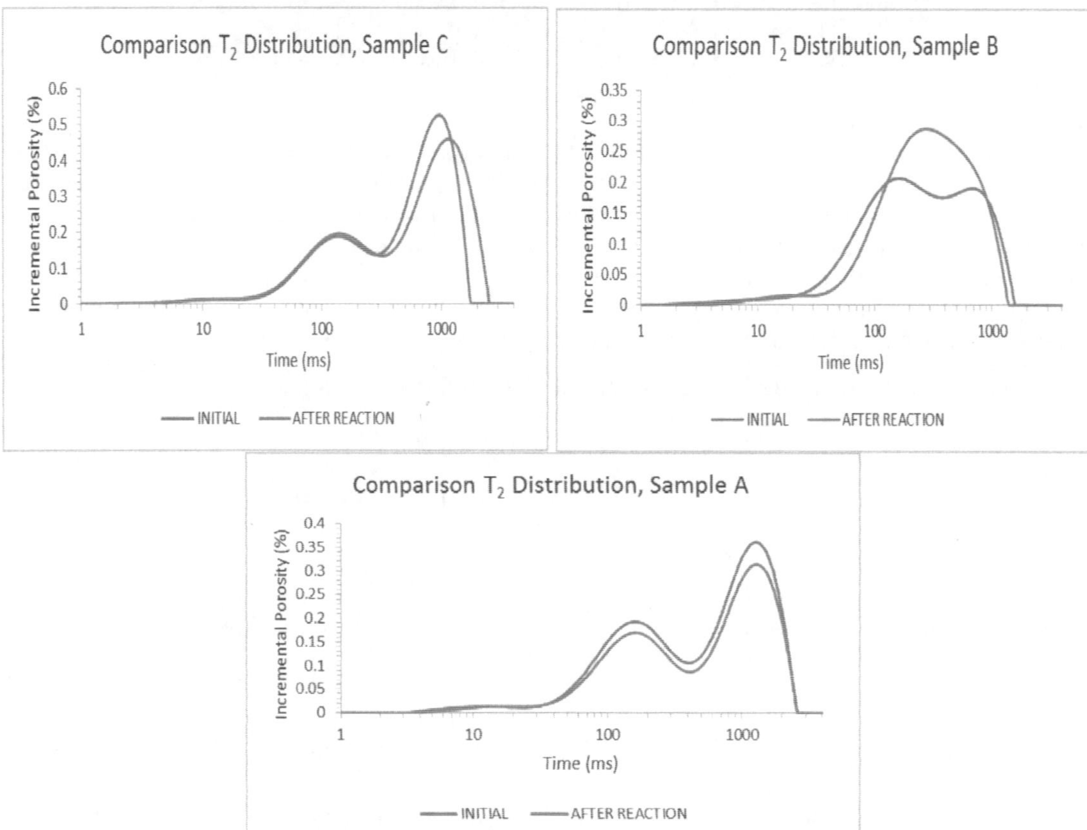

Figure 3. T2 spectrum of carbonate samples

The spectrum in figure 4 clearly represents the increase in the total amount of pore spaces in the sandstone core samples after acidizing, where maximum change in number of pore spaces has been observed in sample B reacted with HEDTA, validating porosity and permeability results (Table 2). The increase in porosity after acidizing can be attributed to the effective action of chelate on the core sample. Sample A reacted with EDTA and there is decrease in porosity which can be attributed to the ineffectiveness of EDTA on sandstone core sample. HEDTA proved to be very effective in creating more spaces in both sandstone and carbonate samples.

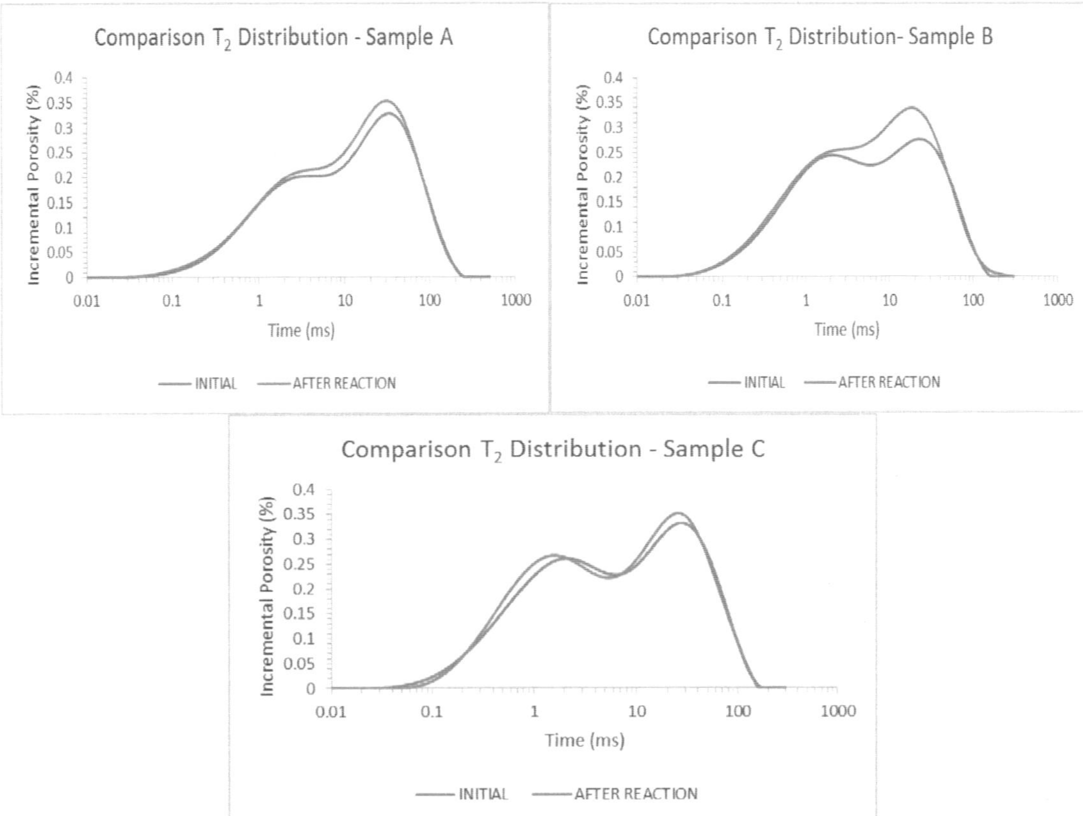

Figure 4. T2 spectrum of sandstone samples.

Table 3 explains the porosity results obtained from NMR analysis. Figures 3 and 4 are based on these results. These results are validating the porosity results obtained previously mentioned in Table 2; where the maximum change in the porosity has been observed when carbonate sand sandstone samples were reacted with HEDTA. The advantage of NMR porosity measurement over general porosity measurement is the knowledge of pore distribution which has been explained and analysed using Figures 3 and 4.

Table 3. Porosity Results using NMR.

	Sample Name	Chelate used	Initial ø %	Final ø %	% change
	Sample A	EDTA	11.75	12.49	6.30
Tight sandstone	Sample B	HEDTA	12.63	14.13	11.90
	Sample C	GLDA	14.02	14.39	2.60
	Sample A	EDTA	18.02	15.59	-0.13
Carbonate	Sample B	HEDTA	14.66	16.02	9.30
	Sample C	GLDA	21.33	22.35	4.80

4.3 Inductively Coupled Plasma (ICP) Analysis

Inductively Coupled Plasma analysis is used to investigate the concentration of different ions dissolved by the acid during the reaction. During the core flooding experiments effluent samples (reacted acid) were collected at regular intervals of time and utilized to determine the concentrations of Si, Ca, Na, K, Fe and Mg present in the reacted chelate. The results presented are in parts per million (ppm). From the analysis presented in Table 4, it can be seen clearly that the maximum amount of each mineral is dissolved when HEDTA has been applied. All samples are collected after 20 minutes of core flood experiments.

Table 4. Concentration of dissolved ions (ppm)

	EDTA		**HEDTA**		**GLDA**	
Element	**Sandstone**	**Carbonate**	**Sandstone**	**Carbonate**	**Sandstone**	**Carbonate**
Sodium	3710	4930	4290	4440	2870	2090
Magnesium	614	198	1370	2570	900	962
Aluminium	0.432	2.27	136	6.65	0.962	39.1
Silicon	4.47	0.954	97.9	0.684	0.1	42.1
Phosphorous	0.1	0.1	2.85	0.1	0.1	0.1
Sulphur	105	111	67.7	133	26.6	17
Potassium	869	1970	313	2310	130	2610
Calcium	1090	178	4600	4320	2640	2010
Titanium	0.1	0.1	0.319	0.150	0.1	0.1
Chromium	0.1	0.1	0.269	0.1	0.1	0.1
Manganese	1.33	0.522	155	5.73	1.6	141
Iron	0.285	7.81	207	54.9	3.28	145
Cobalt	0.1	0.1	0.524	0.121	0.1	1.31
Zinc	31.6	4.32	101	12.1	1.39	22.8

Table 4 represents that HEDTA can dissolve dissolves more amount of sodium, magnesium, aluminium, silicon and calcium. That's why increase in porosity and permeability is quite significant when HEDTA was applied on sandstone samples. While EDTA removed more amount of sodium, sulphur and potassium from carbonate samples as compared to HEDTA and GLDA. GLDA removed more amount of magnesium and calcium from sandstone samples as compared to EDTA. From ICP analysis it can be concluded that HEDTA id effective in removing positive ions as compared to other chelates applied.

5. Conclusion

It can be concluded that HEDTA chelate had significant effect on the porosity and permeability of the core sample and removed positive ions efficiently. Moreover, NMR analysis concluded that fresh large pore spaces has been created using the chelates especially HEDTA results were quite significant. HEDTA can be recommended for tight sandstone and carbonate acidizing.

References

[1] Aboud, R.S., Smith, K.L., Forero Pachon, L. and Kalfayan, L.J., 2007. Society of Petroleum Engineers, "Effective Matrix Acidizing in High-Temperature Environments". SPE 109818-MS
[2] Al-Harthy, S., 2008/2009. Oil Field Review, Options for High-Temperature Well Stimulation. 20(4).
[3] Economides, M.J., Nolte, K.G. 2001. Reservoir Stimulation, Third Edition. Prentice Hall, Ebglewood Cliffs, New Jersey.
[4] Halliburton, 2000a. Carbonate Matrix Acidizing Treatment. Best Practice Series. Halliburton.
[5] Halliburton, 2000b. Effective Sandstone Acidizing, Best Practice Series. Best Practice Series. Halliburton.
[6] Ji, Q., L. Zhou and H. A. Nasr-El-Din (2014). Society of Petroleum Engineers, Acidizing Sandstone Reservoirs Using Fines Control Acid, SPE-169395
[7] Gidley, J.L. 1985. Society of Petroleum Engineers, Acidizing Sandstone Formations: A Detailed Examination of Recent Experience. Paper SPE 14164.
[8] Smith, C. F. and A. R. Hendrickson (1965). Society of Petroleum Engineers, Hydrofluoric Acid Stimulation of Sandstone Reservoirs. SPE-980
[9] M.A. Buijse (Halliburton European Research Centre B.V., 1997, Mechanisms of Wormholing in Carbonate Acidizing, SPE-37283
[10] Alexandre Z.I. Pereira, Mario Germino, Luiz Carlos Paixao, Thiago Judson Lima De Oliveira, Paulo Dore Fernandes, 2012, Society of Petroleum Engineers, Used Approaches for Carbonates Acidizing Offshore Brazil, SPE-151797
[11] A.D. Hill, SPE, Kamy Sepehrnoori, 1994, Society of Petroleum Engineers, Design of the HCl Preflush in Sandstone Acidizing, SPE, and P.Y. Wu: SPE, U. of Texas. SPE-21720
[12] Al-Harbi, B.G., Al-Khaldi, M.H. and AlDossary, K.A., 2011. Society of Petroleum Engineers, "Interactions of Organic-HF Systems with Aluminosilicates: Lab Testing and Field Recommendations". SPE-144100-MS
[13] Mian Umer Shafiq, Hisham Khaled Ben Mahmud and Mohamed Ali Hamid, (2015), IOP Conf. Series: Materials Science and Engineering, Comparison of Buffer Effect of Different Acids During Sandstone Acidizing, 78 (2015).
[14] Mian Umer Shafiq and Hisham Khaled Ben Mahmud, (2016), IOP Conf. Series: Materials Science and Engineering, An Effective Acid Combination for Enhanced Properties and Corrosion Control of Acidizing Sandstone Formation, 121 (2016).
[15] Chunming Xiong, 2010, Society of Petroleum Engineers, Application and Study of Acid Techniques Using Novel Selective Emulsified Acid System, presented at CPS, SPE International Oil and Gas Conference and Exhibition in China, Beijing 8-10 June, 2010, SPE 131216
[16] Yang, F., Nasr-El-Din, H.A. and Al-Harbi, B.M., 2012. Society of Petroleum Engineers, Acidizing Sandstone Reservoirs Using HF and Formic Acids. SPE-150899-MS
[17] Mahmoud, M.A., Nasr-El-Din, H.A., De Wolf, C. and Alex, A., 2011. Society of Petroleum Engineers, Sandstone Acidizing Using A New Class of Chelating Agents. SPE-139815-MS
[18] Frenier, W. et al., 2004. Society of Petroleum Engineers, Hot Oil and Gas Wells Can Be Stimulated Without Acids, SPE-86522-PA
[19] Ali, S.A. et al., 2008. Society of Petroleum Engineers, Stimulation of High-Temperature Sandstone Formations From West Africa With Chelating Agent-Based Fluids, SPE-93805-PA
[20] Mian Umer Shafiq, Hisham Ben Mahmud, Journal of Petroleum Exploration and Production Technology "Sandstone Matrix Acidizing Knowledge and Future Development", 2017, DOI 10.1007/s13202-017-0314-6
[21] McLeod, H. O., Jr., L. B. Ledlow and M. V. Till (1983). Society of Petroleum Engineers, The Planning, Execution, and Evaluation of Acid Treatments in Sandstone Formations, SPE-11931
[22] Shafiq et al, 2016, New Acid Combination for a Successful Sandstone Acidizing, presented in 29th Symposium of Malaysian Chemical Engineers (SOMChE), Miri, Sarawak, Malaysia

[23] Marschall, D.M, 2000, HBVI: An NMR Method to Determine BVI as a Function of Reservoir Capillarity. Presented at the SPWLA 41st Annual Logging Symposium, Dallas, 4–7 June. SPWLA-2000-KK.

IOP Conf. Series: Materials Science and Engineering **217** (2017) 012024 doi:10.1088/1757-899X/217/1/012024

On-off skyhook semi-active control via a magneto-rheological (MR) damper for airfoil-based energy harvesting systems

J N I Pang, H G Harno[1] and V C C Lee[2]

Curtin University Malaysia, CDT 250, 98009, Miri Sarawak, Malaysia.

E-mail:[1]hg.harno@curtin.edu.my [2]vincent@curtin.edu.my

Abstract. An on-off skyhook semi-active control strategy via MR damper is proposed in this study to enhance the performance of a two degree-of-freedom (DOF) airfoil-based energy harvester. For simplicity, only the plunge mode of the airfoil model is examined. NACA0012 airfoil is selected as the vibrating body where energy is harvested and converted into electricity via piezoelectric transduction. Behavioural performance of an actual MR damper is represented numerically with MATLAB/SIMULINK blocks of a conventional Bouc-Wen model. Simulation of the airfoil model is also performed on MATLAB to acquire its dynamic responses. A comparison between passive and semi-actively controlled airfoil systems demonstrates that the proposed strategy is superior in three different aspects – versatility, energy harvesting and sustaining structural integrity.

1. Introduction

Vibration suppression systems are commonly grouped into three types of control methods, namely passive, semi-active and active [1]. A passive system dissipates unwanted vibration in the form of heat without exerting energy from external sources. Due to their simplicity, passive systems are unable to flexibly react to multiple frequencies and parameter changes [2]. In contrast, active vibration systems do not face this problem and are outstanding in controlling unwanted vibration, but with the cost of sensors and actuators. Active vibration systems also require external sources of power, which means greater costs and complication [3]. Over the decades, semi-active control has been more preferable as it preserves the simplicity of a passive system, yet adapts to performance which is to that of an active system, but with minimal amount of external energy supplies [4]. Semi-active control involves time-varying parameters, such as damping and stiffness, to improve system performance [5].

Vibration is known as an oscillatory mechanical phenomenon that exists ubiquitously. Although the presence of vibration is typically unwanted, researchers have proved the feasibility of cultivating vibratory motions as an alternative source of renewable energy [6]. In fact, techniques to energy harvesting via ambient vibrations have been studied comprehensively. Some related applications are electrostatic micro-electro-mechanical systems (MEMS) vibration energy harvester [7] and large-scale energy harvesters [8].

This study aims to numerically study the performance enhancement of an airfoil-based energy harvesting system by introducing the semi-active control via MR damper. The conventional Bouc-Wen model [9] is used to approximately replicate the behavioral performance of an actual MR damper. Then,

IOP Conf. Series: Materials Science and Engineering **217** (2017) 012024 doi:10.1088/1757-899X/217/1/012024

the on-off skyhook semi-active algorithm is incorporated to improve the system performance by controlling aero-elastic vibration of the NACA0012 airfoil. The amount of energy that can be harvested by the semi-active airfoil system is attained and compared against the conventional passive airfoil system via MATLAB. It is demonstrated that the proposed semi-active model has increased the energy harvested in comparison with the passive model. Other benefits that the proposed semi-active model has brought about include versatility and conservation of structural integrity.

2. Numerical Methods

MATLAB/SIMULINK is used in this study to simulate the system's dynamics. Numerical models are explicitly spelled out in following subsections.

2.1. Dynamic model of an airfoil-based energy harvester

In order to quantify the amount of energy harvested in terms of voltage, a piezoelectric coupling intended for energy conversion from mechanical vibration into electricity is attached onto the plunge DOF of the airfoil-based energy harvester. A typical 2-DOF airfoil section coupled with a piezoelectric transducer subjected to an airflow excitation is shown in Figure 1.

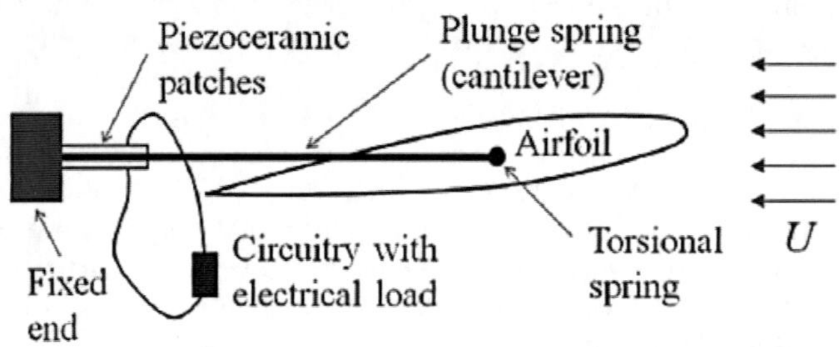

Figure 1. A 2-DOF airfoil section coupled with piezoelectric transducer on the plunge DOF [6].

The equations governing the 2-DOF airfoil section as shown in Figure 1 are expressed in Equations (1) to (3). For the computation of the plunge displacement and output voltage responses, Equations (1) to (3) are re-written in the form of state space equations to be solved via MATLAB's *ode23* integration function.

$$(m + m_e)\ddot{h} + mbx_\alpha\ddot{\alpha} + d_h\dot{h} + k_h h - \frac{\theta}{l} = -L \quad (1)$$

$$mbx_\alpha\ddot{h} + I_\alpha\ddot{\alpha} + d_\alpha\dot{\alpha} + k_\alpha\alpha = M \quad (2)$$

$$C_p\dot{v} + \frac{v}{R_1} + \theta\dot{h} = 0 \quad (3)$$

2.2. On-off skyhook control algorithm

The on-off skyhook control depends on the absolute mass velocity and the damper velocity. When only plunge motion of the semi-active model is concerned, the conditional functions are governed by the plunge displacement to which the airfoil model is confined to move. Since semi-active dampers do not provide energy explicitly, the resulting semi-active damping force must always satisfy the following inequality:

$$F_{sa}(\dot{h} - \dot{\alpha}) \geq 0 \quad (4)$$

where F_{sa} denotes the semi-active damping force to be exerted onto the plunge DOF of the airfoil system and the algorithm describing the on-off skyhook control is expressed as follows:

IOP Conf. Series: Materials Science and Engineering **217** (2017) 012024 doi:10.1088/1757-899X/217/1/012024

$$d_h = \begin{cases} d_{max}, & h > h_{max} \text{ or } h < h_{min} \\ d_{min}, & h \le h_{max} \text{ and } h \ge h_{min} \end{cases} \tag{5}$$

where d_{max} and d_{min} denote the maximum and minimum semi-active damping coefficients in the plunge DOF, respectively; h_{max} and h_{min} denote the maximum and minimum plunge displacements the airfoil system is confined to move, respectively.

In order to maximize the energy harvested, optimal values are obtained and set for both high damping and low damping during the "on" and "off" states of the semi-active controller, respectively. This ensures consistent aero-elastic vibration amplitude of the NACA0012 airfoil to be maintain within the working range. The time-varying damping coefficient of the airfoil-based energy harvester in the plunge DOF is obtained using the following expression:

$$d_h(t) = \frac{F_{sa}(t)}{\dot{h}(t)} \tag{6}$$

Median value of the time-varying damping coefficient $d_h(t)$, which is a constant, is selected as the resulting damping coefficient d_h.

2.3. *Conventional Bouc-Wen model*

Differential equations which constitute the conventional Bouc-Wen model are expressed as follows [10]:

$$F = c_o\dot{x} + k_o(x - x_o) + \alpha z \tag{7}$$

$$\dot{z} = -\gamma|\dot{x}|z|z|^{n-1} - \beta\dot{x}|z|^n + A\dot{x} \tag{8}$$

where c_o, k_o, α, β, γ, x_o, n and A represent parameters describing the non-linear hysteresis of an MR damper [1]. The SIMULINK block of the Bouc-Wen model [10] is built and shown in Figure 2. In order to compute the semi-active damping force for Equation (6), parameters α, β, γ and A in Equations (7) and (8) are set to be 0, 50, 50 and 100, respectively while the data for the remaining four parameters are shown in Table 1.

Table 1. Parameters as function of excitation current.

Current level	c_o	k_o	n	x_o
0 A	5.0743	5.106	1.1678	0
0.05 A	5.8399	8.4644	1.0858	0
0.1 A	6.4582	11.955	1.0017	0
0.2 A	8.6107	23.51	1.0029	0
0.3 A	11.132	42.893	1.0038	0
0.4 A	12.811	60.711	1.0013	0

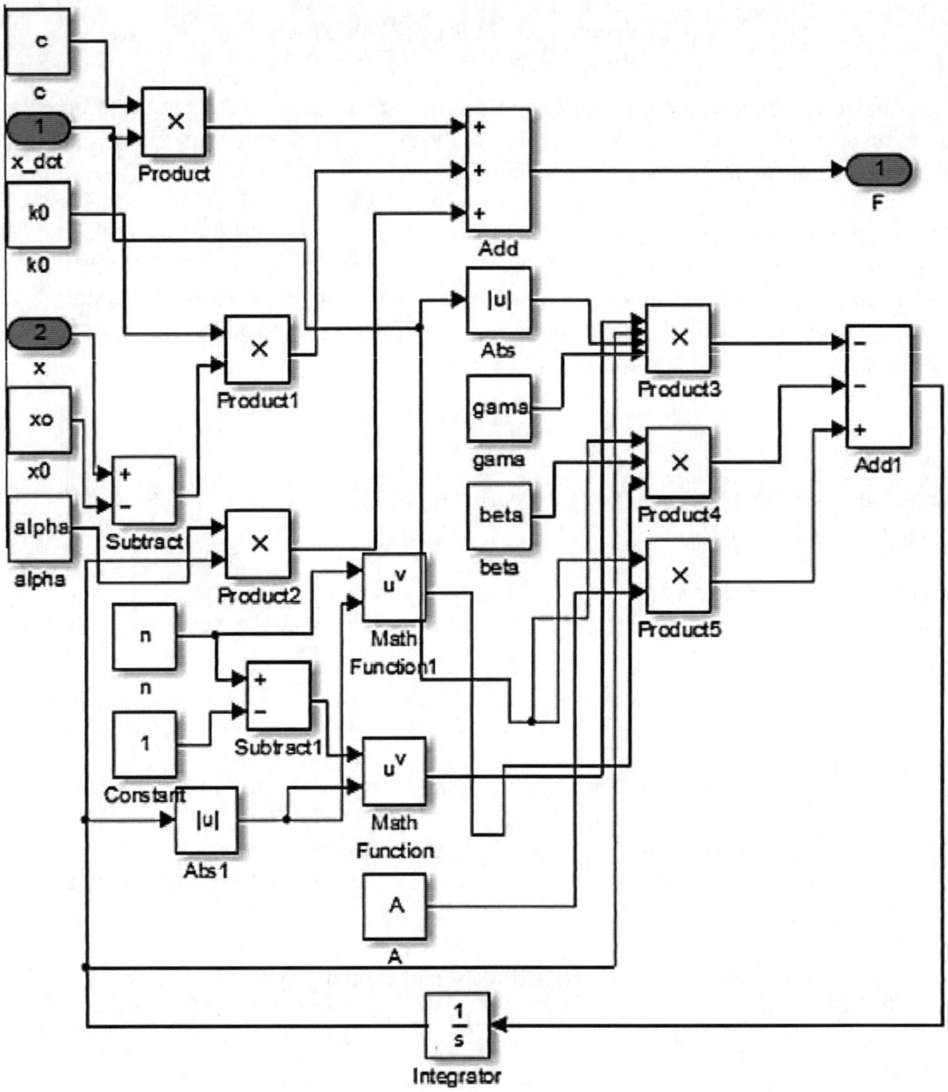

Figure 2. SIMULINK block diagram of the conventional Bouc-Wen model [10].

2.4. Placement of the MR damper

The NACA0012 airfoil is attached onto a cantilever beam which connects it to a fixed point on the right end, as shown in Figure 3. Apparently, the plunge motion is distributed over the entire span length of the cantilever beam. Revisiting Equation (6), the time-varying damping coefficient is a function of the plunge velocity of the airfoil-based energy harvesting system. Therefore, the resulting damping coefficient d_h is also distributed over the whole span length. In order to localize the damping into a particular location to resist only a point, the damping coefficient d_h is multiplied with a dimensionless length. The governing expression is as follows:

$$d_{h_{actual}} = d_h s \tag{9}$$

where s denotes the dimensionless length calculated as follows:

$$s = \frac{a}{200} \tag{10}$$

where a denotes the distance between the MR damper and the fixed point on the right end.

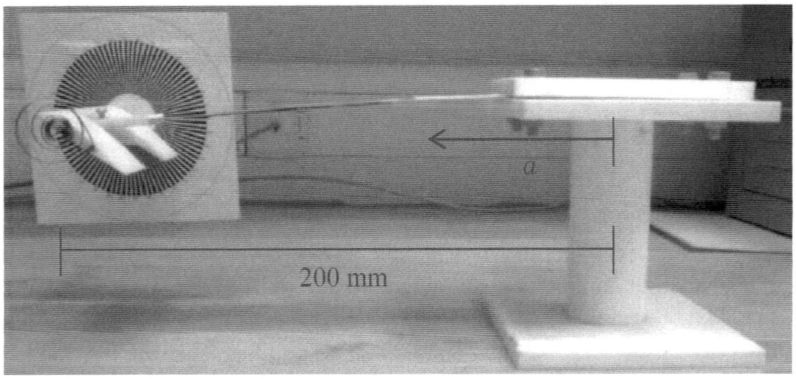

Figure 3. The setup of the complete airfoil-based energy harvester [11].

2.5. Simulation of the semi-active airfoil system

Depending on each placement of the MR damper over the whole span length, different control algorithms as depicted in Equation (5) are implemented. Equation (5) is then integrated into *ode23* to simulate the semi-active airfoil system to acquire its displacement and voltage responses. Controlled simulations for different current inputs are also performed to examine the effects that they have on the resulting output voltage.

3. Results and Discussion

Parametric values of each variable involved in the simulation of the airfoil-based energy harvester are tabulated in Table 2. When the airfoil system is semi-actively controlled, the value of plunge damping coefficient is changed according to the MR damper placement and the inputted current level.

Table 2. Simulation parameters for the airfoil system [11].

Parameter	Parametric value
Airfoil mass (kg)	0.0175
Equivalent mass (kg)	0.0349749
Semi-chord length (m)	0.03
Dimensionless chord-wise offset of elastic axis from centroid	0.2
Beam flexural stiffness (N/m)	77.625
Torsional spring stiffness (Nm/rad)	0.15
Plunge damping coefficient (Ns/m^2)	0
Pitch damping coefficient (Ns/rad)	0
Electromechanical coupling (N/V)	1.55×10^{-3}
Mass moment of inertia (kgm^2)	3.76398×10^{-6}
Equivalent capacitance of piezoelectric layers (F)	113×10^{-9}
Load Resistance (Ω)	570
Initial angle of attack (rad)	-0.8727
Simulation time (s)	5

3.1. Dynamic response of the passive airfoil system

It is apparent that both the plunge displacement and the local voltage responses of the passive airfoil system show beating phenomena as depicted in Figure 4. Beating phenomenon occurs when the ratio between the forcing frequency (ω_f) and the natural frequency (ω_n) is close to one [12]. In this case, $\omega_f/\omega_n \approx 1.19$. This signifies that the airfoil-based energy harvester is vibrating at frequency close to the natural frequency, which also explains why the localized voltage achieved is fairly significant, but at a risk of reaching structural destruction.

Apart from that, the pattern of the local voltage response is also closely synchronized with that of the plunge displacement, except it does not fluctuate as much from t = 2.5 s to t = 5 s. This is due to the mechanism of energy harvesting via a piezoelectric transducer. As the vibrating body of the airfoil system plunges repeatedly, the continuous upwards and downwards motion allows the cantilever beam, which is attached to the energy harvester from the other fixed end, to deform accordingly. This in turn deforms the piezoelectric transducer in the same manner, and generates bi-directional electric current. Hence, this mechanism converts mechanical energy into electricity, and renders the local voltage response to have both positive and negative values for the full duration. Furthermore, the differences between the plunge displacement and the local voltage responses from t = 2.5 s to t = 5 s is due to the nature of piezoelectric material where voltages are extracted within the range of deflection and not instantaneous deflection. In addition, it is also evident that there is no phase difference between the plunge displacement and the local voltage responses. This is because the electric voltage is generated as soon as the piezoelectric material is deformed.

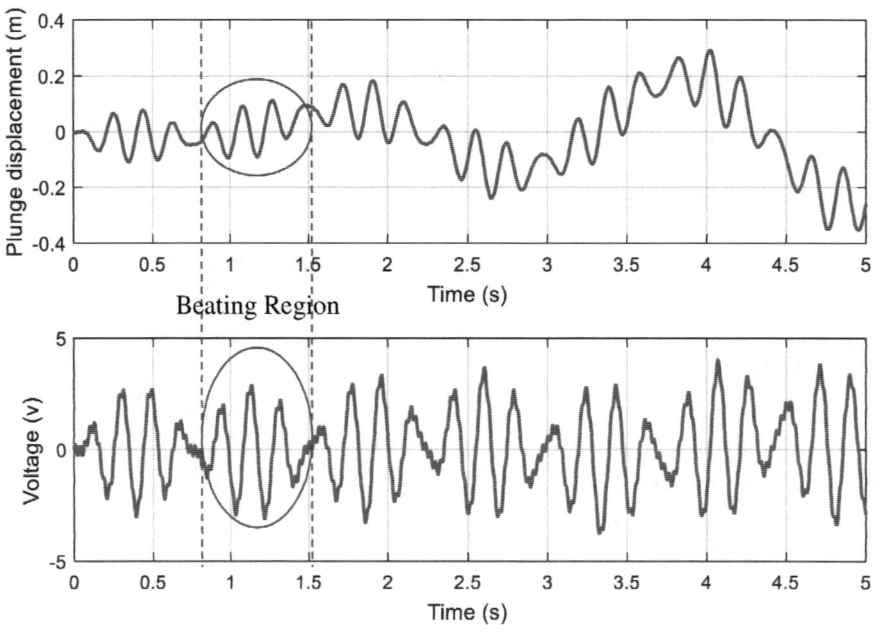

Figure 4. Plunge and voltage responses of the passive airfoil-based energy harvester.

3.2. Dynamic response of the semi-active airfoil system

Inferring from Figures 5 and 6, the incorporation of the semi-active control strategy into the airfoil system has successfully eliminated beating. This is because the natural frequency has been shifted to the extent that the frequency ratio is no longer close to one. In the case where the damping coefficient during the "on" state is 0.64032 Ns/m, i.e. when the input current is maximum, $\omega_f/\omega_d \approx 1.21$, which is slightly further away from the natural frequency as compared to the passive airfoil system. This proves the feasibility of an MR damper for the NACA0012 airfoil in vibration attenuation. The relatively low local voltage from t = 0 s to t = 0.2 s is due to the need for the airfoil system to start up when it is initially subjected to the wind excitation.

Plunge displacement amplitude undergoes suppression each time it exceeds the configured upper and lower bounds, which in this case are h = 0.14 m and h = – 0.19 m, respectively. This is when the skyhook control is activated for the greater damping to be applied to the airfoil system. This proves the viability of the semi-active control in preventing the airfoil system from failing in the midst of cultivating energy. However, it is also visible that the plunge motion has not actually been reduced within the preset bounds. This shows the limitation of the conventional Bouc-Wen model since the provided parametric values, such as α are insufficient and it is apparent that a much larger input current is needed for the on-off skyhook control to provide greater damping. Nevertheless, the maximum plunge displacement has undergone a considerable reduction from 0.3322 m to 0.3205 m, which amounts to about 3.52%. Apart from that, there are also clear signs of minor discontinuities in the plunge displacement and the local voltage. This is the due to the occurrences of chatter and jerk when there are immediate switches between the “on” and “off” states. The output voltage shows very consistent trends except the considerable reduction when the semi-active control is activated. This is unavoidable since the greater the plunge motion, the greater the deformation of the piezoelectric transducer, which is why the local voltage is reduced whenever the vibratory motion of the airfoil in the plunge DOF is suppressed.

Simulations of the semi-active model with different MR damper placements are also performed. It is demonstrated that the closer the MR damper is to the vibrating body, more vibration in the plunge DOF can be suppressed. Concurrently, the output voltage that can be cultivated reduces progressively in correspondence with the plunge displacement. Nevertheless, the responses depict similar trends where the semi-active control becomes impactful especially after the plunge motion of the airfoil model reaches its steady state. For the best illustration of how the input current level can affect the plunge displacement of the airfoil system and the cultivated output voltage, the normalized values of the two parameters are acquired with respect to their initial conditions and plotted against the input current as shown in Figure 7. Normalized values are used because their changes over the available range of current are otherwise, too insignificant to notice.

Inferring from Figure 7, the output voltage response of the semi-active airfoil system is actually decreasing at a higher rate as compared to the plunge displacement response. This shows that keeping the amplitude of the aero-elastic vibration at the maximum and without failing is therefore important in order to cultivate more electricity.

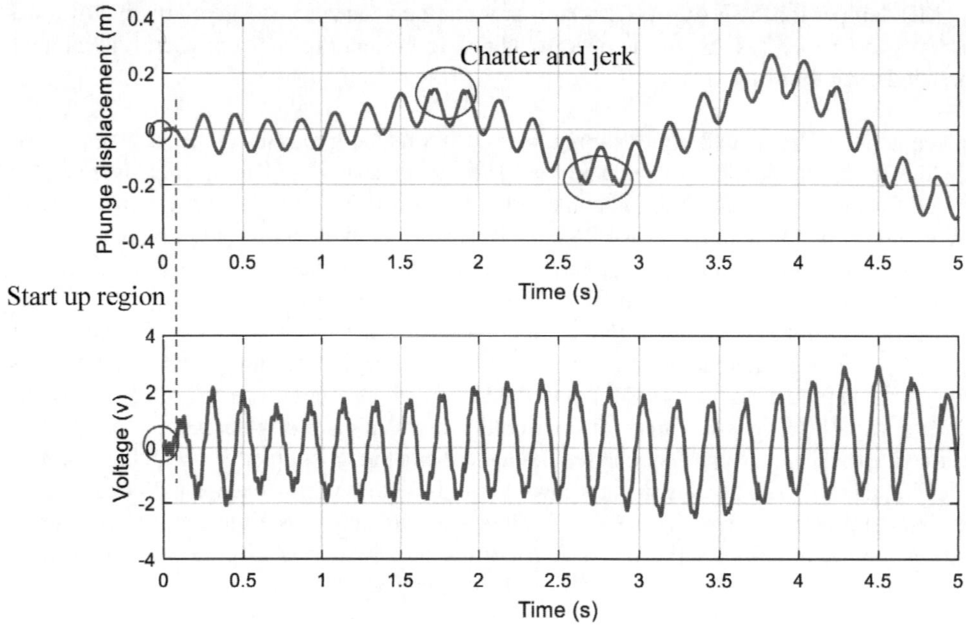

Figure 5. Plunge and voltage responses of the semi-active airfoil-based energy harvester (a = 10 mm).

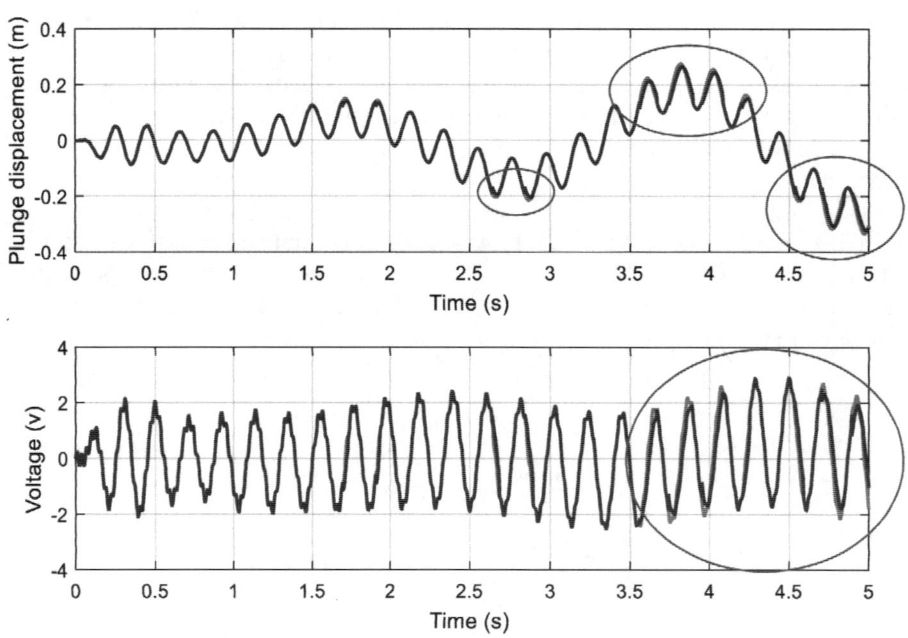

Figure 6. Response comparison between always "off" and semi-active model (a = 10 mm).

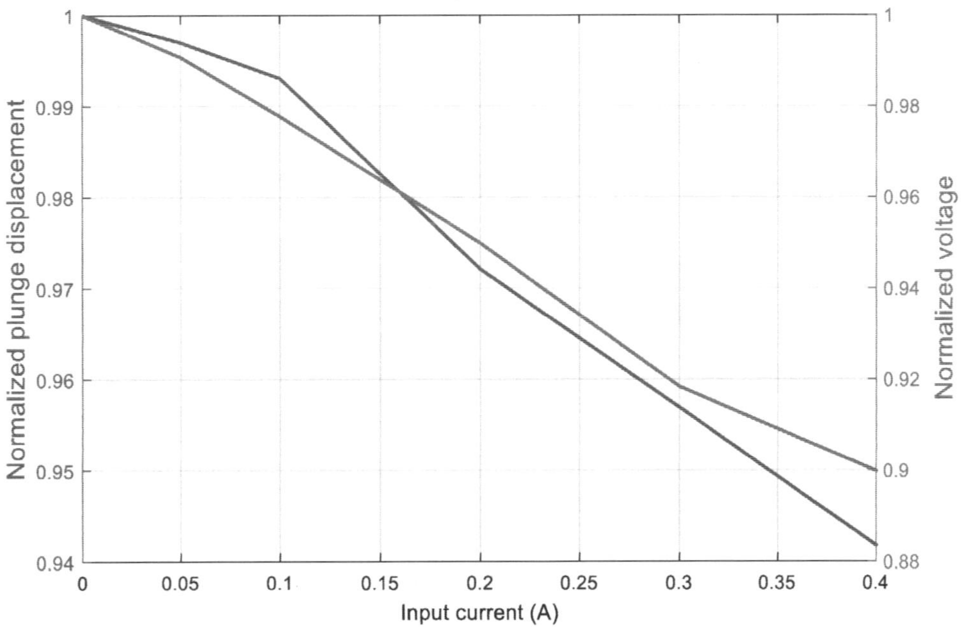

Figure 7. Normalized plunge displacement and voltage versus input current graph.

3.3. Comparison between the passive and semi-active model

The passive airfoil model has reached a maximum plunge displacement of 0.3525 m during its beating peak. At the same instant, the semi-actively controlled airfoil system has managed to achieve a maximum plunge displacement of 0.3205 m. This amounts to a total of 9.08% of displacement reduction, which is significantly favorable for a system which requires vibration suppression. Not only does the elimination of beating prolong the airfoil model's fatigue limit, it also allows a more consistent aero-elastic vibration of the NACA0012 airfoil.

In the case of local voltage response, a maximum of 4.039 V is achieved by the passive system during its beating peak whilst the semi-active airfoil system is only able to hit a maximum of 2.945 V when it peaks at around the fifth second. This brings about voltage reduction of about 27.09%. When observed closely however, it is perceptible that despite the greater voltage that it may acquire, the passive airfoil system spends more than a quarter of its full duration, hovering between 0 to 2 V due to beating. As opposed to this, the semi-active airfoil system peaks rather consistently at approximately 2.4 V. This is proved by acquiring the mean local voltage of the airfoil system for the full duration (from transient to steady state).

Results show that the passive airfoil system has managed to cultivate an output voltage of 45.8 mV whereas the semi-active airfoil system is able to cultivate up to a total of 49.6 mV, amounting to about 8.3% of output voltage increase. This shows that in a long run, the semi-active system is capable of cultivating more energy while being able to suppress vibration simultaneously. Therefore, the semi-active airfoil system is certainly capable of outperforming the passive airfoil system in terms of versatility, vibration suppression, and most importantly energy harvesting.

3.4. Structural integrity

In order to avoid losing structural integrity, systems shall never operate near resonance. Resonance is an undesirable phenomenon which results in the occurrences of extremely large amplitudes, causing structural failure well before a system's fatigue limit is even reached [13]. Resonance can be meddled with damping and is said to occur only when the forcing frequency coincides with a system's natural frequency, i.e. when the frequency ratio equates to unity [13]. Based on the simulation results, the incorporation of the proposed strategy has managed to increase the frequency ratio from 1.19 to 1.21. At first glance, an increment of only 0.02 seems fairly insignificant. However, the elimination of beating demonstrates that even the smallest damping is advantageous at putting a system further away from the realm of resonance. This finding is further validated by examining the variation of a system response with frequency ratio as shown in Figure 8. A damping coefficient of 0.64032 Ns/m which is used for the semi-active airfoil model is equivalent to a damping ratio of approximately 0.16. Inferring from Figure 8, smaller damping ratios result in greater system amplitudes. This implies that an improvement of the frequency ratio from 1.19 to 1.21 is definitely significant for vibration suppression.

Essentially, even without damping, a frequency ratio of 1.19 is already within an acceptable working range. In fact, the plunge displacement response of the passive airfoil system in Figure 4 shows only minor beating. However, for the purpose of maximizing and maintaining consistency of energy harvested, a persistent aero-elastic vibration which occurs not too far away from resonance is desirable. Therefore, the 0.02 increment in frequency ratio is certainly a great achievement. Not only does the proposed strategy eliminate beating, a larger amount of energy has been cultivated. Despite the operation of the airfoil system at above the unity frequency ratio, note that its plunge velocity does not actually pass through resonance since the wind speed that the system is subjected to, is constantly 5.32 m/s and is not required to start up from 0 m/s.

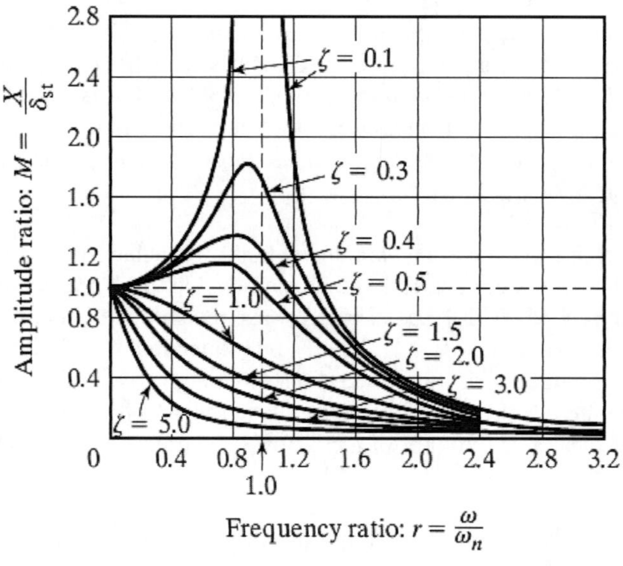

Figure 8. Variation of system response with frequency ratio [12].

4. Conclusion

Numerical results show that a semi-actively controlled airfoil system is more valuable than a passive one in terms of versatility as well as simultaneous energy harvesting and conserving structural integrity. Depending on the scale of a desired application, the MR damper can be placed at different locations along the span length of the cantilever beam and be injected with distinct current levels to provide a range of damping coefficients, $\boldsymbol{d_h}$. It has been demonstrated that the implementation of the semi-active control strategy does not only eliminate beating, but also suppresses up to 9.08% of the maximum plunge displacement of the airfoil system, and thus, prolongs the period of its fatigue life. Moreover, the vibrating body of the airfoil system – NACA0012 airfoil, is also able to vibrate more consistently, which in turn results in a steady and reliable output voltage. Despite its lower maximum voltage magnitude (2.939 V), it is also established that the semi-active model can cultivate more energy in a long run. Unlike the conventional passive model, it does not hover around negligible voltages. Apart from that, an improvement of the frequency ratio from 1.19 to 1.21 is also a significant achievement from two different aspects – protecting the system from losing structural integrity and maintaining the cultivation of a large output voltage. Lastly, the elimination of beating is also noteworthy in terms of terminating thermal stresses and thus, helps conserving the energy storage component for piezoelectric transduction

Nomenclature

Symbol	Description
C_p	Equivalent capacitance of piezoelectric coupling.
I_α	Mass moment of inertia.
R_1	Load resistance.
d_h	Damping coefficient in the plunge DOF.
d_α	Damping coefficient in the pitch DOF.
k_h	Stiffness in the plunge DOF.
k_α	Stiffness in the pitch DOF.
m_e	Effective fixture mass.
x_α	Dimensionless chord-wise offset of the elastic axis from the centroid.
$h, \dot{h}, \ddot{h}$	Displacement, velocity and acceleration in the plunge DOF.
C	Centroid of the airfoil section.
L	Lifting force.
M	Moment.
b	Semi-chord length of the airfoil section.
l	Length of span.
m	Mass of the airfoil.
v	Voltage across the resistant load.
$\alpha, \dot{\alpha}, \ddot{\alpha}$	Displacement, velocity and acceleration in the pitch DOF.

References

[1] Yao G Z, Yap F F, Chen G, Li W and Yeo S H 2002 *Mechatronics* **12** (7) p 963-973.
[2] Gawthrop P J, Neild S A and Wagg D J 2012 *J Intel Mat Syst Str* **23** (18) p 2103-2116.
[3] Liu Y, Waters T P, and Brennan M J 2005 *J Sound Vib* **280** (1) p 21-39.
[4] Di Monaco F, Tehrani M G, Elliott S J, Bonisoli , and Tornincasa S 2013 *J Sound Vi* **332** (23) p 6033-6043.
[5] Dong X M, Yu M, Liao C R and Chen W M 2010 *Nonlinear Dynam* **59** (3) p 433-453.
[6] De Marqui Jr C and Erteuk A 2013 *Advances in Energy Harvesting Methods* (New York: Springer) p 269-294.
[7] Wong Z J, Yan J, Soga K and Seshia A 2009 *A multi-degree-of-freedom electrostatic MEMS power harvester.*
[8] Tang X and Zuo Z 2009 *ASME 2009 International Mechanical Engineering Congress and Exposition* (American Society of Mechanical Engineers) p 885-896.
[9] Dyke S J and Spencer B F 1997 *Intelligent Information Systems, 1997. IIS'97. Proceedings* (IEEE) p 580-584
[10] Peng G R, Li W H, Du H, Deng H X and Alici G 2014 *Appl Math Model* **38** (15) p 3763-3773.
[11] Lai K H 2015 *On the Design and Analysis of an Airfoil-Based Energy Harvester* (Curtin University, Malaysia).
[12] Palm W J 2006 *Mechanical Vibration* (New Jersey: John Wiley & Sons).
[13] Rao S S 1990 *Mechanical Vibration* (New York: Addison-Wesley).

Evaluation of the Passive Cooling Strategies for Pei Min Sport Complex

K S Yam[1]*, W L Yem[1], V C C Lee[1]
[1]*Curtin University Malaysia*, Miri, Sarawak, Malaysia.

E-mail: yamks@curtin.edu.my

Abstract. This paper presents a modelling study on the evaluation of the passive cooling strategies for Pei Min sport complex at Miri. The squash centre has experienced excessively high temperature during peak hours that results in complains from the users. We discussed several passive cooling mechanisms and proposed four strategies for the sport centre. Thermal energy simulations were performed on these strategies using OpenStudio to evaluate their impact on the hourly temperature profile within the building. It was found that the peak temperature during the noon was significantly reduced when conductive material was applied at the lower surface of the roof, and the top of the roof was coated with white paint. However, insulating the roof also leads to weaker heat dispersion from the building which lower the rate of temperature drop in the late afternoon. Partitioning the roof was found to have similar effect as insulating roof. Air infiltration is essential for promoting air movement and regulating the temperature within the building. It was found the complex already have sufficient opening for the full effect of air infiltration.

1. Introduction

In today's worldwide trend towards green energy, there is an increasing demand on the new or existing buildings in the commercial, industrial and residential sectors to be energy efficient. While new buildings can be designed to be energy efficient, meeting this requirement for old buildings is more challenging due to the building design constraints. For this reason, retrofitting or refurbishing existing buildings in order to minimize the building's energy consumption is becoming increasingly common these day. Numerous study has reported significant reduction in energy consumption and added environmental benefit through proper retrofitting [6-8]. Invariably, refurbishing and retrofitting building has seen intensive research recently that result in numerous development of effective modelling approaches [e.g. 1-3], and strategy, as summarised in these recently published review articles [4-5].

The main challenge faced in retrofitting of existing buildings are the numerous uncertainties that needs to be considered, such as the building design, climate change, human behaviour change, government policy change, etc. In addition, large building generally has extended layout design with a complex network of subsystems that are highly interactive. Dealing with the uncertainties and theses system interaction causes the selection of retrofit technologies becomes very complex. On top of the technical challenge, there are the financial limitations and payback period that needs to be considered. Consequently, in the early stages of a retrofitting studies, integrated approach that can combine the all

* To whom any correspondence should be addressed.

IOP Conf. Series: Materials Science and Engineering **217** (2017) 012025 doi:10.1088/1757-899X/217/1/012025

essential components of a building (building information, energy balance, and economical analysis) is increasingly becoming a necessity in order to effectively size the influence of various refurbishing measures [e.g. 9-10]. There is already various modelling software of this kind been developed. The assessment on the pros and cons of these software was conducted recently [11]. OpenStudio is one of the few building simulation software that is open-source, and is capable of performing energy simulation, and design optimisation [12]. It is employed here to evaluate the performance of the proposed passive cooling strategies in reducing the heat problem experienced at Pei Min Squash Court in Miri.

2. Problem Statement

The squash centre in Miri have a gross area of (25000 × 45000) mm2. Inside the complex, it has nine squash courts, one admin room, one activity room, two toilets and one sports equipment. Each squash court that includes a wall partition had a dimension of (10000 × 6600) mm2. The material of the roof is corrugated sheet made from galvanized steel, and the wall is brick concrete. The unusual high temperature inside the squash court during the afternoon period have led to constant complains from the players. There are several factors that contributes to the overheating in the centre:

- Miri experiences hot and humid weather annually, especially during the dry season that happens from March to September where the volume of rainfall is significantly lower, as shown in Fig. 1.
- The high sunlight intensity during dry season that increases the radiation heat on the surface of sport complex, coupled with the conduction and convection of heat into building environment cause the increase of the temperature
- The lack of proper and efficient ventilation results in heat being trapped inside the complex.

The management had proposed to installed two air-conditions in two enclosed squash court for competition usage. However, the remaining squash court without air-conditioners will require alternative low-budget solutions. The solution should effectively reduce the temperature within the centre which will increase the thermal comfort of the players. The ease of the implementation is another important factor that needs to be considered. Complex equipment generally leads to higher implementation cost.

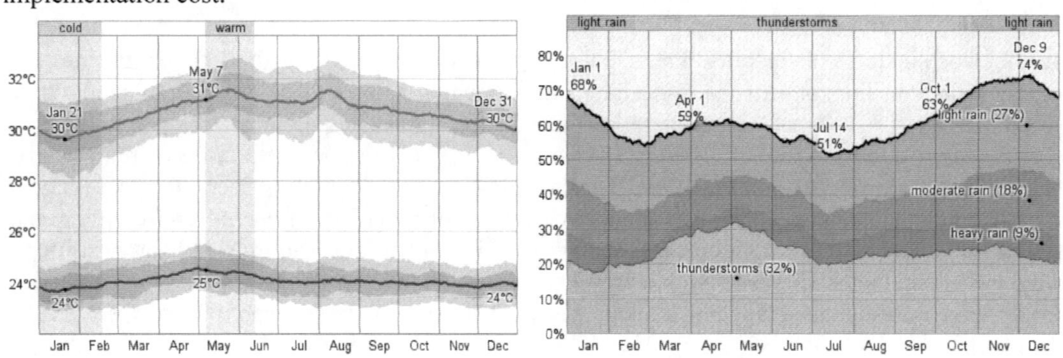

Figure 1. (left) The daily average low (blue) and high (red) temperature. (right) The fraction of days in which various types of precipitation are observed. [15]

3. Passive Cooling Mechanism

3.1 Solar Reflective Insulation

Solar reflective insulation is a barrier material which resists/blocks/reflects radiant heat energy to prevent its transfer through the boundary between two spaces which are at different temperatures. Radiation transfers heat via invisible electromagnetic wave. When sunlight is shined on a roof, the roof absorbs a portion of the radiation heat and reflect the rest back into the atmosphere. Since the roof itself emits radiation heat, the amount of radiation absorbed by the roof depends also on the radiation emittance of the roof. A good reflective material should have high reflection, high emittance and low absorption rate. Cool Roof Rating councils introduced the following definitions for evaluating the reflective performance of roof:

- Solar absorptance, which refers to proportion of the overall solar radiation incidence that is absorbed by the material.
- Infrared emittance, which refers to emittance of heat of the material in form of infrared.

The level of the absorptance and emittance depends on the colour coating of the external surfaces, surface roughness, and the thermal properties of the material. As shown in Fig. 2, light colour coated material such as the white plaster and white paints have both high reflectance and infrared emissivity. Aluminium foil has high reflector but low emissivity. Dark asphalt has good emissivity but very poor reflectance.

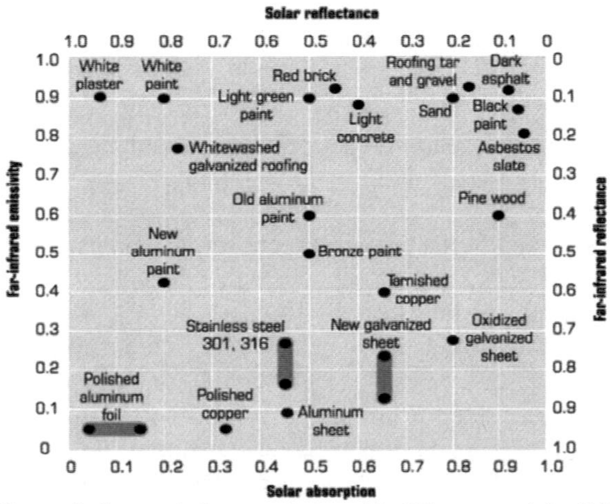

Figure 2. Spectral characteristics of building materials. [13]

3.2 Thermal Conductive Insulation

Thermal insulation involves the use of a material or combination of materials that has low thermal conductivity for reducing the rate of heat flow in and out of a building. These materials are often placed near to the surface of the heat entry. There four type of insulations as shown in Fig. 3. The selection of these materials depends on several factors, i.e. the durability of the material, safety reason, the environment impacts, and the cost to benefit ratio. For instance, fibrous materials such as glass wool fibres may not be an ideal choice for building with high humidity because the thermal conductivity of the material increases when its moisture content increases. On the other hand, organic material such as polyurethane, have low moisture absorption rate, are cheap, and have high thermal resistance, but emits chorofluorocarbons (CFC) over its life cycles which has detrimental effect on the ozone, and also have higher flammability index.

3.3 Air Infiltration / Natural ventilation

Temperature difference between the indoor air and outdoor air causes the pressure difference across the building envelope and moves air in and out from the building through the openings. Depending internal structure of the building, and the climatic condition, air movement can reduce the temperature and moisture level inside the building, and improve the thermal comfortability of the occupants. Ignoring the internal structural resistance, wind speed of 2.5m/s or less can generate an exterior wind pressure of 1 or 2 Pa, and higher speed of 10m/s can generate pressure of 25 Pa or more. In addition to the natural ventilation, the operation of mechanical equipment, ventilation systems, and local exhaust fans further influences the net flow of air into and out of the building.

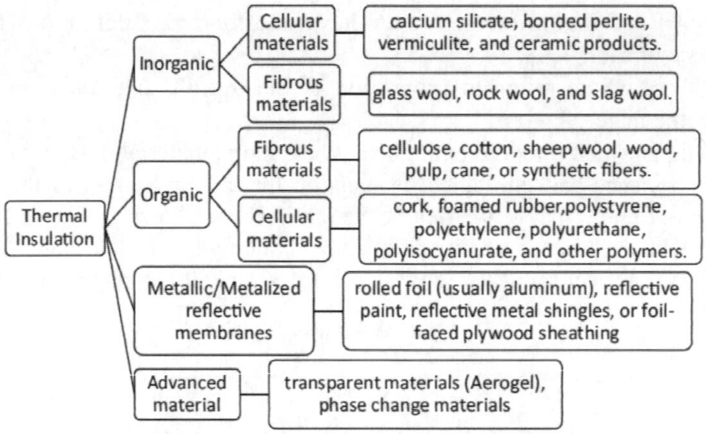

Figure 3. Different types of thermal insulation [14]

4. Solution Strategy

The cooling mechanisms discussed in the previous section was employed on the sport complex. Four scenarios were created as listed in Table 1. Case A considers a layer of thermal conductive resistance material (glass wool with thermal conductivity of 0.04W/mK) overlain with a thin layer of aluminium foil to be installed underneath the roof of the sport complex to reduce the heat transfer from the roof into the building. Three thicknesses in the range of 0.1 to 0.3m will be considered to evaluate the effect of the thickness. Case B considers white paint (solar absorptance of 0.1) to be included on the roof as the solar radiation reflector, in addition to the installation of the insulation material in Case A. Case C considers the roof to be partitioned using a layer of wooden platform in order to create a closed airspace between the roof and the inner section of the building, as shown in Fig. 5. The airspace functions as thermal insulation in addition to a layer of glass wool on top of the wooden platform. Lastly, Case D considers the size of the openings that exists in the original building to be increased by 25%. The purpose is to investigate whether enhancing the air movement within the building can reduce the temperature in the building.

Table 1. Evaluation of passive cooling strategy for the sport centre

Scenario	Description
Case A	Insulation materials (glass wool) overlain with aluminium foil are installed underneath the roof. Three thickness are considered, i.e. 0.1m, 0.2m, and 0.3m.
Case B	In addition to the installation of the insulation material (0.1m) underneath the roof, white reflective coating is painted on top of the roof.
Case C	The roof is partitioned by installing thin layer of wood. A layer of insulation material and aluminium foil is placed on top of the partition.
Case D	The opening at the wall is increased by 25% to increase the rate of infiltration.

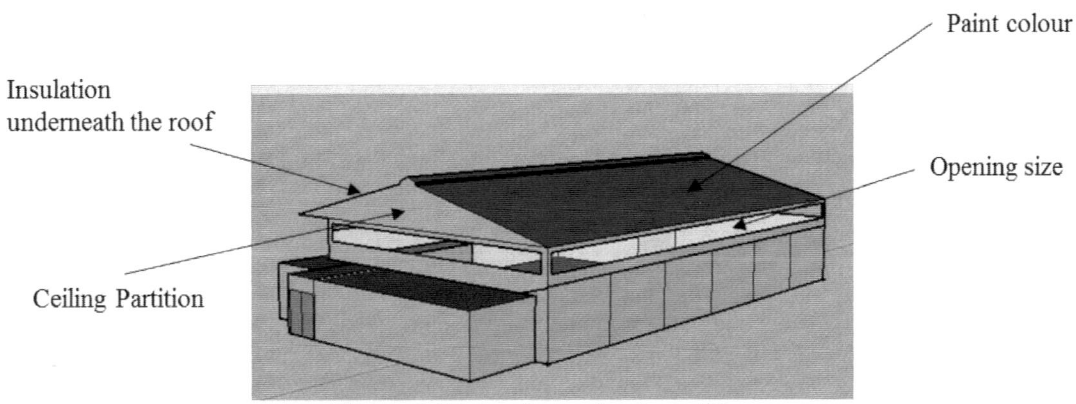

Figure. 4. Strategy for cooling the building

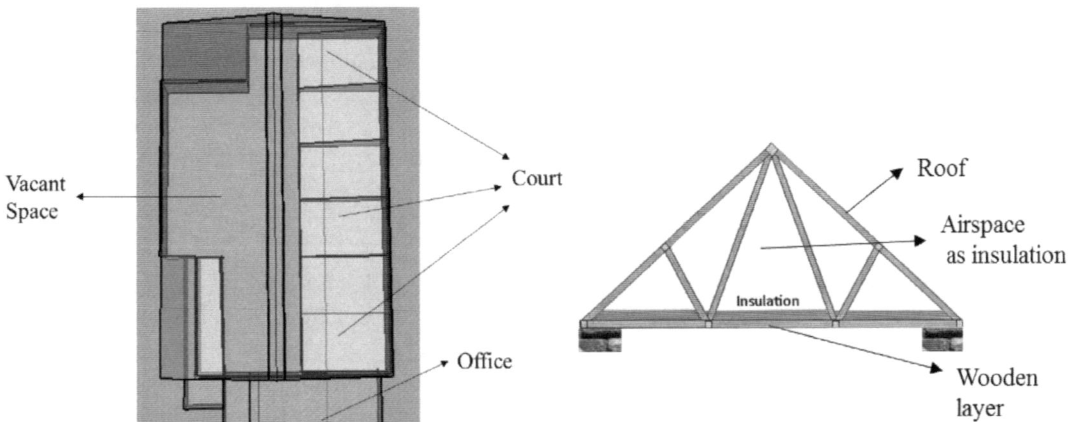

Figure 5. *(left)* Interior layout of the sport complex. *(right)* Roof partition with insulation.

5. Thermal-Energy Simulation

The strategy outlined in previous section was evaluated using OpenStudio, a building energy software that is derived from EnergyPlus (released by National Laboratory of the U.S. Department of Energy) for performing thermal and energy balance simulation on buildings. OpenStudio employs heat balance-based solution of radiant, convection and conduction effects to determine the temperature, and thermal comfort of thermal zones prescribed in the buildings. In addition, the solutions take into account the weather effect that includes the solar radiation, daylight illuminance, and precipitation, in addition to the standard parameters of temperature, humidity, pressure, wind speed and direction. Therefore, sub-hourly interaction between the thermal zones and the environment can be modelled. The following steps have been taken in the process to obtain the energy consumption and thermal behaviour of the sport complex:

i. A 3D computer model of the sport complex was created using SketchUp according to the dimension given in the layout plan.
ii. The model is then imported into the OpenStudio for the thermal energy balance analysis.
iii. A thermal zone was created for the airspace inside the building. Since all the courts and the majority of the vacant space share the airspace, only one single thermal zone was created.
iv. The thermal properties of the components of building (roof, and wall) and number of occupants was assigned.

IOP Conf. Series: Materials Science and Engineering **217** (2017) 012025 doi:10.1088/1757-899X/217/1/012025

v. The weather file of Brunei Seri Begawan (the location nearest to Miri that is available in the database) in 2015 was incorporated into the model. One of the hottest days in 2015 (22-May-2015) was selected. The peak temperature reaches 33°C and the average wind speed was low (7 km/h).
vi. Simulations were performed to predict the hourly temperature profile inside the building. Various cooling strategy discussed was implemented into OpenStudio and the predicted hourly temperature profile was compared.

6. Result & Discussions

Fig. 6(a) shows that the temperature in the original building rises steadily from 6am until it reaches the peak of 34 °C at 2pm on 22 May 2015. The temperature drops thereafter to 30 °C at 8pm. Since the peak usage hours of the sport centre is between 2pm and 8pm, the discussions on the result will be focussed on this period of time.

The result of case A is shown in Fig. 6(a). The installation of thermal insulation under the roof was able to reduce the overall temperature and most importantly, the peak temperature at noon by approximately 1.5 degrees. Unfortunately, the installation of insulation also allows the building to retain the heat better and therefore reduce the rate of temperature drop after the peak at 2pm. As a result, no significant temperature reduction was observed from 2pm to 8pm. Increasing the thickness twice or three times unfortunately does not yield any significant temperature reduction. The insulation of 0.1m is sufficiently thick for maximum conductive insulation.

Fig. 6(b) shows the result of the Case B. By coating the roof with white paint, the temperature is further reduced. The temperature was significantly reduced in the period of 12pm to 4pm, where the peak temperature was decreased by approximately 3 degrees to 31 °C. Similar to Case A, temperature reduction is lesser after 4pm due to the heat retention by the roof insulation. The effect of partitioning the roof into closed airspace insulation produces similar effect as the installation of insulation underneath the roof. As shown in Fig. 6(c), the predicted peak temperature was reduced by approximately two degrees. The temperature reduction also become lesser after 4pm due to heat retention in the roof partition. Since installing roof partition incurs greater cost but does not yield any significant improvement, Case B is more favourable approach.

Lastly, the result on the effect of infiltration was (case D) shown in Fig. 6(d). Opening was shown to be essential for cooling the building. If the opening in the original design was closed, significant temperature rise was predicted. However, the original opening size was sufficient because increasing the opening size by 25% does not yield much difference to the temperature.

7. Conclusion

This work presents a modelling study on evaluating a set of passive cooling approaches for Pei Min sport complex using OpenStudio. The proposed approaches are based on thermal conductive insulation, solar reflective insulation, air space insulation and air infiltration. The following conclusions were made:

1. The effective cooling was achieved when glass wool and aluminium foil was installed at the lower surface of the roof and the top surface of the roof is coated with white paint. The peak temperature was reduced by three degrees using this method. However, insulation also retain heat better which lower the rate of temperature drop in the late noon.
2. Partitioning the roof into closed airspace for thermal insulation yields the same impact as installing thermal insulation at the lower surface of the roof.
3. Air infiltration is essential for regulating the temperature of the building and the air movement within the building. The sport complex already has sufficient opening for the full effect of the air infiltration.

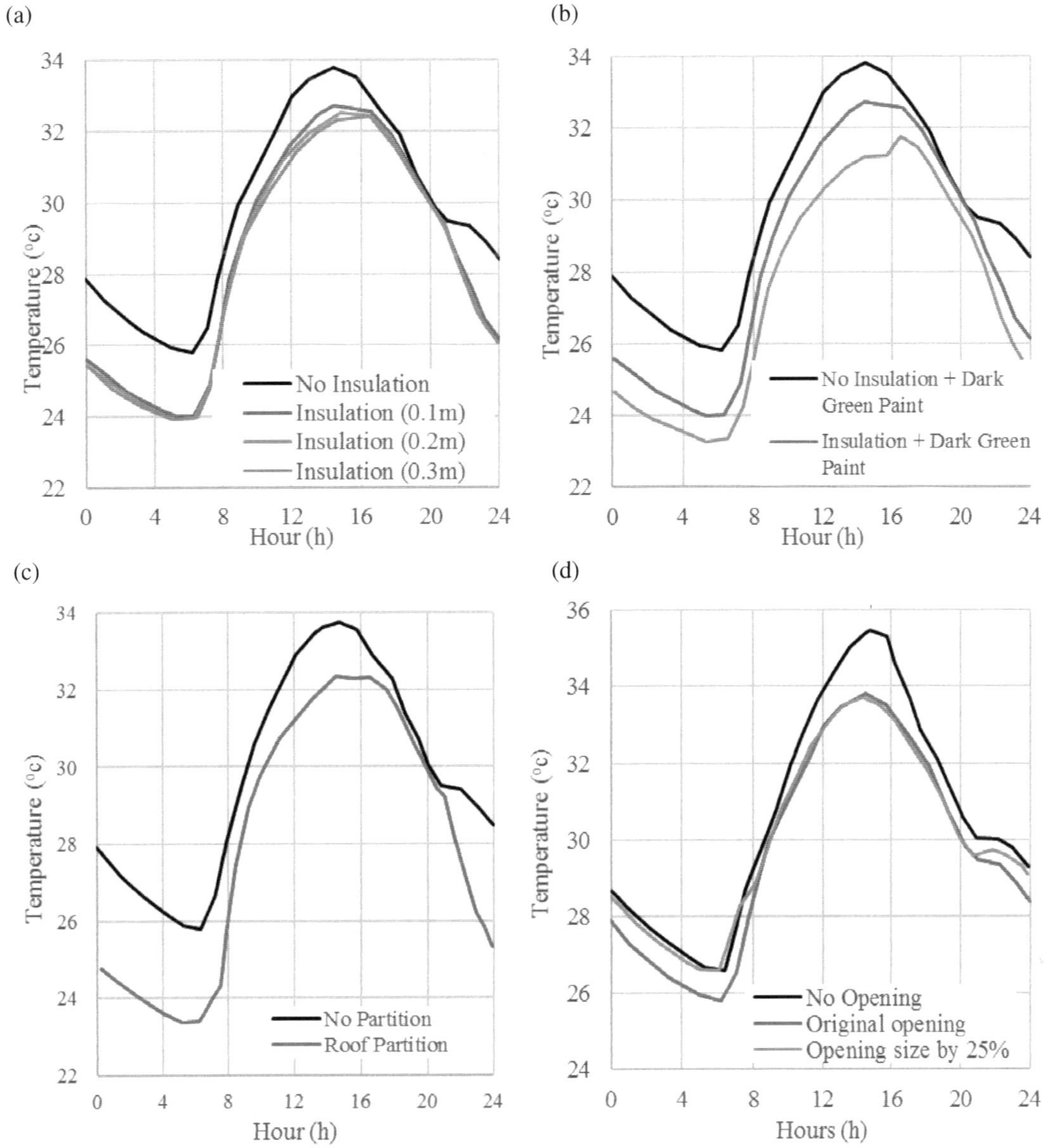

Figure 6. Prediction on the temperature profile within the squash centre with various passive cooling strategies on 22-May-2015 using OpenStudio, (a) Case A, (b) Case B, (c) Case C, (d) Case D.

8. Reference

[1] Ma, Z. and Wang, S., 2009. Building energy research in Hong Kong: a review. Renewable and Sustainable Energy Reviews, 13(8), pp.1870-1883.
[2] Flourentzou, F. and Roulet, C.A., 2002. Elaboration of retrofit scenarios. Energy and Buildings, 34(2), pp.185-192.
[3] Chidiac, S.E., Catania, E.J.C., Morofsky, E. and Foo, S., 2011. Effectiveness of single and multiple energy retrofit measures on the energy consumption of office buildings. Energy, 36(8), pp.5037-5052.
[4] Ma, Z., Cooper, P., Daly, D. and Ledo, L., 2012. Existing building retrofits: Methodology and state-of-the-art. Energy and buildings, 55, pp.889-902.

[5] Pacheco, R., Ordóñez, J. and Martínez, G., 2012. Energy efficient design of building: A review. Renewable and Sustainable Energy Reviews, 16(6), pp.3559-3573.

[6] Ardente, F., Beccali, M., Cellura, M. and Mistretta, M., 2011. Energy and environmental benefits in public buildings as a result of retrofit actions. Renewable and Sustainable Energy Reviews, 15(1), pp.460-470.

[7] Hestnes, A.G. and Kofoed, N.U., 2002. Effective retrofitting scenarios for energy efficiency and comfort: results of the design and evaluation activities within the OFFICE project. Building and Environment, 37(6), pp.569-574.

[8] Chidiac, S.E., Catania, E.J.C., Morofsky, E. and Foo, S., 2011. Effectiveness of single and multiple energy retrofit measures on the energy consumption of office buildings. Energy, 36(8), pp.5037-5052.

[9] Gourlis, G. and Kovacic, I., 2016. A study on building performance analysis for energy retrofit of existing industrial facilities. Applied Energy, 184, pp.1389-1399.

[10] Kensek, K. and Noble, D., 2014. Building information modeling: BIM in current and future practice. John Wiley & Sons.

[11] Attia, S., 2011. State of the art of existing early design simulation tools for net zero energy buildings: a comparison of ten tools. Architectuet climat.

[12] Cole, W.J., Hale, E.T. and Edgar, T.F., 2013, June. Building energy model reduction for model predictive control using OpenStudio. In American Control Conference (ACC), 2013 (pp. 449-454). IEEE.

[13] Center, F.S.E., 2008. Florida Solar Energy Center. Retrieved September, 25, p.2008.

[14] Sadineni, S.B., Madala, S. and Boehm, R.F., 2011. Passive building energy savings: A review of building envelope components. Renewable and Sustainable Energy Reviews, 15(8), pp.3617-3631.

[15] "Average Weather For Miri, Malaysia - Weatherspark". *Weatherspark.com*. N.p., 2017. Web. 29 Mar. 2017.

IOP Conf. Series: Materials Science and Engineering **217** (2017) 012026 doi:10.1088/1757-899X/217/1/012026

Comparative microstructure study of oil palm fruit bunch fibre, mesocarp and kernels after microwave pre-treatment

Jessie S.L. Chang[1], Y.S. Chan[2,*], M.C. Law[1], C.P. Leo[3]

[1] Department of Mechanical Engineering, Faculty of Engineering and Science, Curtin University Malaysia, CDT 250, 98009, Miri, Sarawak, Malaysia
[2] Department of Chemical Engineering, Faculty of Engineering and Science, Curtin University Malaysia, CDT 250, 98009, Miri, Sarawak, Malaysia
[3] School of Chemical Engineering, Engineering Campus, Universiti Sains Malaysia, Seri Ampangan, 14300 Nibong Tebal, S.P.S. Pulau Pinang, Malaysia.

* Email: chanyensan@curtin.edu.my

Abstract. The implementation of microwave technology in palm oil processing offers numerous advantages; besides elimination of polluted palm oil mill effluent, it also reduces energy consumption, processing time and space. However, microwave exposure could damage a material's microstructure which affected the quality of fruit that can be related to its physical structure including the texture and appearance. In this work, empty fruit bunches, mesocarp and kernel was microwave dried and their respective microstructures were examined. The microwave pretreatments were conducted at 100W and 200W and the microstructure investigation of both treated and untreated samples were evaluated using scanning electron microscope. The micrographs demonstrated that microwave does not significantly influence kernel and mesocarp but noticeable change was found on the empty fruit bunches where the sizes of the granular starch were reduced and a small portion of the silica bodies were disrupted. From the experimental data, the microwave irradiation was shown to be efficiently applied on empty fruit bunches followed by mesocarp and kernel as significant weight loss and size reduction was observed after the microwave treatments. The current work showed that microwave treatment did not change the physical surfaces of samples but sample shrinkage is observed.

1. Introduction

Moisture is a major constituent in fresh food and it serves as a good microwave energy absorber. The moisture contains in the food greatly influences the amount of microwave absorption during heating due to the dipolar nature. In microwave field, dipole molecules orient themselves with the electric field of the microwave. This rotation leads to molecular friction and heat generation, which is known as the dielectric polarization [1].

During the microwave irradiation, energy is transport volumetrically to the food through interaction between microwaves, polar water molecules and charged ions in food. The absorption and the conversion of microwave energy to heat depends on the dielectric nature of the food. Dielectric constant and dielectric loss is a measurement of dielectric properties of a food, which predict the amount of electromagnetic energy coupled into food, and the ability of food to convert the coupled energy into heat respectively [2-4]. At present, microwave drying has been widely used in fruits processing [3, 5, 6]. In comparison with conventional drying method, microwave gives higher thermal efficiency, better energy savings, shorter drying time, and lower risk of overheating [7]. Microwave pre-treatment of the fresh fruit bunch (FFB) and its fresh fruits have been studied extensively. Microwave pre-treatment is capable of providing a rapid sterilization process than the conventional process. Sterilization is a process which aims to promote detachment of fruits from FFB and also to inhibit the enzymatic activity of lipase in fruits. The conventional sterilization process requires a large amount of steam at 140 °C for 75 – 90 min and often results in production of polluting waste water, i.e., palm oil mill effluent (POME). The microwave pre-treatment which is a dry, clean and yet encourages fruits softening and promotes fruits detachment from FFB [8, 9] is a great option in replacing the conventional steam sterilization. Furthermore, the quality of palm and kernel oil yielded from a microwave treated oil palm fruits was found to be superior than the commercial crude palm oil such as low free fatty acid (FFA) attribute to reduce microbiological activity [8-10], longer storage duration [11], low moisture content and improved chemical quality [9, 12].

During the microwave drying process, moisture content from food is removed at the temperature below boiling point. The drying process not only affects the moisture content of the food, but also changes its physical, mechanical, and chemical properties [13], such as reduction in dimensions [14], increment in bulk density[15] and decrement in mechanical strength[15, 16], and changes in color [16, 17]. In microwave drying process, Izli and Isik [18] reported that high microwave exposure could damage the materials' microstructure which affected the quality of fruit. Moreover, the quality of dried fruits is dependent in changes or modification of its physical structure including the texture and appearance. Dimension of the texture and appearance of several food grains and mineral in SEM images have been measured [19, 20] and the analysis of the geometric dimensions, morphological features and orientation of components can be identified [21].

Numerous literatures on the effects of the microwave treatment on the quality, chemical and physical properties of the EFB and the oil palm fruits have been conducted. Nevertheless there is limited work on the influences of microwave drying process on its microstructure particularly on the raw EFB fibers, mesocarp and kernel. As such, this comparative microstructures studies on the morphology of EFB, mesocarp and kernel aim to reveal the influences of microwave irradiation on their microstructures.

2. Materials and methods

2.1. Material preparation
The fresh palm fruit bunches (*Elaeis guineensis*) were obtained from United Palm Oil Sdn Bhd, Pulau Pinang, a local oil palm mill in Nibong Tebal, Penang, Malaysia. The unprocessed empty fruit bunches (EFB) were collected after the palm fruits detachment from the fruit bunches. The EFB were cut into 21mm x 13mm x 5.5mm slides by using a cutter and a hacksaw (Stanley 15-265). Palm fruits were cut to obtain its mesocarp and kernel. Mesocarp samples with dimensions of 20mm x 10mm x 4mm and kernel samples with mean diameter of 12mm were used in the experiments. The samples were stored in fridge prior pre-treatment process.

2.2. Microwave treatment
Samples were treated at 100W and 200W for two minutes in 2455MHz microwave (CEM's Microwave Assisted Reaction System 6). Microwave treatment duration of two minutes was chosen as better quality of kernel oil and palm oil from oil palm fruits were obtained [9] whereas 100W and 200 W microwave power selection was due to the limitation of domestic microwave. The weight and size of the samples before and after treatment were measured using an electronic balance (AA-160 Denmer) and a vernier caliper. Samples then were grinded and placed in a desiccator before SEM analysis.

2.3. Scanning electron microscope (SEM) analysis

The morphological structures of the raw and pre-treated samples were observed through a scanning electron microscope (TM3000 Tabletop Microscope). Each sample was mounted on a stub using double-sided adhesive carbon tabs before coated with gold in a sputter coater (Quorum SC7620) in order to prevent charging of the specimen with electron beam. Images were taken at an accelerating voltage of 15kV with magnifications of 500x to 5000x. Particle size analysis was conducted using the ImageJ.

3. Results and discussion

3.1. Weight loss and size reduction

Initial moisture content and the dielectric properties of EFB, mesocarp and kernel were shown in Table 1. It showed that microwave pre-treatment has a greater effect on weight loss in EFB followed by the mesocarp and the kernel [22]. As the microwave effect contributes to the loss of moisture of the samples is due to the dielectric properties, higher dielectric constant of EFB indicates higher ability in energy storage. Hence, a higher dielectric loss of EFB leads to greater microwave energy dissipation. Thus, more heat is generated and the evaporation of the moisture content within the EFB is enhanced [4]. According to Table 1, lower dielectric constant of the mesocarp and kernel lead to lower heat absorption.

Table 1. Properties of raw EFB, mesocarp fiber and kernel

	Initial Moisture Content (kg/kg)	Dielectric constant at 2.45GHz, ε'	Dielectric loss 2.45GHz, ε''
EFB [23]	0.69	57.4	18.6
Mesocarp pulp [8, 24]	0.37	11.5	4.2
Kernel [8, 25]	0.23	11.7	5.8

During microwave pre-treatment, weight losses of the samples were observed due to the lost of moisture within the samples and later led to sample shrinkage. Weight loss and size reduction of the samples after microwave pre-treatment were shown in Table 2.

Table 2. Percentage of weight loss and size reduction of the EFB, mesocarp and kernel after microwave pre-treatment

Sample	Weight loss, %		Size reduction, %	
	100W	200W	100W	200W
EFB	11.4	26.8	22.5	34.6
Mesocarp	6.0	18.6	2.6	2.9
Kernel	1.5	3.0	0.6	1.2

The percentage of sample weight loss (W) and size reduction (S) are defined as:

$$\%W = \frac{W_i - W_x}{W_i} \quad (1)$$

$$\%S = \frac{V_i - V_x}{V_i} \quad (2)$$

where W_i and W_x are the initial and final weight of the sample and V_i and V_x are the initial volume of the sample.

The weight lost of the EFB, mesocarp and kernel were 11.4%, 6.0% and 1.5% respectively after expose to 100W of microwave power while 26.8%, 18.6% and 6.0% respectively after 200 W. A higher microwave power exposure leads to greater dielectric heating effect on the sample.

Moreover, decreased in the size of EFB, mesocarp and kernel samples were also observed after pre-treatment. At the same exposure time and power, the size reduction of the EFB is much greater as compared to mesocarp and kernel. Size reduction occurred because the sample is unable to support its own weight and hence collapses in the absence of moisture [26]. As amount of moisture lost in the sample is directly proportional to its weight lost [22], more moisture loss in the EFB results in a higher size reduction in comparison with the kernel and mesocarp. The size of treated EFB, mesocarp and kernel were reduced by 22.5 - 34.6 %, 2.6 - 2.9 % and 0.6 - 1.2 % respectively as shown in Table 2.

EFB contain high composition of volatiles components; components which formed into gas phases during heating process. The high compositions of flammable volatile components in the EFB had caused it to burn and ignite at a lower temperature [27]. It was also noticed that raw EFB was burnt when exposed to microwave power level of more than 200W. This is possibly due to excessive heat generated within the EFB which led to overheating.

3.2. Changes of microstructure of EFB

Figure 1 shows the surface structures of the raw and treated EFB under 1000x and 5000x magnifications. It is noticed that the raw EFB having a similar surface structure as reported by other researchers [28-32]. In the micrographs, pores that are round in shape were found on the fiber strand. The presence of pores in the raw and treated EFB shows that EFB is a fibrous material [24]. In addition, the granular starch was also found in the fiber strand.

From micrograph, it is shown that the round-shaped granular starch remains deposited after the microwave treatment, but the size of granular starch was reduced after the treatment (Table 3). The size of granular starch was reduced from 0.115 mm to 0.009 mm and 0.006 mm after exposure of 100 W and 200 W respectively. Smaller size of the granular starch is desirable as it could facilitate the removal of starch. The removal of starch could increase the surface area of the fibre for biomass treatment [30, 33]. In the current sterilization process, starch presents in fruit and EFB is removed through hydrolysis, where it is broken down by supplied wet steam [34]. Hence, it is believed that microwave treatment could be used to replace the conventional steam sterilization which produces undesired palm oil mill effluents.

The raw EFB had a rough surface with a great number of spiky and round-shaped globular protrusions at the fiber strands. Previous studies identified the globular protrusions as silica bodies [30]. The attached silica bodies in circular crater were distributed uniformly along the surface of the strand. Silica bodies were observed to remain attached at the surface after the microwave treatment. Silica bodies are undesirable in industrial processes as they are inert to most chemicals. In addition, they are hard and could only be removed by alkali treatment or by mechanically or at high temperature [30]. Figure 1 showed that a small portion of the silica bodies of EFB treated at 200W was disrupted. Table 3 shows the diameter of the silica body, and circular crater before and after microwave drying process. The drying process has increased the diameter of circular crater from 0.112 mm to 0.134 mm and 0.171 mm after exposed to microwave treatment for two minutes. Increased in the diameter of circular crater could facilitate the removal of silica bodies from the crater and hence result in better fiber-matrix interfacial adhesion of the EFB for further composite fabrication. As shown in Figure 1(b) and 1(c), microwave treatment has the potential to remove and disrupt the silica bodies. Despite on the reduction of the circular crater, the diameter of the silica body remains constant after exposed to microwave treatment.

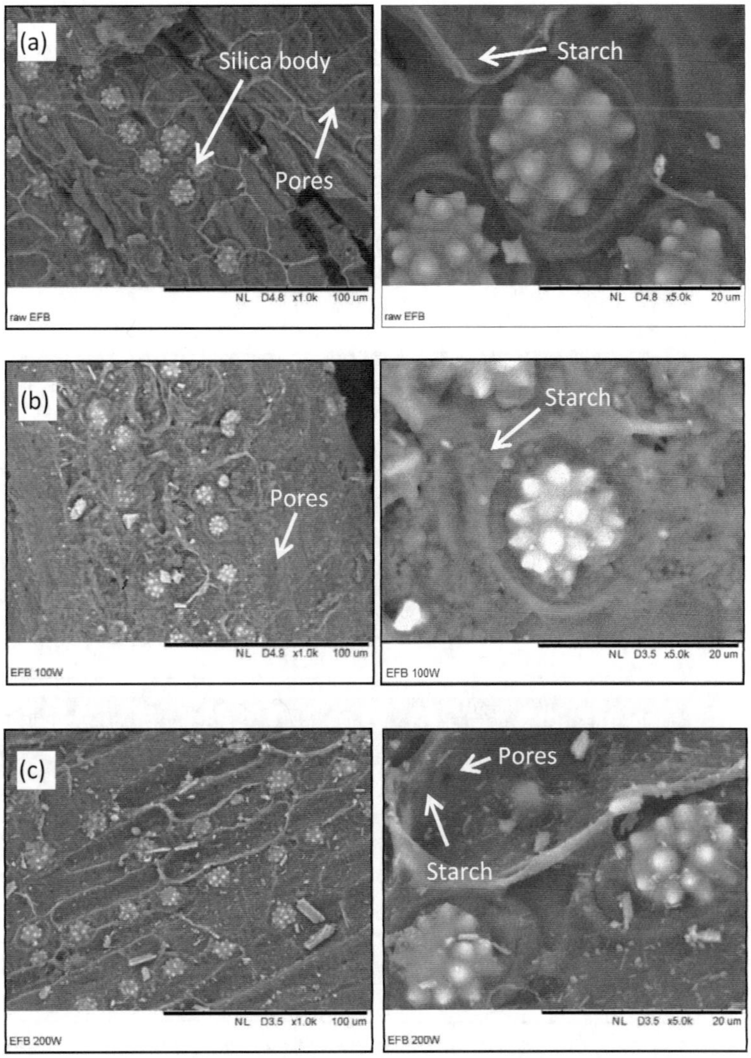

Figure 1. SEM micrographs at 1000x and 5000x magnifications of (a) raw and microwave treated EFB samples at a power level of (b) 100W and (c) 200W.

Table 3. Diameter of the silica bodies, circular crater and granular starch in the raw and microwave treated EFB.

Sample	Diameter of granular starch (mm)	Diameter of circular crater (mm)	Diameter of the silica body in the circular crater (mm)
Raw EFB	0.115	0.112	0.088
Treated EFB at 100W	0.009	0.134	0.087
Treated EFB at 200W	0.006	0.171	0.089

3.3. Changes in microstructure of kernel

Figure 2 shows the surface structure of raw and treated palm kernel. The surface structure of raw and treated kernel under a magnification of 1500x showed similar observations by Zaidul et al. [35]. It has clearly showed the oil glands and cell structures on the surface of kernel. The oil glands are enveloped by cell structure. There is no significant change on the oil gland and cell structure of the treated kernels (Figure 2b and 2c) as compared to the raw kernel (Figure 2a). However, it noticed that the surface of the treated kernel (Figure 2b and 2c) shrunk due to moisture lost. The shrinkage, however, had increased the width of cell structure, as shown in Table 4. The width of cell structure of the raw kernel had increased from 0.145 mm to 0.263 mm and 0.388 mm after microwave pre-treatment at 100 W and 200 W respectively. However, these have no impact on the appearance of the oil gland on the surface. Furthermore, oil were found deposited on the surface of the treated kernel, and this had led to the observed increased of the cell structure.

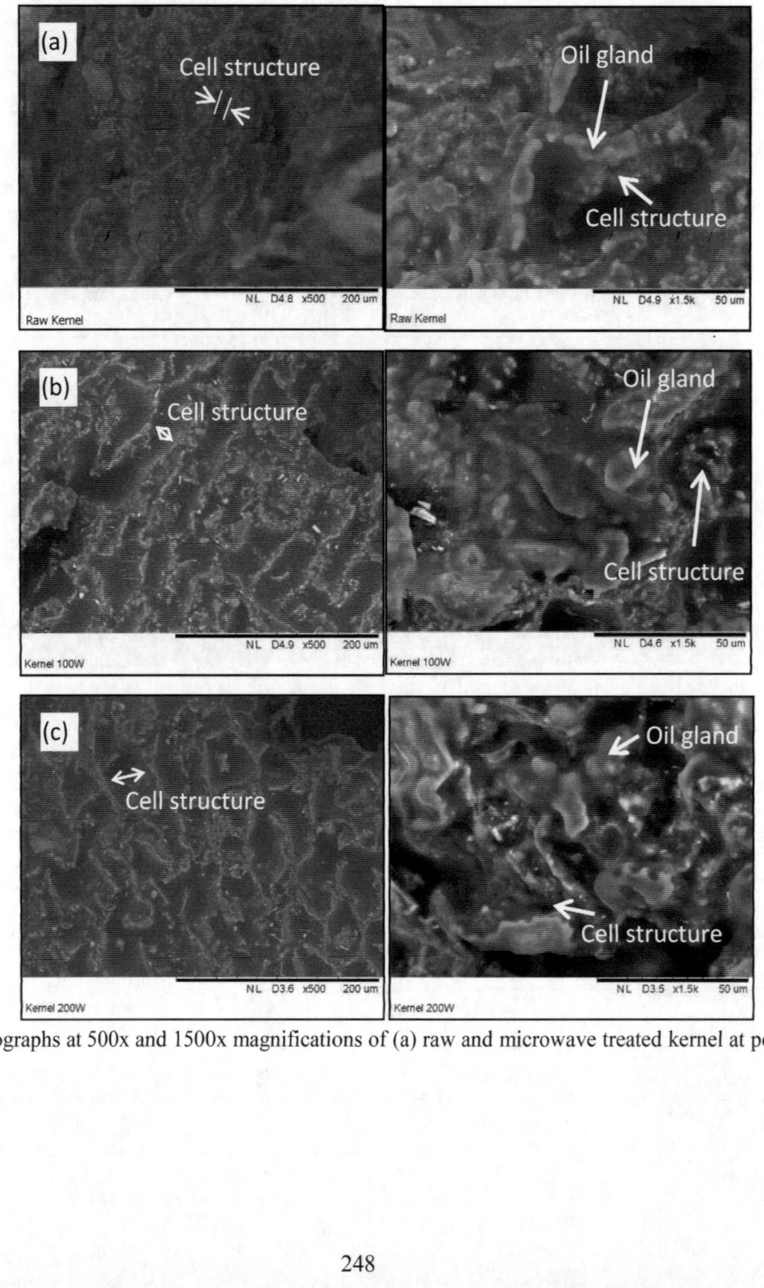

Figure 2. SEM micrographs at 500x and 1500x magnifications of (a) raw and microwave treated kernel at power level of (b) 100W and (c) 200W.

Table 4. Width of cell structure of the raw and microwave treated kernel

Sample	Width of cell structure (mm)
Raw kernel	0.145
Treated kernel at 100W	0.263
Treated kernel at 200W	0.388

3.4. Changes in the microstructure of mesocarps

The microstructures of the fresh and treated mesocarps under a 500x magnification are shown in Figure 3. The morphology of the mesocarp pulp shows a wrinkled shaped texture. Similar morphology were observed by Ho et al. [36]. Oil glands were found deposited on the surface.

No significant changes on the oil gland were also observed in treated mesocarp as compared with the raw mesocarp. However, the moisture lost after the microwave treatment resulted in shrinkage to the mesocarp is noticable. The shrinkage, however, increases the width of cell structure of the mesocarp which as shown in Table 5. The width of cell structure of the raw mesocarp had increased from 0.128 mm to 0.143 mm and 0.1698 mm after microwave exposure at 100 W and 200 W respectively. Surface of the treated mesocarp too was found deposited with oil, as a result of microwave treatment. Yielded oil had led to the observed increased of the cell structure.

Figure 3: SEM micrographs at x500 magnifications of (a) raw and microwave treated mesocarp at power level of (b) 100W and (c) 200W.

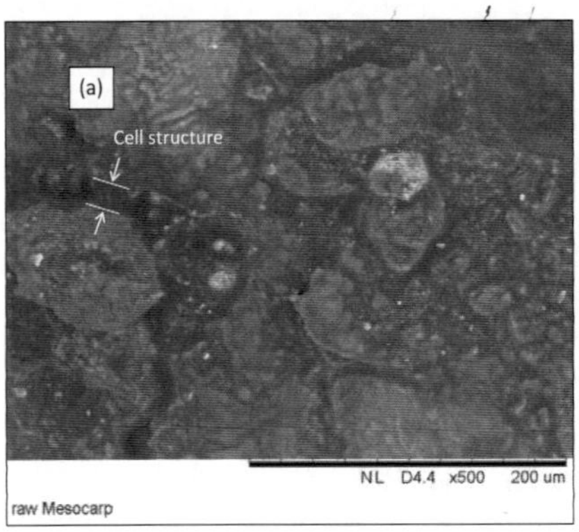

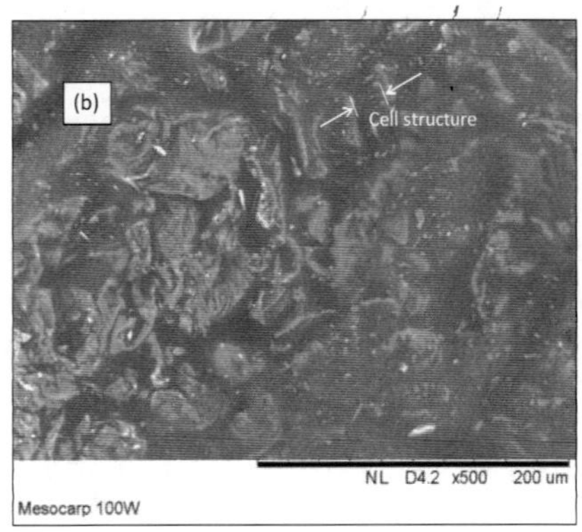

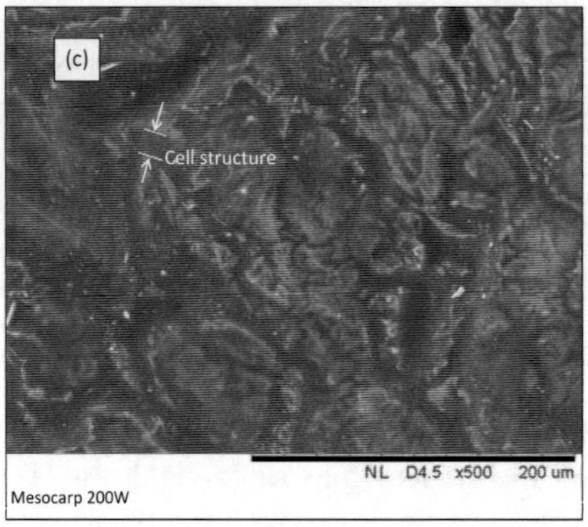

Figure 3. SEM micrographs at 500x magnification of (a) raw and microwave treated mesocarp at a power level of (b) 100W and (c) 200W.

Table 5. Width of cell struture of the raw and microwave heated mesocarp.

Sample	Width of cell structure (mm)
Raw mesocarp	0.128
Treated mesocarp at 100W	0.143
Treated mesocarp at 200W	0.169

4. Conclusion

In spite of microwave irradiation method gives a faster process time and reduces energy consumptions, microwave exposure could damage microstructure of fruit and subsequently affect the fruit's quality. However, limited works report on their texture and appearance after microwave drying process. SEM has shown the appearance change of the microwave treated samples as compared to those before treatment. The following conclusions can be made from the current study:

- High moisture content and high dielectric properties result in a high efficiency of microwave drying treatment. EFB experiences a significant effect on weight loss and size reduction, followed by the mesocarp and kernel. The weights were reduced by 11.4 - 26.8%, 6.0 - 18.6% and 1.5 - 3.0% for treated EFB, mesocarp and kernel respectively. Meanwhile, the sizes reduction of 22.5 - 34.6%, 2.6 - 2.9% and 0.6 - 1.2% was observed on the treated EFB, mesocarp and kernel respectively.
- SEM micrographs confirmed that surface of EFB had visible changes after the microwave drying process but had no significant effect on the kernel and mesocarp. In comparison to the raw samples, insignificant changes of the physical structure on the dried kernel and mesocarp had ensure that the quality has a minimal deterioration [37]. The oil glands deposited on the surface of treated kernels and mesocarps were not influenced by the microwave treatment. However, the microwave treatment on EFB had reduced the size of granular starch, increased the crater surrounded silica bodies and disrupted the silica bodies, as these are beneficial for the further EFB biomass treatment.
- Effects of microwave power on the drying process were investigated on the EFB, mesocarp and kernel. Despite greater reductions of the weight and size of the samples after treated at higher microwave power, the morphologies of the EFB, mesocarp and kernel were only slightly different.

In this study, domestic microwave oven at 2450MHz is used as the main sources, this leads to difficulties in scaling up the results to industrial-scale application. Further studies should use larger microwave systems with higher frequency based on industrial level microwave heating equipment.

XRD analysis is also recommended in future to provide quantitative support on the identification of silica body.

5. Acknowledgement

This work was supported by the Fundamental Research Grant Scheme (FRGS/2/2014/TK06/CURTIN/03/1). The authors acknowledge United Palm Oil Sdn Bhd, Pulau Pinang, Malaysia for providing fresh fruit bunch and Curtin Sarawak for financial support.

References

[1] Anwar J, Shafique U, Waheed-uz-Zamana, Rehman R, Salman M, Dar A, Anzano JM, Ashraf U, Ashraf S. Microwave chemistry: Effect of ions on dielectric heating in microwave ovens. *Arabian J. Chem.* 2015;**8**:100-4.
[2] Mudgett RE. Microwave properties and heating characteristics of foods. Food Technology. 1986;40:84-93.
[3] Umesh Hebbar H, Rastogi NK. Chapter 12 - microwave heating of fluid foods. In: Valdramidis PJCKTP, editor. Novel thermal and non-thermal technologies for fluid foods. San Diego: Academic Press; 2012. p. 369-409.
[4] Magee TRA, McMinn WAM, Farrell G, Topley L, Al-Degs YS, Walker GM, Khraisheh M. Moisture and temperature dependence of the dielectric properties of pharmaceutical powders. *J. Therm. Anal. Calorim.* 2013;**111**:2157-64.
[5] Law CL, Chen HHH, Mujumdar AS. Food technologies: Drying. In: Motarjemi Y, editor. Encyclopedia of food safety. Waltham: Academic Press; 2014. p. 156-67.
[6] Sabarez HT. 4 - modelling of drying processes for food materials. In: Fryer SBKJ, editor. Modeling food processing operations: Woodhead Publishing; 2015. p. 95-127.
[7] Atong D, Ratanadecho P, Vongpradubchai S. Drying of a slip casting for tableware product using microwave continuous belt dryer. *Drying Technol.* 2006;**24**(5):589-94.
[8] Chow MC, Ma AN. Processing of fresh palm fruits using microwaves. *J. Microwave Power Electromagn. Energy*. 2007;**40**(3):165-73.
[9] Cheng SF, Mohd NL, Chuah CH. Microwave pretreatment: A clean and dry method for palm oil production. *Ind. Crops. Prod.* 2011;**34**(1):967-71.
[10] Nokkaew R, Punsuvon V. Sterilization of oil palm fruits by microwave heating for replacing steam treatment in palm oil mill process. Advanced Materials Research. 2014;**1025-1026**:470-5.
[11] Umudee I, Chongcheawchamnan M, Kiatweerasakul M, Tongurai C. Sterilization of oil palm fresh fruit using microwave technique. International Journal of Chemical Engineering and Applications. 2013;4(3):111-3.
[12] Sukaribin N, Khalid K. Effectiveness of sterilisation of oil palm bunch using microwave technology. *Ind. Crops. Prod.* 2009;**30**(2):179-83.
[13] Barbosa-Canovas GV, Vega-Mercado H. Dehydration of foods. New York, United States of America1996.
[14] Li H, Pordesimo LO, Igathinathane C, Vinyard B. Physical property effects on drying of chile peppers. *Int. J. Food Prop.* 2009;**12**:316-30.
[15] Karababa E, Coskuner Y. Moisture dependent physical properties of dry sweet corn kernels. *Int. J. Food Prop.* 2007;10:549-60.
[16] Abasi S, Minaei S. Effect of drying temperature on mechanical properties of dried corn. *Drying Technol.* 2014;**32**:774-80.
[17] Guiné RPF. Influence of drying method on some physical and chemical properties of pears. International Journal of Fruit Science. 2011;**11**:245-55.
[18] Izli N, Isik E. Color and microstructure properties of tomatoes dried by microwave, convective, and microwave-convective methods. *Int. J. Food Prop.* 2015;**18**(2):241-9.

[19] Igathinathane C, Pordesimo LO, Batchelor WD. Major orthogonal dimensions measurement of food grains by machine vision using imagej. *Food Res Int.* 2009;**42**:76-84.

[20] Pollastri S, Gualtieri AF, Gualtieri ML, Hanuskova M, Cavallo A, Gaudino G. The zeta potential of mineral fibres. *J Hazard. Mater.* 2014;**276**:469-79.

[21] Mur A, Purcell C, Soong Y, Crandall D, McLendon TR, Haljasma IV, Warzinski R, Kutchko B, Kennedy S, Harbert W. Integration of core sample velocity measurements into a 4d seismic survey and analysis of SEM and CT images to obtain pore scale properties. *Energy Procedia.* 2011;**4**:3676-83.

[22] Dincer I. Moisture loss from wood products during drying—part ii: Surface moisture content distributions. *Energy Sources*. 2007;**20**:77-83.

[23] Omar R, Idris A, Yunus R, Khalid K, Aida Isma MI. Characterization of empty fruit bunch for microwave-assisted pyrolysis. *Fuel*. 2011;**90**(4):1536-44.

[24] Sabil KM, Aziz MA, Lal B, Uemura Y. Effects of torrefaction on the physiochemical properties of oil palm empty fruit bunches, mesocarp fiber and kernel shell. *Biomass Bioenergy*. 2013;**56**:351-60.

[25] Kok S, Ong-Abdullah M, Ee GC, Namasivayam P. Comparison of nutrient composition in kernel of tenera and clonal materials of oil palm (elaeis guineensis jacq.). *Food Chem.* 2011;**129**:1343-7.

[26] Witrowa-Rajchert D, Rząca M. Effect of drying method on the microstructure and physical properties of dried apples. *Drying Technol.* 2009;**27**(7-8):903-9.

[27] Abdullah N, Sulaiman F. The properties of the washed empty fruit bunches of oil palm. Journal of Physical Science. 2013;**24**(2):117-37.

[28] Norul Izani MA, Paridah MT, Anwar UMK, Mohd Nor MY, H'ng PS. Effects of fiber treatment on morphology, tensile and thermogravimetric analysis of oil palm empty fruit bunches fibers. *Composites Part B* 2013;**45**(1):1251-7.

[29] Nasri NS, Hamza UD, Abdulkadir A, Ismail SN, Ahmed MM, editors. Utilization of sustainable palm empty fruit bunch sorbents for carbon dioxide capture. 6th International Conference on Process Systems Engineering (PSE ASIA); 2013; Kuala Lumpur.

[30] Law KN, Daud WRW, Ghazali A. Morphological and chemical nature of fiber strands of oil palm empty-fruit bunch (opefb). BioResources. 2007;**2**(3):351-62.

[31] Baharuddin AS, Sulaiman A, Kim DH, Mokhtar MN, Hassan MA, Wakisaka M, Shirai Y, Nishida H. Selective component degradation of oil palm empty fruit bunches (opefb) using high-pressure steam. *Biomass Bioenergy*. 2013;**55**:268-75.

[32] Shinoj S, Visvanathan R, Panigrahi S, Kochubabu M. Oil palm fiber (opf) and its composites: A review. *Ind. Crops Prod.* 2011;**33**(1):7-22.

[33] Zheng Y, Pan Z, Zhang R. Overview of biomass pretreatment for cellulosic ethanol production *Int. J. Agric Biol Eng.* 2009;**2**(3):51-67.

[34] Pradeepkumar T, Jyothibhaskar BS, Satheesan KN. Management of horticultural crops. Peter KV, editor. New Delhi: New India Publishing; 2008.

[35] Zaidul ISM, Norulaini NAN, Omar AKM, Sato Y, Smith Jr RL. Separation of palm kernel oil from palm kernel with supercritical carbon dioxide using pressure swing technique. *J. Food Eng.* 2007;**81**(2):419-28.

[36] Ho LS, Nair A, Yusof HM, Kulaveerasingam H, Jangi MS. Morphometry of lipid bodies in embryo, kernel and mesocarp of oil palm: Its relationship to yield. American Journal of Plant Sciences. 2014;**5**:1163-73.

[37] Kong F, Singh RP. 2 - chemical deterioration and physical instability of foods and beverages. In: Kilcast D, Subramaniam P, editors. Food and beverage stability and shelf life: Woodhead Publishing; 2011. p. 29-62.

IOP Conf. Series: Materials Science and Engineering **217** (2017) 012027 doi:10.1088/1757-899X/217/1/012027

Factors affecting the immobilization of fungal biomass on CNT as a biosorbent for textile dyes removal

Ibrahim Adebayo Bello [1], Nassereldeen, A, Kabbashi[1,2*], Md Zahangir Alam[1,2], Ma'an F Alkhatib[1,2], Fatin Nabilah Murad[1]

[1]Department of Biotechnology Engineering, International Islamic University Malaysia Gombak, 50728, Malaysia
[2] Bioenvironmental Engineering Research Centre (BERC), Kuliyyah of Engineering, IIUM

*Email: nasreldin@iium.edu.my

Abstract. Effluents from dye and textile industries are highly contaminated and toxic to the environment. High concentration of non-biodegradable compounds contributes to increased biochemical oxygen demand (BOD) and chemical oxygen demand (COD) of the wastewater bodies. Dyes found in wastewater from textile industries are carcinogenic, mutagenic or teratogenic. Biological processes involving certain bacteria, fungi and activated carbon have been employed in treating wastewater. These methods are either inefficient or ineffective. These complexities necessitates search for new approaches that will offset all the shortcomings of the present solutions to the challenges faced with textile wastewater management. This study produced a new biosorbent by the immobilization of fungal biomass on carbon nanotubes. The new biosorbent is called "carbon nanotubes immobilized biomass (CNTIB)" which was produced by immobilization technique. A potential fungal strain, *Aspergillus niger* was selected on the basis of biomass production. It was found out in this studies that fungal biomass were better produced in acidic medium. *Aspergillus niger* was immobilized on carbon nanotubes. One-factor-at-a time (OFAT) was employed to determine the effect of different factors on the immobilization of fungal biomass on carbon nanotubes and optimum levels at which the three selected parameters (pH, culture time and agitation rate) would perform. Findings from OFAT showed that the optimum conditions for immobilization are a pH of 5, agitation rate of 150rpm and a culture time of 5 days.

1. Introduction

Immobilization is a process whereby cells are attached or ensnared in a torpid support. Usually, the substrate is converted to a product by the immobilized cells. There is often a retention of the immobilized cells which is continuously utilized in the bioreactor. Over the years, it has been shown that in many processes, using immobilized cells bring about more efficiency when compared to suspended cells [1]. In this research, fungal biomass was immobilized on carbon nanotubes for the production of a novel biosorbent; carbon nanotubes immobilized biomass for the removal of textile dyes.

IOP Conf. Series: Materials Science and Engineering **217** (2017) 012027 doi:10.1088/1757-899X/217/1/012027

2. Experimental

Aspergillus niger and Penicillum sp were cultivated in a liquid medium using a shake flask method. The liquid medium was 98 ml of (17g/1000ml) malt extract, (0.1-0.5) g of carbon nanotubes and 2mls of inoculum. The malt extract solution was prepared by dissolving 17g of malt extract powder into 1000mls distilled water. It was placed on a magnetic stirrer and allowed to make a uniform mixture. Varying quantities of carbon nanotubes were measured and poured into the liquid media in the Erlenmeyer conical flask. The pH of each samples were recorded and then autoclaved at 121°C for 2-3 hours. The liquid medium was allowed to cool down to room temperature after autoclaving. The inoculum, 2% (2mls) of each fungus was used to inoculate the liquid medium and transferred to the incubator shaker at 150 rpm for 3 days. Immobilized activated carbon was filtered with whatman filter paper (100mm) and dried at 105^{0}C in a drier. The weight of dried biomass was recorded, crushed and then stored in a dry container.

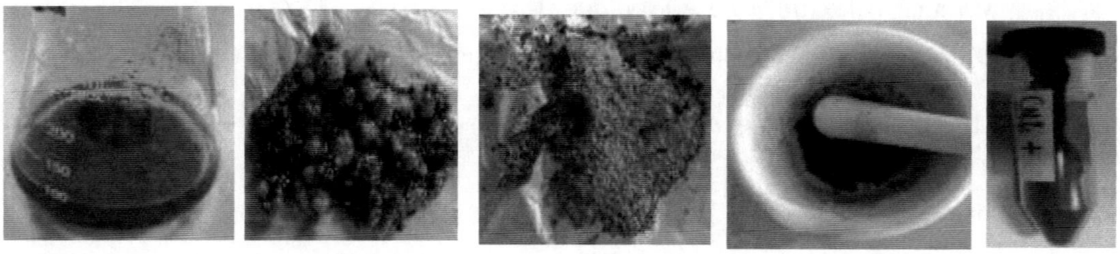

(A) (B) (C) (D) (E)

Figure 1 (A) immobilized carbon nanotubes by Aspergillus niger in culture broth (B) filtered CNTIB (C) dried CNTIB (D) crushed CNTIB (E) stored CNTIB

The growth of fungus (*Aspergillus niger*) was observed after two days as it grew into pellicles. The pellicles increased in diameter subsequently. After the third day of growth (72 hours), the carbon nanotubes immobilized biomass was harvested by filtering through a 150μm filter. It was then washed thoroughly with distilled water to ensure that CNTs were tightly attached to the biomass. The filtrate which contained the unattached CNTs was passed through the pump, washed thoroughly with deionized water to remove the growth medium sticking on to its surface and filtered to get a clear solution. The CNTIB and unattached CNTs were transferred to the drier for 24 hours at a temperature of 100°C. The weight of dried CNTIB and CNTs were taken. The mass of biomass and mass of attached CNTs were also determined. The percentage of carbon nanotubes attached to fungal biomass was then calculated thus:

$$M_{uCNT} = M^0_{CNT} - M^1_{CNT}$$

$$M^1_{CNT} = (M^0_{CNT} - M_{uCNT})$$

$$M_B = \left(M_{CNTIB} - M_{uCNT}\right)$$

$$E_{B\rightarrow CNT} = \frac{M_B}{M_{CNTIB}} \times 100$$

Where

M_B = Mass of biomass

M_{uCNT} = Mass of unattached Carbon nanotubes

M^{0}_{CNT} = Initial mass of carbon nanotubes

M^{1}_{CNT} = Mass of attached carbon nanotubes

M_{CNTIB} = Mass of carbon nanotubes immobilized biomass

$E_{B \to CNT}$ = Efficiency of biomass immobilization on CNT

3. Results and discussion

3.1. *Effect of pH on biomass production*

The effect of pH on *Aspergillus niger* for the immobilization on carbon nanotubes was studied. The mass of biomass produced and percentage of carbon nanotubes attached after immobilization were also determined.

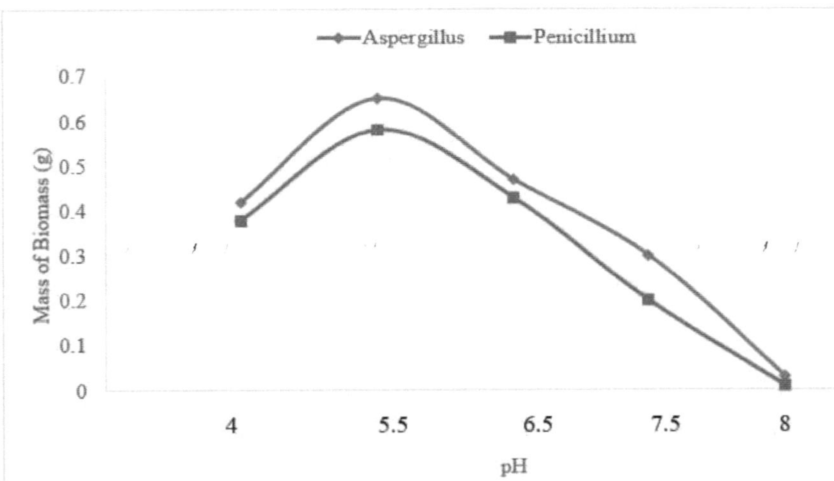

Figure 2: Effect of pH on the biomass production at temperature 32^{0}C, 150 rpm and 72 hours culture time by *Aspergillus niger*.

The effect of pH was studied on two major types of fungal strains with respect to carbon nanotubes immobilization. As shown in Figure 2, the mass of biomass production was plotted against various pH ranging from 4 to 8. The mass of biomass produced was determined after 3 days of culturing. From the result, it is shown that at pH 4, there was a gradual increase in the mass of biomass produced from 0.42g and 0.38g for *Aspergillus niger* and *Penicillum sp* respectively. At about pH 5.5, the biomass of *Aspergillus niger* and *Penicillum sp* were highest, producing a biomass of 0.65g and 0.58g respectively. Further increase in pH resulted into a decline in production of biomass.

Based on results from the experiment, *Aspergillus niger* was chosen as the fungal strain to be used in the immobilization process. A pH of 5.5 was also observed to be the optimum for biomass production because most of the biomass was produced at that point. It was also observed that growth of fungal biomass was best in acidic media.

The effect of pH was studied on the germination and growth of seven xerophilic fungi - *Eurotium rubrum, E. repens, Wallemia sebi, Aspergillus penicillioides, Penicillium roqueforti, Chrysosporium xerophilum* and *Xeromyces bisporus*. They were grown at varying temperatures and on media with varying pH values. From the results of the experiment, it was found out that all fungal strains grew faster under acidic than neutral or alkaline pH conditions [2].

Furthermore, according to [3], in the studies of the effects of operational parameters including pH on maturity of *Aspergillus niger* and *Spirogyra sp* in aqueous solution on dye removal, the results showed that the maximum performance of fungal biomass was obtained at pH 3 (acidic). [5] also suggested an acidic pH of 3 for growth of fungal biomass. The optimum pH for the biomass growth of *P. ostreatus,* in the determination of biodegradation products from sulfonated dyes was also suggested at 5 [1].

3.2. Effect of agitation rate

The percentage of biomass attached to carbon nanotubes was examined. The agitation rates were varied from 100 i.e., immobilization increased with corresponding increase in agitation. From the graph, immobilization increased with an increase in agitation rate. Agitation of 100 – 150 rpm gave corresponding increase in immobilization. After agitation of 150 rpm, further increase to 180-200 rpm led to a fall in percentage immobilization. The immobilization was highest (85.1%) at 150 rpm for *Aspergillus niger*. In the study of the influence of agitation rate on the performance of a stirred anaerobic sequencing batch reactor containing immobilized biomass, [6] found out that the use of agitation increased the efficiency of the reactor and enabled reduction of the total cycle time. Moreover, it was found that the biomass of Aspergillus niger was obtained at 150 rpm in accordance with 5 days culturing time [4]. The study by [7] on a white rot fungus, *Pleurotus ostreatus* was also conducted at an agitation rate of 150 rpm.

The effect of agitation rate on the immobilization of fungal biomass on carbon nanotubes is shown in figure 3.

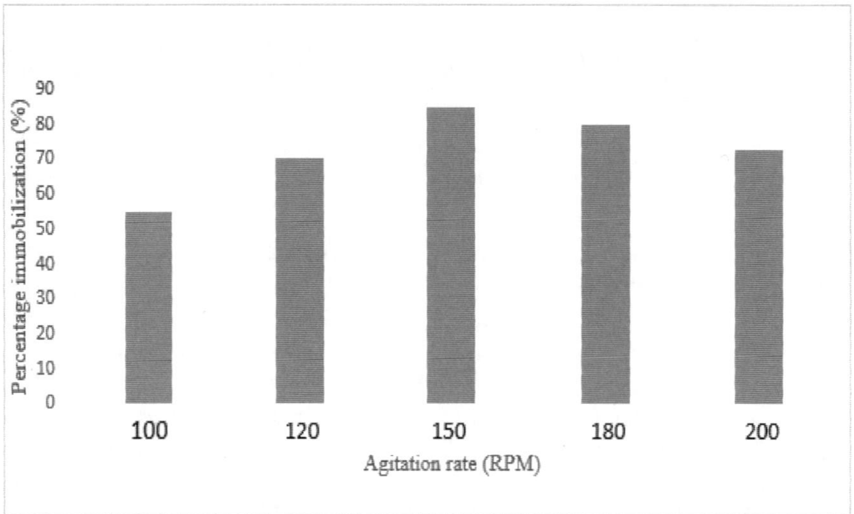

Figure 3: Effect of Agitation rate on the immobilization of carbon nanotubes at temperature 32^{0}C, 150 rpm and 72 hours culture time by *Aspergillus niger*

3.3 Effect of pH on immobilization of CNTs

The effect of pH was studied on *Aspergillus niger for* the immobilization of its biomass on carbon nanotubes as shown in Figure 4.

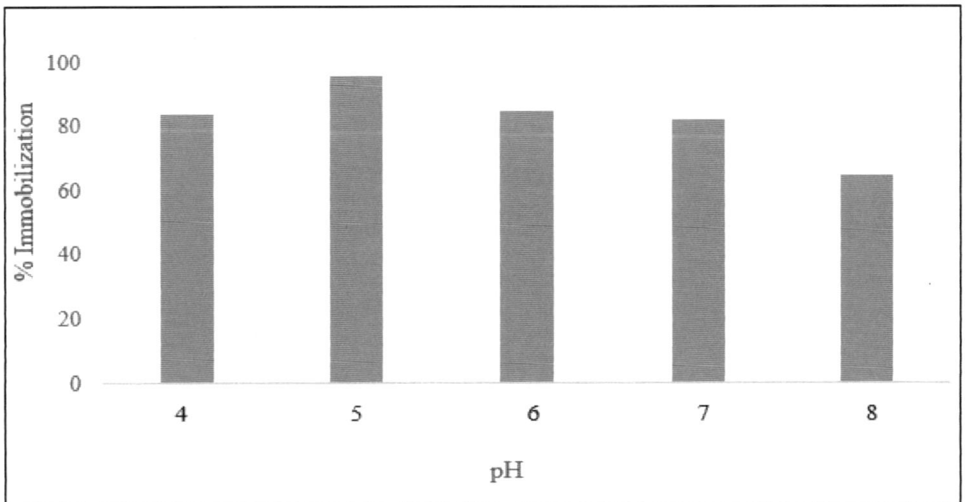

Figure 4 Effect of pH on the immobilization of carbon nanotubes at temperature 32^{0}C, 150 rpm and 72 hours culture time

The percentage of biomass immobilized on carbon nanotubes was determined. The pH ranges were varied from 4 to 8. After 72 hours of culturing, the percentage immobilization was determined. At pH 5, the carbon nanotubes showed the highest percentage immobilization in which the strain of *Aspergillus niger* showed about 96% attachment. Further increase in pH from 5-8 brought about a decrease in immobilization. This suggests acidic medium as the best medium for growth and immobilization of fungal biomass. Various studies have shown that the acidic media is the most favorable for fungal biomass growth. *T. versicolor*, *P. ostreatus*, *P. sajor-caju* and *P. chrysosporium* strains were studied by some researchers and they suggested the optimum pH of 4-5 for these fungal biomass growth [5]. Based on the results observed, pH 5 was selected for immobilization of fungal biomass onto carbon nanotubes for the production of the biosorbent, carbon nanotubes immobilized biomass.

3.4 Effect of time on Immobilization of Aspergillus niger

The effect of culture time on the immobilization of carbon nanotubes was studied as shown in Figure 5.

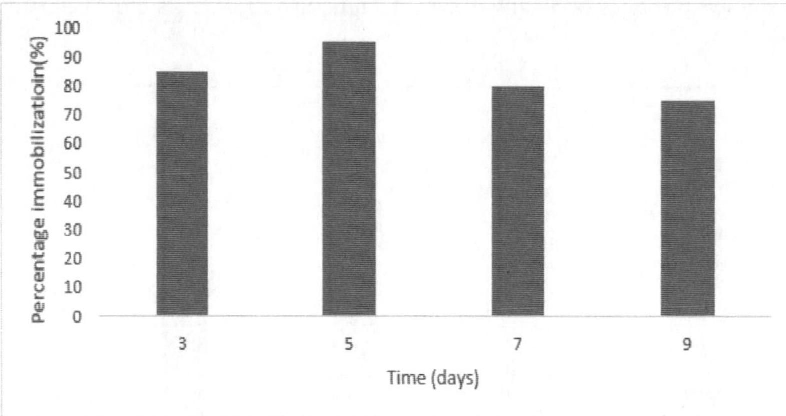

Figure 5: Effect of culture time on the immobilization of carbon nanotubes at temperature 32^0C, 150 rpm and pH of 5

The culture time was varied for 3, 5, 7 and 9 days. From the result, it was shown that the highest immobilization was achieved on the 5th day (120 hours culture time). The 5 day culturing time was also supported by [4] in his research which involved fungal chitosan production from potato processing wastewater and its characterization.

4. Conclusion

The results obtained in this study showed that *Aspergillus niger* was able to immobilize its biomass on carbon by pallet formation. The effective pH was pH 5, agitation rate was 150 rpm and culture time was found to be 5 days for optimum immobilization of biomass on carbon nanotubes. The highest biomass production in this research for *Aspergillus nige*r was found to be 89% while the highest immobilization was at 96%.

Acknowledgement
The authors acknowledge the financial assistant for this project from FRGS 2015, to all technicians and staff of BTE and BERC.

References

[1] Brodelius, P., & Nilsson, K. (1980). Entrapment of plant cells in different matrices: a comparative study. *Febs Letters*, *122*(2), 312–316.

[2] Gock, M., Hocking, A., Pitt, J., & Poulos, P. (2003). Influence of temperature, water activity and pH on growth of some xerophilic fungi. *International Journal of Food Microbiology*, *81*(1), 11–19.

[3] Khalaf, M. (2008). Biosorption of reactive dye from textile wastewater by non-viable biomass of Aspergillus niger and Spirogyra sp. *Bioresource Technology*, *99*(14), 6631–6634.

[4] Khalaf, S., & Khalaf, M. (2005). Fungal chitosan production from potato processing wastewater and its characterization. *Isotope and Radiation Research*, *37*(5), 1465–1475.

[5] Nilsson, I., Möller, A., & Mattiasson, B. (2006). Decolorization of synthetic and real textile wastewater by the use of white-rot fungi. *Enzyme and Microbial Technology*, *38*(1), 94–100.

[6] Ratusznei, S., & Rodrigues, J. (2001). Influence of agitation rate on the performance of a stirred anaerobic sequencing batch reactor containing immobilized biomass. *Water Science and Technology*, *44*(4), 305–312.

[7] Zhao, X., Lu, Y., & Hardin, I. (2005). Determination of biodegradation products from sulfonated dyes by Pleurotus ostreatususing capillary electrophoresis coupled with mass spectrometry. *Biotechnology Letters*, *27*(1), 69–72.

IOP Conf. Series: Materials Science and Engineering **217** (2017) 012028 doi:10.1088/1757-899X/217/1/012028

Computational investigation of fluid flow and heat transfer of an economizer by porous medium approach

C Rajesh Babu [1], P Kumar [2], G Rajamohan [1]

[1]Department of Mechanical Engineering, Curtin University, CDT 250, 98009 Miri, Malaysia
[2]Department of Chemical Engineering, Curtin University, CDT 250, 98009 Miri, Malaysia

E-mail: rajesh.babu@postgrad.curtin.edu.my

Abstract. Computation of fluid flow and heat transfer in an economizer is simulated by a porous medium approach, with plain tubes having a horizontal in-line arrangement and cross flow arrangement in a coal-fired thermal power plant. The economizer is a thermal mechanical device that captures waste heat from the thermal exhaust flue gasses through heat transfer surfaces to preheat boiler feed water. In order to evaluate the fluid flow and heat transfer on tubes, a numerical analysis on heat transfer performance is carried out on an 110 t/h MCR (Maximum continuous rating) boiler unit. In this study, thermal performance is investigated using the computational fluid dynamics (CFD) simulation using ANSYS FLUENT. The fouling factor ε and the overall heat transfer coefficient ψ are employed to evaluate the fluid flow and heat transfer. The model demands significant computational details for geometric modeling, grid generation, and numerical calculations to evaluate the thermal performance of an economizer. The simulation results show that the overall heat transfer coefficient 37.76 W/ (m^2K) and economizer coil side pressure drop of 0.2 (kg/cm^2) are found to be conformity within the tolerable limits when compared with existing industrial economizer data.

1. Introduction

Due to the widespread utilization of heat exchangers in various industrial applications. Burning coal for electricity generation creates products of combustion which contain solid particles (fly-ash). These small solid particles are carried by the hot (flue) gas flow through the heat exchanger in a power utility boiler. The transported fly-ash can cause severe erosion to the heat exchanger boiler tubes, especially in the economizer. Such solid particles in suspension may also significantly influence the heat transfer characteristics of boiler heat exchangers [1].

Conventional thinking about how energy is extracted, converted, and consumed is being challenged by rising concerns about the environmental impacts of power generation on land, water, air quality, and climate. Currently, about half of our electricity is generated by coal-fired plants, which are responsible for more than 80% of carbon emissions from the power sector [2]. A comprehensive carbon management policy should both address ways to reduce emissions through a change in the fundamental energy supply structure which will take significant time and consecutively improve the performance of the existing infrastructure [2]. One such option to improve the existing energy infrastructure is to increase the thermal efficiency of existing coal-fired power plants. It has become important to improve the heat transfer characteristics of tubes in view of the energy situation [3].

The main equipment for recovering heat from exhaust flue gas is economizer in a coal-fired power plant. The economizer is a mechanical device used to preheat boiler feed water before it enters the boiler drum by utilizing the waste heat captured from boiler flue gasses [4]. The demand for reducing primary energy claims an improvement of the efficiency of heat exchangers, which contributes to the reduction of the CO_2-production [5].

There are plenty of techniques to increase heat transfer in tubular heat exchangers, such as rough surfaces, extended surfaces, surface vibration, fluid vibration, and so on. But porous media have a large contact surface with fluids, which can enhance the heat transfer process. A porous medium not only changes the condition of the flow field and makes the thermal boundary layer thinner, but also its thermal conductivity is usually higher than that of the studied fluid. Thus, adding a kind of highly conductive porous insert to a heat exchanger can effectively enhance its heat transfer characteristics and, consequently, we will be able to save the material used to construct the device and the space occupied by it [6]. The numerical investigation carried out in this study is a contribution to the research done for the heat transfer enhancement using porous media.

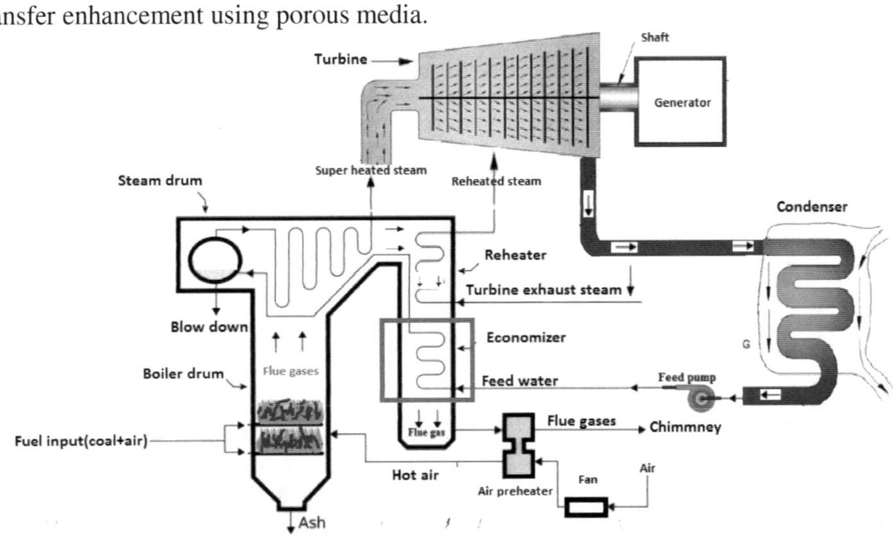

Figure 1. Economizer in a coal-fired Boiler (adapted from [7]).

Fig 1. shows the location of economizer in a coal-fired boiler. The economizer is typically the last water cooled heat transfer surface upstream of the air preheater. Flue gasses are the combustion exhaust gasses produced at power plants consist of mostly nitrogen, carbon dioxide, water vapor, soot carbon monoxide etc [8]. Frantisek et.al [9] investigated the flow and temperature fields in an economizer for the inline and staggered tube arrangement through numerical analysis. Patankar and Spalding [10] was perhaps the first one capable of calculating simultaneously the thermal fields in both fluids and in the tube metal and can be regarded as the framework from which successive contributions were developed.

1.1 Physical model, governing equations and numerical method

A schematic of the economizer is depicted in Fig 2.while the associated computational domain shown in Fig 3.some detailed geometric parameters for conducting simulations are tabulated in Table1 and Table 2.The physical properties of flue gasses and feed water are tabulated in Table 3 and Table 4.As the flue gasses flows across the inline bare tube economizer is actually a quite complex interaction amid flow filed and obstacles. Hence some assumptions are made in the following to simplify the calculation:

1. Geometry is simplified into three different zones (Fluid, porous zone 1 and porous zone 2
2 .All the water tubes are eliminated to reduce the complexity of the computational process.
3. All the heat lost from the flue gas is taken away by the water domain. This indicates that the losses in the models (heat loss to the surrounding atmosphere, thermal resistance due to fouling etc) are neglected.
4. Secondary fluid (water) is assumed to be stationary without any flow conditions being given. Water domain is modeled separate parallel to the flue gas domain (Dual Cell Method).
5. Viscous and inertial resistance for the porous zones is calculated based on the pressure drop recorded for the economizer.
6. Temperatures are fixed for the water domain based on calculating the heat released from the flue gas

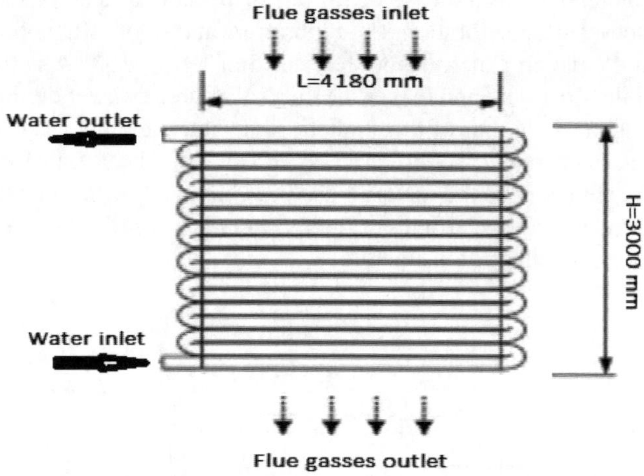

Figure 2. Schematic diagram of multipass counter cross-flow economizer (own study)

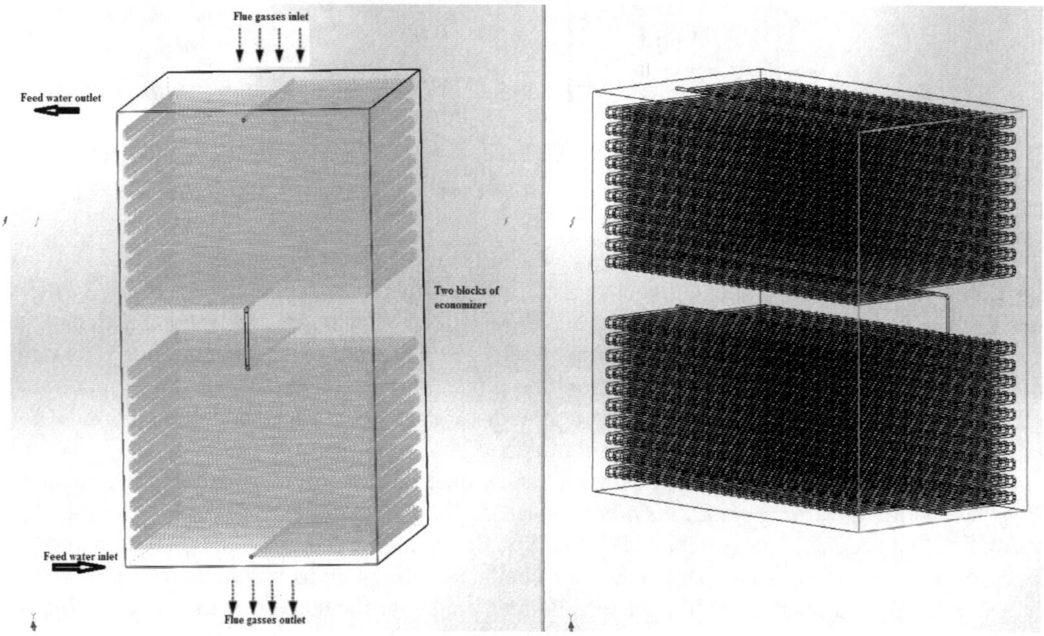

Figure 3. Geometrical details of multipass counter cross-flow heat exchanger water flow circuit inside the heat exchanger N_{row}=74

Table 1. Detail geometric parameters of the economizer

d_o(mm)	d_i(mm)	P_L(mm)	P_T(mm)	n_t	N_{row}	Tube material
55.3	50.8	95	135	20	35	SA 210 GRA1

Remarks: d_i = Tube inside diameter; d_o = tube outside diameter; P_L = longitudinal tube pitch; P_T = transverse tube pitch; n_t = number of tubes in row; N_{row} = number of tube rows.
Notes: Tube layouts of the heat exchanger are the in-line layout. (SA 210 GRA1)

Table 2. Economizer specifications

Parameters	Values
Type of economizer	Plain tube, horizontal in-line
Type of flow	Multipass counter cross-flow
Location of the economizer	Upstream of air preheater
Total heating surface of an economizer	1915 m^2
Economizer duty	4.902Mmkcal/hr
Total number of blocks	2
Economizer coil side pressure drop	0.2 kg/cm^2
Economizer casing thickness	6 mm

1.2 Data reduction and interpretation

1.2.1 Governing heat transfer equations

In this research, the effect of fly ash deposit on the heat exchanger performance is studied via thermal resistance of fouling. The following is the method for finding the thermal resistance due to fouling.

The heat transfer rate of a heat exchanger (Q) can be calculated from water side as

$$Q = m_w(h_{w2} - h_{w1}) \tag{1}$$

The heat transfer rate of the combustion gas is defined by

$$Q = m_g c_{pg}(T_{g1} - T_{g2}) \tag{2}$$

Heat conduction through the tube wall is known as

$$Q = kA_{tot}\Delta T_{ln} \tag{3}$$

Logarithmic mean temperature difference between input and output of the heat exchanger

$$\Delta T_{ln} = \frac{(T_{g1}-T_{w2})-(T_{g2}-T_{w1})}{ln\frac{T_{g1}-T_{w2}}{T_{g2}-T_{w1}}} \tag{4}$$

The total thermal resistance can be calculated by

$$\frac{1}{UA} = \frac{1}{h_i A_i} + \frac{\ln(d_o/d_i)}{2\pi k_t L} + \frac{1}{\eta_o h_o A_o} \tag{5}$$

For multipass counter cross-flow with (N_{row}=37 or α)

$$\varepsilon_c = \frac{1-e^{-NTUA(1-C_A)}}{1-C_A e^{-NTUA(1-C_A)}} \tag{6}$$

Basic law governing the flow of fluids through porous media is Darcy law

$$Q = \frac{-kA}{\mu}\frac{(p_b-p_a)}{L} \tag{7}$$

Mass conservation

$$\frac{d(\rho q)}{dx} + \frac{d(\rho q)}{dy} + \frac{d(\rho q)}{dz} = \frac{-d(\rho\varphi)}{dt} \quad (8)$$

1.2.2 Governing equations of the fluid flow and heat transport

Continuity equation

$$\frac{\partial \rho}{\partial t} = \frac{\partial(\rho u_i)}{\partial x_i} = 0 \quad (9)$$

Momentum equation

$$\rho\left[\frac{\partial u_j}{\partial t} + u_i\frac{\partial u_j}{\partial x_i}\right] = -\frac{\partial p}{\partial x_j} - \frac{\partial \tau_{ij}}{\partial x_i} + \rho g_j \quad (10)$$

Where p is the static pressure, τ is the stress tensor and ρ g is the gravitational body force.

$$\tau = \mu\left[(\nabla v + \nabla v^T) - \frac{2}{3}\nabla v\, I\right] \quad (11)$$

Where μ is the molecular viscosity, I is the unit tensor and the second term on the right side is the effect of volume dilation.

Energy equation

$$\rho\left[\frac{\partial h}{\partial t} + u_i\frac{\partial h}{\partial x_j}\right] = \frac{\partial}{\partial x_i}\left[\lambda\frac{\partial y}{\partial x}\right] - \tau_{ij}\frac{\partial u_j}{\partial x_i} + \frac{\partial p}{\partial t} + u_i\frac{\partial p}{\partial x_i} \quad (12)$$

1.2.3 Porous medium formulation

Momentum equations for porous media

Porous media are modeled by the addition of a momentum source term to the standard fluid flow equations. The source term is composed of two parts: a viscous loss term and an inertial loss term.

$$Si = -\left(\sum Dij\mu vj + \sum Cij\frac{1}{2}\rho|v|vj\right) \quad (13)$$

Where Si is source term V is the magnitude of the velocity and D and C are matrices.

To recover the case of simple homogeneous porous media

$$Si = -\left(\frac{\mu}{\alpha}Vi + C2\frac{1}{2}\rho|v|vj\right) \quad (14)$$

Where α is the permeability and C2 is the inertial resistance.

Fluent allows the source term to be modeled as power law of the velocity magnitude

$$Si = -C_o|V|^{C1} = -C_o|V|^{(C1-1)}vj \quad (15)$$

Where C_0 and C_1 are user defined empirical coefficients.

Table 3. Flue gasses specifications

Parameters	Values
Flue gas temperature entering economizer	370°C
Flue gas temperature leaving economizer	259°C
Flue gas pressure at economizer inlet	458.95 pa
Flue gas pressure at economizer outlet	615.85pa
Flue gas velocity	5.61 m/s
Flue gas density	1.31 kg/m^3
Flue gas mass flow in economizer	12280kg/m^2s

IOP Conf. Series: Materials Science and Engineering **217** (2017) 012028 doi:10.1088/1757-899X/217/1/012028

Table 4. Feed water specifications

Parameters	Values
Feed water temperature economizer inlet	210°C
Feed water temperature economizer outlet	260.9°C
Feed water pressure economizer inlet	135.1 kg/cm^2
Feed water velocity inside the tubes	0.31 m/s
Total number of parallel paths on water side	74
Water flow to Economiser	110 t/h MCR (Maximum continuous rating)

2. Methodology

The basic steps involved in solving any CFD problem are as follows:

- Identification of flow domain.
- Geometry Modeling.
- Grid generation.
- Specification of boundary conditions and initial conditions.
- Selection of solver parameters and convergence criteria.
- Results and post processing.

Figure 4. shows the overall view of the CFD approach.

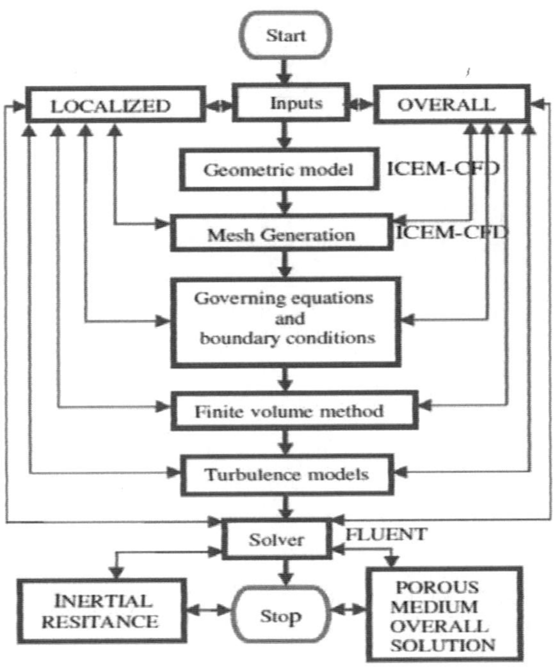

Figure 4. CFD approach

2.1 Identification of flow domain

Before constructing modeling, it is required to understand the exact flow domain of an economizer. In the model, water flows through tubes and the surrounding space was flowing with flue gasses. Beside symmetry conditions, velocity inlet and pressure outlet were defined for both the zones. Convection and conduction heat transfer was considered and the suitable mesh was created for this purpose. A schematic of the economizer for multipass counter cross-flow inline bare tube arrangement is depicted Fig.5

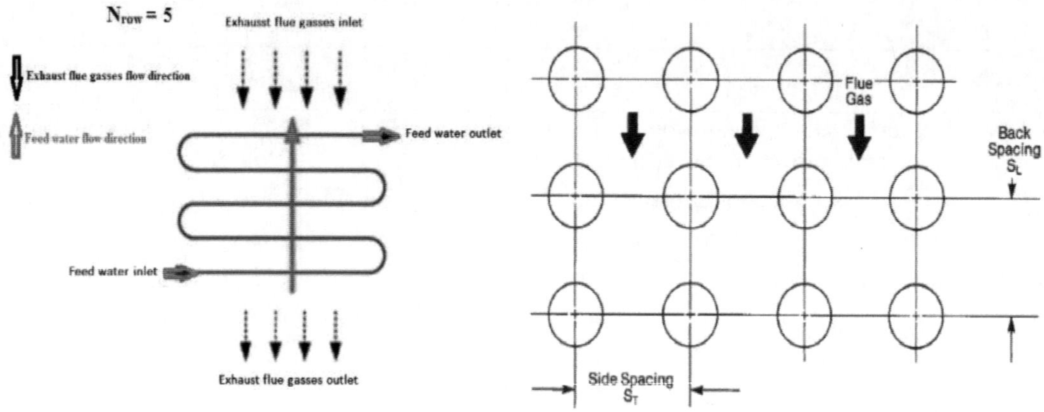

Figure 5. Schematic diagrams of the economizer for multipass counter cross-flow inline bare tube arrangement (N_{row}=5)

2.2 Geometric modeling

The software that is used for generating the geometry and meshing is decided based on nature and complexity of the geometry. The most common and reliable economizer design is the bare tube, in-line horizontal multipass counter cross-flow economizer as shown in Figure. 6. When a coal is fired, the fly ash (flue gasses) creates a high fouling and erosion environment. The advantage of the bare tube, the inline arrangement was it minimizes the erosion and trapping of ash. It also easiest geometry to kept clean by soot blowers. The computational domain of the inline bare tube economizer is shown in Fig 6 and the physical properties of porous media are tabulated in Table 5. and Fig 7

.

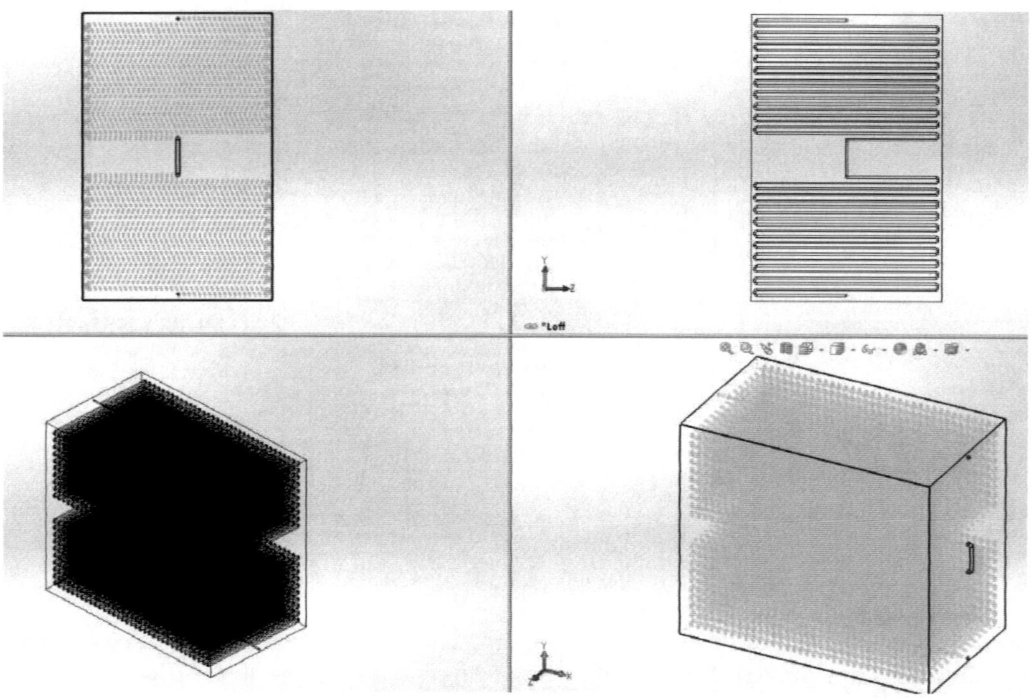

Figure 6. Solid geometric model of an economizer with four viewports.

Table 5. Boundary conditions for porous media

Parameters	Values
Porous media length	3m
Total heating surface area	1915 m^2
Volume of the porous zone	85.48 m^3
Overall heat transfer coefficient	37.76 kcal/ m^2 hr deg C

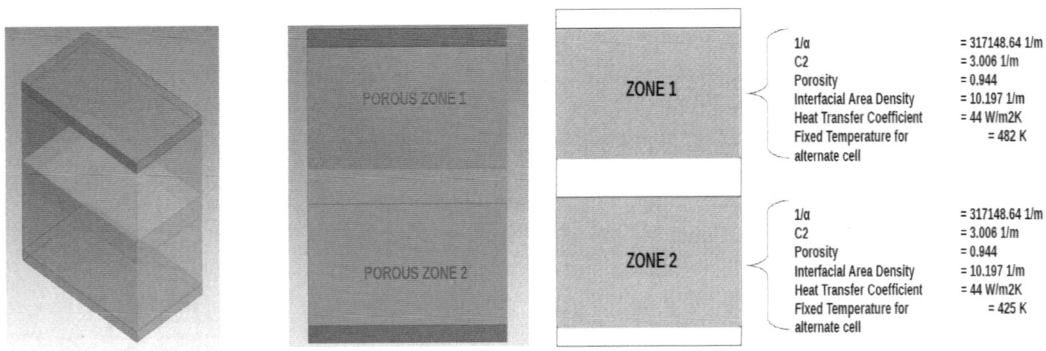

Figure. 7. Geometric details of a model and Boundary conditions for porous zones 1 & 2

2.3 Grid generation.

Meshing is an integral part of the CFD analysis process. The mesh influences the accuracy, convergence, and speed of the solution. For the purpose of flow domain discretization, The economizer model considered for 3D grid generation. The mesh used in this study is shown in Fig 8. The grid dependency checked and the quality was 1, max skewness was $1E^{-10}$, aspect ratio-1.114 and grid having 238896 cells. The grid generated with dual cell method.

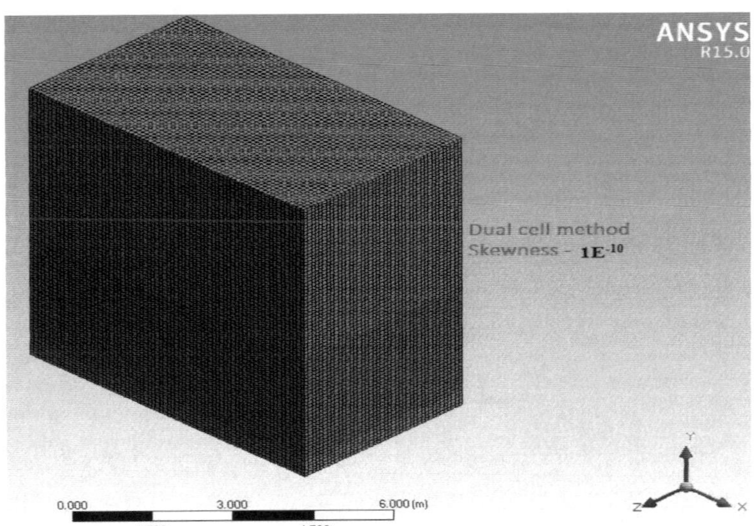

Figure 8. Grid generation with dual cell method

IOP Conf. Series: Materials Science and Engineering **217** (2017) 012028 doi:10.1088/1757-899X/217/1/012028

2.4 Mesh details

Table 6. Mesh statistics

Parameters	Value	Limit
Min orthogonal quality	1	>0.2
Max skewness	$1E^{-10}$	<0.8
Aspect ratio	1.1142	<70
Number of cells	238896	

2.5 Solver

To avoid excessive computational demands of full monolith modelling, this paper is an attempt to apply and demonstrate the porous medium approach for predicting the flow and heat transfer characteristics. The incompressible steady state form of governing equations for mass conservation, momentum balance, and energy conservation together with the equations for modelling turbulence are solved using Ansys Fluent 17.1.The pressure correlation approach using the SIMPLE algorithm is adopted.

3. Results and Discussions

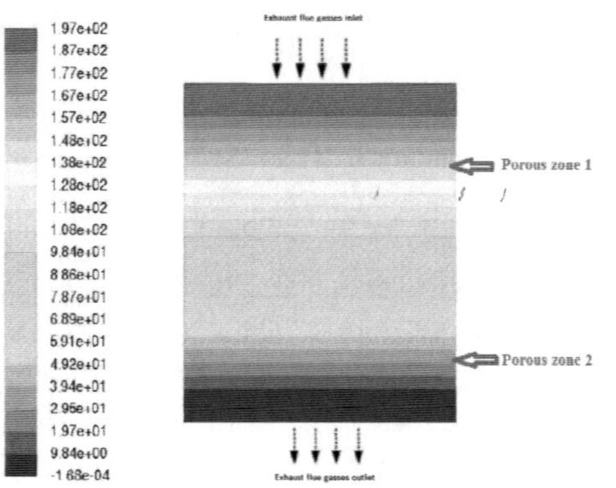

Figure 9. Variation of contour pressure plots

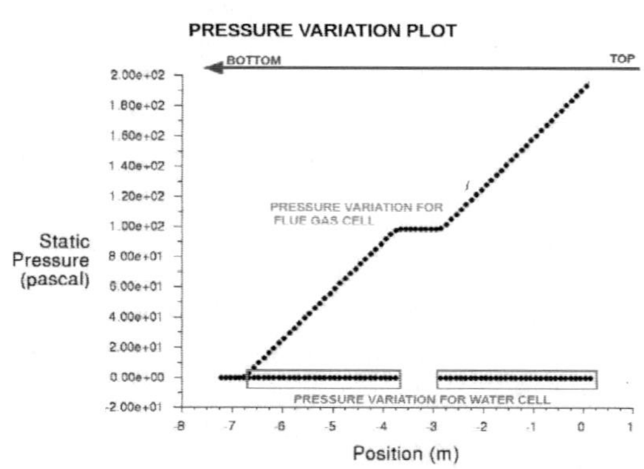

Figure 10. Pressure drop variation of flue gasses

Figures 9 and 10 shows the contours of the pressure of the flow from top to bottom. The variation of pressure across the porous zones 1 and 2 decreasing from 202.21 Pa to 5.45 Pa.

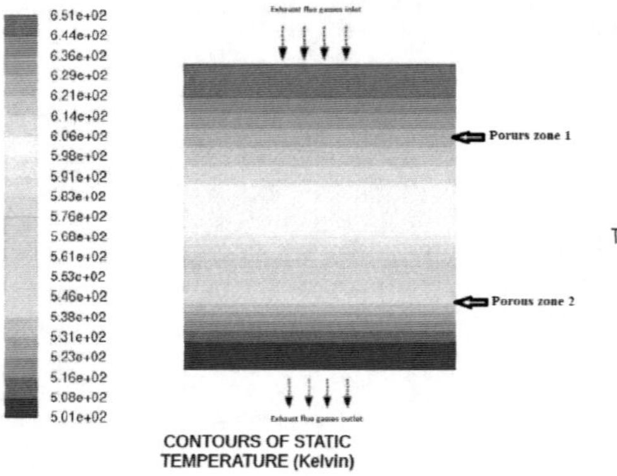

Figure 11. Variation of contour temperature plot

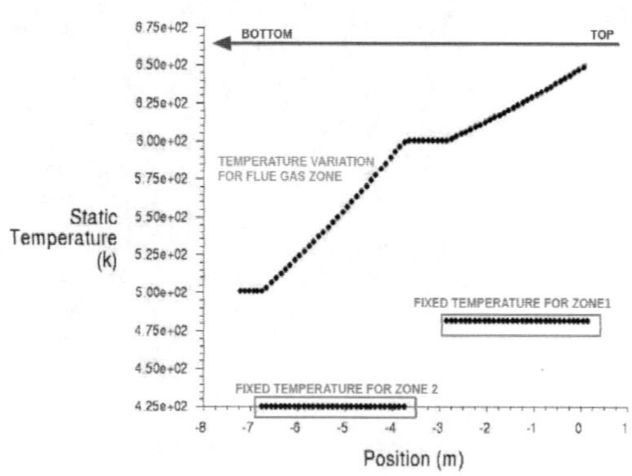

Figure 12. Temperature drop variation of flue gasses

Figures 11 and 12 shows the contours of the temperature of the flow from top to bottom. The variation of temperature across the porous zones 1 and 2 decreasing from 651°K to 501°K. with a flue gas velocity of 5.61 m/s, the flow direction is from top to bottom.

4. Conclusions

The study numerically examines the geometric parameters of a multipass counter cross-flow economizer on the flue gas side performance of a 37-row bare tube inline arrangement economizer. The main purpose of this article was to show that CFD can be a helpful tool in analyzing the economizer. The gas is observed to be uniformly circulated at the heating surface. The predicted values of gas velocity, temperature, and pressure distributions through the heat transfer bundles indicate the acceptable distribution of flow in the heat transfer.

Numerical simulation has been conducted for a porosity range of 0.94. The results obtained show that the amount of heat transfer from the flue gasses to water from 651°K to 501°K with the heat transfer area of 1915 m^2 with a flue gas velocity of 5.61 m/s, and pressure variation from 202.21 Pa to 5.45 Pa. The forming of foulant is mostly on the front part of the bare tubes. The developed empirical model could be used to predict the thermal resistance due to the fouling. The simulation results show that the overall heat transfer coefficient 37.76 W/ (m^2K) and economizer coil side pressure drop of 0.2 (kg/cm^2) are found to be conformity within the tolerable limits when compared with existing industrial economizer data.

References

[1] Tu, J., C. Fletcher, and M. Behnia, *NUMERICAL MODELLING OF THREE-DIMENSIONAL FLY-ASH FLOW IN POWER UTILITY BOILERS.* International journal for numerical methods in fluids, 1997. **24**(8): p. 787-806.

[2] Ross Brindle, J.E., Amanda Greene, *Improving the Thermal Efficiency of Coal-Fired Power Plants in the United States*. 2010: Sheraton Baltimore City Center Baltimore, MD.

[3] Kareem, Z.S., et al., *Passive heat transfer enhancement review in corrugation.* Experimental Thermal and Fluid Science, 2015. **68**: p. 22-38.

[4] Pongsoi, P., S. Pikulkajorn, and S. Wongwises, *Heat transfer and flow characteristics of spiral fin-and-tube heat exchangers: A review.* International Journal of Heat and Mass Transfer, 2014. **79**: p. 417-431.

[5] Hofmann, R., F. Frasz, and K. Ponweiser. *Experimental Analysis of Enhanced Heat Transfer and Pressure-Drop of Serrated Finned-Tube Bundles With Different Fin Geometries.* in *Proceedings of the 5th WSEAS International Conference on Heat and Mass Transfer (HMT08), WSEAS*. 2008.

[6] Rad, S.E., H. Afshin, and B. Farhanieh, *Heat Transfer Enhancement in Shell-and-Tube Heat Exchangers Using Porous Media.* Heat Transfer Engineering, 2015. **36**(3): p. 262-277.

[7] Beychok, M. *Steam generator.* 2013; Available from: http://en.citizendium.org/wiki/Steam_generator.

[8] Nuntaphan, A. and T. Kiatsiriroat, *Thermal behavior of spiral fin-and-tube heat exchanger having fly ash deposit.* Experimental thermal and fluid science, 2007. **31**(8): p. 1103-1109.

[9] Frantisek BRUMERCIK, D.S., Michal LUKAC,Aleksander NIEOCZYM, Stawomir WIERZBICKI, *Modeling of flow and temperature filed in an economizer.* Applied computer science, 2016. **12,no.2**: p. 63-73.

[10] Patankar, S. and D. Spalding, *A calculation procedure for the transient and steady-state behavior of shell-and-tube heat exchangers*. 1974: McGraw-Hill, New York.

Acknowledgment

I would like to thank all my thesis committee members and I would like to express my gratitude to the Curtin University Malaysia for providing Curtin Sarawak Postgraduate Research Scholarship (CSPRS).

Distribution network connection pricing framework and methodology: identification of areas of improvement for Sarawak Energy Berhad Connection Charges Guidelines through modified delphi method

J K Tan[1,2] and N Abas[1]

[1] UTM Razak School of Engineering and Advanced Technology, Universiti Tekonologi Malaysia, Jalan Sultan Yahya Petra, Kuala Lumpur, Malaysia
[2] Sarawak Energy Berhad, Kuching, Sarawak, Malaysia

E-mail: jookok@gmail.com

Abstract: Complaints on issues and matters related to connection charges have been very common for electricity supply utility companies around the world including Sarawak Energy Berhad. In order to identify the areas that can be improved, a mixed method of exploratory research involving qualitative and quantitative methods have been designed and undertaken rather than a single method of survey. This will ensure a more comprehensive and detailed understanding of the issues from various target groups. The method is designed under three phases, employing Modified Delphi Technique for phase 1 through a series of stake holder engagements, online and offline survey questionnaires to be filled by internal wiring contractors for phase 2 whilst under phase 3, case studies shall be carried out on the issues identified from phase 1 and phase 2 of the study. This paper presented the findings from the Modified Delphi Technique. The findings revealed that there are areas of improvement for Sarawak Energy Berhad connection guidelines in term of differentiation of dedicated and shared assets which leads to unfairness to the connecting customers, inconsistency and non-transparent in charging. The findings of Modified Delphi Technique shall be used for implementation of phase 2 and phase 3 of the study.

1. Introduction

Complaints on connection charges issues related to electricity supply are prevalent throughout the world for electricity supply utility companies such as Sarawak Energy Berhad (SEB), as this involves dollars and cents that customers pay to get connected to the distribution electrical network.

Some of the common issues raised are mostly related to the pricing, technical problem or the related application process and procedures, which has direct impact to the connecting customers.

In order to gain a more understanding on this topic and to explore any deficiency in SEB connection charges, a mixed method of exploratory research was employed, in view of the fact that connection charges is a technical issue but it is not totally an engineering problem. The conventional use of single survey method may not be adequate to fully reflect the true picture of the topic under study. An extensive review on mixed method could be found by studies published by Plano Clark et. al.[1].

2. Proposed Research

The study was undertaken via reviewing and understanding connection charges policy of SEB. Identification of areas for improvement on SEB connection charges guidelines were undertaken in phases through dialogues, engagement, questionnaires and case study for various groups of customers. This paper will discuss the results/findings based on the surveys done using the Modified Delphi Method.

3. Literature Review

3.1 Typical Connection Frame Work Model

Figure 1 shows the supply connection of a customer from the distribution intake substation for SEB. It is also the typical connection frame work model used for computation of the connection charges and the policies for other utilities.

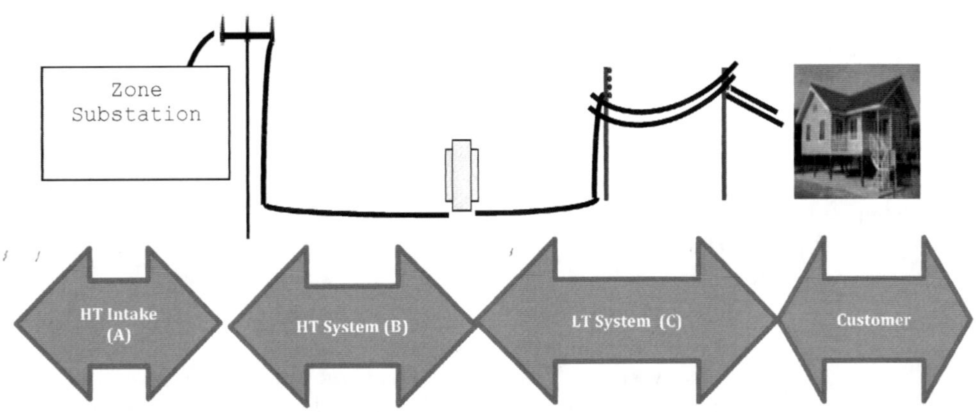

Figure 1. Typical distribution network connection frame work model for computation of connection charges

It is made up of the following chargeable sectors:

Development of High Tension (HT) intake (33KV substation): Covers the development of major distribution substation, typically at 33KV and with voltage transformation to 11KV before distributing to various customers as indicated in Figure 1 as "A".

Development of High Tension (HT) system: Covers the development of HT system which includes the underground cables, overhead lines, transformers and all related accessories as indicated in Figure 1 as "B".

Development of Low Tension (LT) system: Covers the development of LT system which includes the underground cables, overhead lines and all related accessories as indicated in Figure 1 as "C".

3.2 SEB Connection Charges Guidelines

Table 1 summarized the key components of current connection charges guidelines of SEB based on components spelt out in figure 1, extracted from their connection charges guidelines [2].

Item	Description
HT Intake (33KV substation)	Case to case basis. No clear guidelines on the charging methodology
HT System	Covers the: - Connection to existing HT network - Extension of HT network - Installation of HT transformer Housing and shop houses development: - Capacity charge of RM500/KVA + Actual HT costs (if more than 1km from the nearest HT connection point) Single residential, commercial and industrial premise: - Capacity charge of RM500/KVA + Actual HT charge for HT cost that exceed 300% of the capacity charge + Actual HT costs (if more than 1km from the nearest HT connection point) Government and temporary supply: - Estimated project actual cost
LT System	Estimated actual project cost
Salient feature	Key Features: - A fixed RM500 / KVA capacity chare imposed on all customers regardless of if HT development is required - Subsidy (up to 300% of capacity charge paid) on HT development cost for single premise customers and unlimited subsidy cap (Differences between total HT project cost and capacity charge paid) for development projects - Inconsistency in charging HT intake (33KV substation) construction

Table 1. Key components of current charging methodology of SEB as per extracted from their connection charges guidelines

3.3 Generic Characteristics of Connection Charges

Connection charges cover all categories of customers that require connection of supply to utilities' network, at various voltage level. In a study by Mark Cox [3], fourteen (14) countries covered under his study revealed a few issues regarding their respective connection charges policies which include: inappropriately designed rules, complex procedures, in-transparent and high connection costs. All these issues can create barriers for the realization of new connections, especially to household customers. G Strbac, J Mutale and D. Pudjianto [4] also raised the same concerns regarding the connection charges from the perspective of the regulator, network companies and the customers.

In order to deliver the most appropriate network pricing (tariff and connection charges) to customers of electricity, and to ensure that the utilities is able to operate the system efficiently as well as are able to sustain their business on long term basis, it is necessary to ensure that during computation or revision of network pricing, it shall be able to meet the utility's financial objectives and to address the issues faced by the utility from time to time.

G Strbac, J Mutale and D. Pudjianto [4] presented in their working report to Office of Gas and Electricity Market (OFGEM) on the fundamental objectives for computation of network pricing that will address the issues faced by the regulators, network operators as well as the customer.

In their report, it is mentioned that in order to achieve effective economic efficiency, the computed network pricing shall distinguish between different user categories and at different locations at which a customer is connected in order to avoid cross subsidies between customers.

On the other hand, from the utility company point of view, the network pricing shall also reflect an efficient network investment and to avoid over investment that will affect the company's financial standing. A network pricing based on shallow approach will trigger additional investment from the utility companies hoping that the upfront investment can be recovered through long term collection from the customers through tariff. Likewise, a deep approach will trigger collection of fund from the customer to develop the system for future use.

From the regulator point of view, excessive network charges are always the main culprit of complaints. Though lower the network prices would solve the problem, the utility companies may not be able to generate enough revenue to invest in network operation and development thus affecting the service quality to the customers. With the appropriate network pricing computed, it will enable the company to deliver the required revenue to ensure company sustainability in operating, maintaining and upgrading the assets.

However, besides considering the network pricing based on the financial standing of the company, G Strbac, J Mutale and D. Pudjianto [4] mentioned that it is necessary to ensure that the network pricing is stable and predictable. This is an important aspect that influence a connecting customer's decision. A right balance must however be struck between price stability and flexibility allowing prices to respond to changing situations.

In addition to this, the network price shall also be transparent, auditable and consistent [5][4] to allow the interested connecting customer to understand the structure of the fee and to make necessary decision for connection to the existing system.

Guy Nicholson, an independent specialist in the Grid Integration of renewable Energy, who has carried out design and costing studies for 8 out of 13 UK offshore windfarms commented that the connection charges may not need to be low but it must be reasonably charged based on the Use of System (UoS) [5]. It is also supported by G Strbac, J Mutale and D. Pudjianto [4] where it is mentioned that the connection charges should be charged based on the capacity used or applied for. This is to ensure that all customers are fairly treated.

Mark Cox [3], in his working paper to OFGEM commented that in order to determine an appropriate framework to sustain for long term distribution charging, a few charging principles shall be considered which are all in line with [4][5], which can be summarized as below:

- Cost reflectivity
- Simplicity
- Transparency
- Predictability

3.4 The Delphi Technique

The Delphi Technique is a widely accepted method for gathering opinions from experts in particular area. It was originally created in 1950 by the RAND Corporation [6]. It is a process of gathering information and opinions from the respondents within their area of expertise through several iterations of communication with the respondents.

It is carried out by gathering information in the first iteration of communication. The information is consolidated and to be used to generate more detailed questionnaires in the subsequent round of communication. The iteration process will continue until convergence is achieved.

Typically, the Delphi Techniques comprises of four rounds of iterations of communication and the conventional method of communication used are through questionnaire to aggregate the opinions of a group of individual. There are now many different types of Delphi Technique. For this study, the modified Delphi Technique was used. Instead of conducting the survey through questionnaires, the modified Delphi Technique will comprise of engagements, dialogue sessions and questionnaires.

In theory, the process of the Delphi can be optimized with more iterations until a perfect consensus is achieved. However, some researchers have found that three iterations are adequate for data collecting and reaching a consensus [7]. It is thus decided that three (3) iterations for this study. Three iterations of the Delphi Technique included face-to-face engagements and dialogues, written questionnaires seeking for opinions based on the summarized, controlled feedback from the engagements/dialogues and literature review.

Undertaking surveys through Delphi Technique provides several advantages: It offers a platform for investigator to have at least two to four iterations of opinion seeking in a controlled manner also giving the respondents extensive opportunities to provide feedback and to review and to reconsider their initial opinion generated in the previous rounds. In addition to this, Delphi Technique can reduce the effect of "noise"; that is, extraneous or error driven information.

4. Research Methodology

4.1 Target Participants

Representatives who are fluent and familiar with SEB CCG were drawn from various government departments and non-government agencies to make up "the Delphi panel". These departments and agencies are also the ones who have raised concerns over the SEB CCG to Minister of Public Utility (MPU) and have attended the lab session conducted by MPU [8]. The departments and agencies invited are as below:

- Ministry of Housing (MOH) Sarawak
  - Agency that is fluent in any policy dealing with various housing and real estate development
  - Represented by the 2 active officer in charge
- Ministry of Industrial Development (MID) Sarawak
  - Agency that is dealing with all the large, medium and small industrial estate
  - Represented by the 2 active officer in charge
- Association of Consulting Engineers Malaysia (ACEM) Sarawak Branch
  - Association dealing with all the potential connecting customers and to provide technical and financial advices
  - Represented by 4 members of ACEM who are also the active electrical consultants dealing with various electricity connecting customers
- Sarawak Housing And Real Estate Developer Association (SHEDA)
  - Association dealing and ensure the interest of housing and real estate developers is well taken care
  - Represented by the 2 members of active council members

The survey shall be conducted in four different locations, namely Kuching, Sibu, Bintulu and Miri, being the four (4) main city/town of Sarawak. This made up of a total of 40 respondents for the survey.

The usual panel in a Delphi study is typically less than 50 [9]. More subjects will ensure diversity in the reference groups and to ensure that the judgements will not be biased and compromised.

The forty (40) respondents were provided with details of this research and invited to participate in the survey.

4.2 Procedures for Implementation
The three (3) rounds of the modified Delphi Technique were coordinated using email and face-to-face engagements to introduce the study and its goals, to answer questions and to invite participation in the study. All respondents were maintained anonymous throughout the whole study.

Round One (R1) of Delphi implementation was undertaken through engagements in the form open-ended dialogue sessions with an explanation of the study's purpose. It is an information gathering session and all respondents are free to give comments and feedback on any issues they faced with SEB CCG.

This first round of Delphi Technique aimed to sought the Delphi panel's views on the preliminary findings from MPU lab session on their concerns over SEB CCG. After receiving the responses from the Delphi panel, the collected information is analysed individually using content analysis for themes and overlap. A consolidated summary of the findings shall be used as the basis for ***Round Two (R2)*** questionnaires formulation.

In ***Round Two (R2)***, statements from R1 were used to compose questionnaire requesting the panel to review the summarized items discussed in R1. All the panels were asked to reconsider and to revisit the information feedback in R1 to ensure all relevant issues are covered. This round was done through face-to-face engagement. The panel was then asked to rate the impact of each issue and to establish priorities for if the issue can only be solved one-by-one using a 4-point scale, ranging from 1 (least impact/priority) to 4 (high impact/priority). The respondents are also offered the opportunity to comment on any items shall it was missed out during R1 engagement.

The impact level as well as the priority of each issue raised were computed and ranked. The summary of response from R2 was transmitted to respondents for the final consideration in R3.

In ***Final Round (R3)***, the respondents received a summary of responses from R2 through presentation for their final consideration of the issues raised in SEB CCG. All respondents are given another opportunity to make further clarifications of any issues before a consensus was reached. The respondents were also asked to give comments on how would the issues can be solved and what are the strategies that can be taken on implementing SEB CCG.

The returned responses of R3 were summarized to reach for a consensus. The high impacts issues that should be prioritized were identified. The potentials solution proposed by the respondents were also summarized and categorized into different categories. Focus area for the model construction is identified.

***4.3** Implementation Plans and Timelines*
The whole Delphi Process shall be carried out in four main stations, one month apart for each station. one months upon completion of the last engagement sessions, R2 questionnaires shall be sent out to the relevant respondents to feedback within a month. R3 questionnaires shall then be sent out 1 month after all the R2 questionnaires were all returned. The whole process shall take up to maximum of 6 months.

Out of the 40 participants, 35 of them completed all three rounds of the modified Delphi method, for a response rate of 87.5%.

5. Results and Findings

The Delphi Technique was modified to operate in three rounds to collect data to identify the gap/deficiency in SEB connection charges guidelines. The result of the Delphi Technique is as shown in Table 2:

Rank	Description
1	**Inconsistent treatment of charging:** Different customers applying for the same energy demand but was charged differently
2	**Inconsistency in the charging:** The connecting cost varies from station to station with the same scope of work or connecting to same number of living units
3	**Difficult to budget for connection charges:** The connection cost per living unit come out to different values. This makes it difficult for developers to budget for connection cost.
4	**Positioning of substation:** Developer is often requested to position the substation at a prime location within a development which significantly taken up the opportunity cost of the whole development.
5	**Inconsistent charging of shared and dedicated asset:** Some customers pay full cost of the network development whilst some customers pay portion of the network construction cost
6	**Manipulating of applied load:** SEB connection charging policies allow for customers to pay less when applying for higher energy demand
7	**Different customers make different contribution to shared HT System Development (11KV and below):** RM500/KVA is imposed on all customers regardless of if any augmentation is required on HT system
8	**Land for substation:** The land required or constructing distribution substation and zone substation affect the overall project development cost
9	**Change of Policies/Regulations:** SEB often changes policies without giving adequate lead time for implementation makes for an uncertain market to the developer
10	**Complexity and transparency:** Clients are in the opinion that SESCO connection charges guidelines are being too complex and not transparent to the customers
11	**Channel for appeal:** There is no proper contact to lodge a complaint on any "high" connecting cost imposed by the company
12	**Time frames for approvals of application:** Typical application for electricity supply takes a long time, typically up to 6 months.
13	**Inconsistent / Unclear process for assessing application:** For any application, there is no clear guidelines on how the company assesses an application.
14	**Cost increases driven by SEB:** The price of connecting customers to electricity has been going up over time and it is due to change of SEB policy on electrical designs

15	**No clear guidelines on assisting disadvantaged customer:** Disadvantaged customers pay high connection cost to be connected to SEB network

Table 2. The compiled result of the Delphi Technique.

The issues/limitations computed were categorised into three categories as shown in Table 3 below. For the paper, the study will emphasis on issues related to pricing methodology.

Categories	Description
Processes and procedures	• Inconsistent / Unclear process for assessing application • Time frames for approvals of application • Channel for appeal • No clear guidelines on assisting disadvantaged customer
Pricing methodology	• Inconsistent treatment of charging • Inconsistency in the charging • Difficult to budget for connection charges • Different customers make different contribution to shared HT System Development (11KV and below) • Inconsistent charging of shared and dedicated asset • Manipulating of applied load • Complexity
Technical standards	• Cost increases driven by SEB • Positioning of substation • Change of Policies/Regulations • Land for substation

Table 3. The results of Delphi Technique are grouped into three categories.

a) Inconsistent treatment of charging to different customer

It is found that different customer categories of the same applied load would be charged different amount of connection charge. In general, housing/show houses developers are receiving higher subsidy than single premise, industrial and residential customer. For example, assume a housing development and an industrial customer require exactly the same load on the system (and they require exactly the same HT assets to be constructed to serve that load), if the industrial customer breaches the 300% subsidy threshold, then it bears the full HT costs above that threshold, yet in the same circumstances, the housing development would not (Unlimited subsidy)

b) Inconsistency in the charging

The low tension (LT) portion of SEB connection charges guidelines is based on the estimated actual cost of the construction. This cost may vary between stations for the same scope of work to be carried out. This is mainly due to the fact that the estimated cost may be carried out by different planners. In addition to this, the work charges and the material cost may vary from one station to another stations.

c) Difficult to budget for connection charges

With reference to item b) above, which makes the connection charges to be incurred to the customers vary, this makes the developers of housings or shop houses difficult to budget for the connection charges for each premise. This is more difficult to be achieved with the change of technical standards, which makes the per living unit to vary from RM3,000.00 to RM14,000.00.

d) Different customers make different contribution to shared HT System Development (11KV and below)

For charges on HT system development, a standard charge of RM500/KVA is imposed to ALL customers requesting for connection to the system regardless of whether any HT network augmentation is required. It creates unfairness of charging whereby for small customers that does not require any augmentation to the upstream system or requires no dedicated HT assets, would also have to contribute RM500/KVA. On the other hand, a development which requires connection to the existing HT network and extension of HT network up to 1km dedicated to their development would also pay the same. This indirectly cause the cross subsidy between customers whereby the small customers would subsidise the big customers.

e) Inconsistent treatment of HT/LT assets (Shared and dedicated assets)

HT assets, regardless of whether developed to serve a single development (dedicated) or multiple developments (Shared) are covered by the RM500/KVA (+ additional cost funded by SEB) whereas all LT assets are treated as dedicated assets (and all funded directly by the connecting customer) thus inconsistency in the charging of shared/dedicated asset.

f) Manipulation of KVA applied for

The 300% subsidy rule introduced in SEB connection charges guidelines may encourage some industrial and commercial customers to arbitrarily increase their "required" nominated electrical load (KVA) to reduce their connection charges. This will eventually increase SEB's contribution whilst at the same time leading to the development of oversized asset in the distribution system thus reduction in asset utilization factor.

g) Complexity

Due to the inconsistency of the charging between station and the cost per living unit varies significantly, it is no doubt that it has created the perception that SEB connection charges guidelines is too complicated to be understood and the charging methodology is not transparent to public.

6. Summary and Discussion

The 7 issues/limitations listed under pricing methodology were analysed and it was clearly shown that SEB connection charges guidelines has not demonstrated the four basic criteria of network supply connection models as below, as mentioned in Section 3.3:

- Cost reflectivity
- Simplicity
- Transparency
- Predictability

Revised SEB connection charges guidelines shall comprise of the following fundamental principles:

- Simple and easy to understand charging principle
- Transparency of charging via introduction of schedule of rate for the commonly used scope of work
- Fairer connection charges through "Pay-for-what-you-use" principle
- Apply "least-cost means of providing electricity services to end customers
- Differentiation of charging for dedicated and shared assets
- Equity of charging – Customers require similar assets should impose similar costs, avoiding cross subsidy between customers

Following the above fundamental principle as the frame work, additional work needs to be carried out via:

- Case study to determine and to verify the deficiency of the current connection charges policy

- Financial modelling and financial impact study on the current connection charges policy to determine the current and future financial position of continuing to practice the current policy

7. Reference

[1] V. L. Plano Clark, C. a. Huddleston-Casas, S. L. Churchill, D. O'Neil Green, and a. L. Garrett, "Mixed Methods Approaches in Family Science Research," *J. Fam. Issues*, vol. 29, no. 11, pp. 1543–1566, 2008.
[2] S. E. B. Distribution Services Division, "New Connection Charges Guidelines," vol. 2008, pp. 1–6, 2008.
[3] M. Cox, "Structure of electricity distribution charges: Consultation on the longer term charging framework," 2005.
[4] G. Strbac, J. Mutale, and D. Pudjianto, "Methodology for pricing of distribution networks with distributed generation," *18th Int. Conf. Exhib. Electr. Distrib.*, no. June, pp. 6–9, 2005.
[5] G. Nicholson, "Structure of Charges Workshop Distributed Generation Issues."
[6] D. H. G. A.L. Delbecq, A. H. Van de Ven, *Group Techniques for Program Planning: A guide to nominal group and delphi processes*. United States of America: Scott, Faresman and Company, 1975.
[7] B. R. S. R.L. Custer, J.A. Scarcella, "The modified Delphi technique: A rotational modification," *J. Vocaional Tech. Educ.*, vol. 15(2), pp. 1–10, 1999.
[8] M. of P. Utility, "Lab For SEB/SESCO Connection Charges," in *Lab For SEB/SESCO Connection Charges, 19 - 21 March 2012, Four Points Hotel Kuching*, 2012.
[9] J. W. A. B.R. Witkin, *Planning and conducting needs assessment: A practical guide*. Thousand Oaks, CA: Sage Publications, Inc., 1995.

Acknowledgement
The authors would like to thank Sarawak Energy Berhad for the support, who provided insight and expertise that greatly assisted the research and acknowledged Universiti Technology Malaysia and MOHE for supporting this research.

IOP Conf. Series: Materials Science and Engineering **217** (2017) 012030 doi:10.1088/1757-899X/217/1/012030

Electricity Breakdown Management for Sarawak Energy: Use of Condition-Based Equipment for Detection of Defective Insulator

J K Tan[1,2] and N Abas[1]

[1] UTM Razak School of Engineering and Advanced Technology, Universiti Tekonologi Malaysia, Jalan Sultan Yahya Petra, Kuala Lumpur, Malaysia
[2] Sarawak Energy Berhad, Kuching, Sarawak, Malaysia

E-mail: jookok@gmail.com

Abstract: Managing electricity breakdown is vital since an outage causes economic losses for customers and the utility companies. However, electricity breakdown is unavoidable due to some internal or external factors beyond our control. Electricity breakdown on overhead lines tend occur more frequently because it is prone to external disturbances such as animal, overgrown vegetation and defective pole top accessories. In Sarawak Energy Berhad (SEB), majority of the network are composed of overhead lines and hence, is more prone to failure. Conventional method of equipment inspection and fault finding are not effective to quickly identify the root cause of failure. SEB has engaged the use of corona discharge camera as condition-based monitoring equipment to carry out condition based inspection on the line in order to diagnose the condition of the lines prior to failure. Experimental testing has been carried out to determine the correlation between the corona discharge count and the level of defect on line insulator. The result shall be tabulated and will be used as reference for future scanning and diagnostic on any defect on the lines.

1. Introduction

The world's requirement for a more reliable electricity supply has been ever increasing, nevertheless electricity supply utilities have not been able to effectively reduce service disruptions due to unplanned outages on lines. The causes of unplanned outages were studied and are categorized as animal, weather, vegetation, human intervention and unknown [1]. Various approaches and initiatives have been undertaken to mitigate and to reduce service downtime due to unplanned outages on lines.

Sarawak Energy Berhad (SEB), a utility company in south east region, is responsible for the generation, transmission and distribution of electricity for the state of Sarawak, Malaysia [2]. While majority of its systems are made up of overhead lines, SEB strives to minimize the occurrences of unintended outage on lines. In order to improve the reliability of the lines and to effectively identify the root cause of the outages, SEB is looking into improving company maintenance philosophy from reactive maintenance and time based maintenance to predictive maintenance. This is mainly due to the fact that it has been found that equipment failed before the next scheduled maintenance date.

This paper discusses the use of corona discharge camera as part of the diagnostic maintenance on overhead lines to detect any abnormalities prior to outages.

2. Proposed Research

The study shall be undertaken via experiments to determine the effectiveness of the corona discharge camera in detecting any abnormalities in the overhead lines.

3. SEB Distribution System

Due to the geographical nature of Sarawak, and the customers are sparsely populated, 75% of the distribution system is made up of overhead lines with a total circuit length of 24,031.49KM [3]. Table 1 shows the breakdown statistic for SEB from year 2014 to 2016 [4] based on circuit category. It is noticed that unplanned interruption due to overhead lines is increasing for the last three years up to 69.27% at year 2016.

Circuit Category	2014	2015	2016
Overhead lines	***62.61%***	***62.72%***	***69.27%***
Underground cable	9.79%	8.40%	9.21%
Equipment failure	5.45%	6.76%	5.04%
Others	22.15%	22.13%	16.48%
Total:	100%	100%	100%

Table 1. Breakdown statistics (in %) by circuit category from 2014 - 2016

Table 2 shows the contribution of unplanned outage on overhead lines on SAIDI for the year 2014 to 2016 [4]. Breakdown due to overhead lines has contributed to over 50% of the total distribution SAIDI.

Circuit Category	2014	2015	2016
Breakdown on overhead lines	***66***	***59***	***58***
Distribution SEB	123	103	87
(%)	53.7%	57.3%	66.7%

Table 2. SAIDI contribution from outages on overhead lines from 2014 – 2016

Based on the statistics in table 1 and 2, it is clearly shown that the performance of overhead lines circuit has significant impact on the company performance. Further analysis and data gathering were undertaken on SEB breakdown statistics, emphasising on identifying the root cause of the unplanned outage on overhead lines. Table 3 shows the root cause of all outages from the year 2014 to year 2016 [4].

Analysis results in table 3 show that unplanned outage on overhead lines due to transient fault contributed to over 40% of the total outages. In view of the significant impact of unplanned outages due to transient fault (Cause unknown) on overhead lines, it is necessary to relook into various maintenance methodologies to reduce this figure and to improve the line performance.

Cause of outages	**Year**		
	2014	**2015**	**2016**
Excavation	1.09%	1.74%	2.60%
Public tree clearing	1.36%	1.19%	1.73%
Vehicle	1.77%	2.43%	2.83%
Aerial cable fault	3.52%	2.61%	2.62%
Cable fault	3.91%	3.53%	3.57%
Equipment failure	5.45%	6.76%	5.04%
Lightning	**15.81%**	**16.89%**	**17.61%**
Transient	**37.41%**	**38.18%**	**42.41%**
Vegetation	4.29%	4.42%	2.83%
Protection non-discrimination	0.02%	0.05%	0.05%
Mis-operation	0.09%	0.17%	0.05%
Animal	7.56%	6.61%	7.43%

Table 3. Cause of outages for SEB from 2014 – 2016

3.1 Diagnostic Maintenance

Diagnostic maintenance was first proposed in 1978 by the Electric Power Research Institute (EPRI), also known as predictive maintenance [5]. This maintenance is done through state monitoring tools, diagnostic equipment and health level to determine the condition of the equipment and the best time to carry our maintenance and repair. There are various types of diagnostic maintenance tools used for determining the condition of the overhead lines as well as to detect the location of fault on the overhead lines. In most European countries, there is also ongoing reform of the maintenance system towards condition-based maintenance.

There are various diagnostic maintenance tools currently in use. The first and the most common one used by many utilities is the overhead lines infrared thermography scanning camera. It is put into use since 2010 through which a number of major defects were detected easily [6]. On the other hand, corona discharge scanning has also been widely used for detection of fault and to determine the condition of the equipment on the line.

Comparing to corona discharge scanning, infrared thermography scanning has been used in a wide variety of fields such as medical, military and electrical equipment diagnostic. Well-established Delta-T acceptance criteria from the International Testing Association Maintenance Testing Specifications, 1997 (NETA MTS-1997) has also been set up to ease on evaluation of site findings before any action (repair maintenance works) needs to be taken [7]. However, no similar acceptance criteria on corona discharge count that has been set up.

Sarawak Energy Berhad has purchased one (1) set of CoroCAM 504, a corona discharge camera developed by South Africa's Eskom Transmission. This CoroCAM is able to detect corona discharges released from an equipment and to quantify the number of photons reaching the camera into a value called photon count.

Scanning was carried out on frequently tripped feeders and lots of data were gathered. All the pin insulator detected with counts were shut down for inspection but no abnormality was found. It is thus necessary to find out any correlation between the photo count and the severity of insulation failure within an equipment in order to carry out "just-in-time" maintenance. The findings will be able to provide a better understanding and be able to interpret the scanning result accurately and effectively.

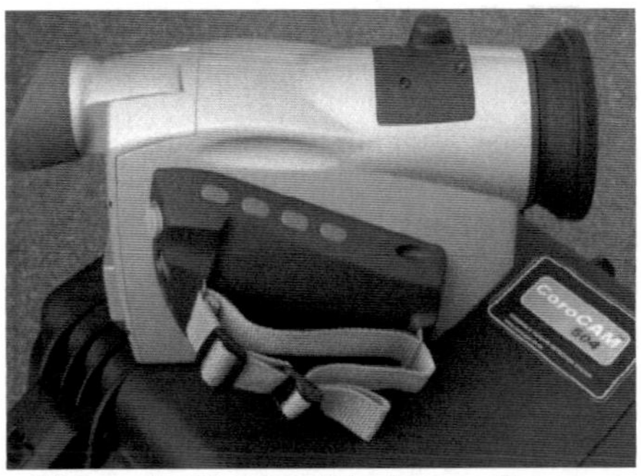

Figure 1. Corona Discharge Camera CoroCAM 504.

Figure 2. Scanned image on SEB overhead lines extracted from CoroCAM 504

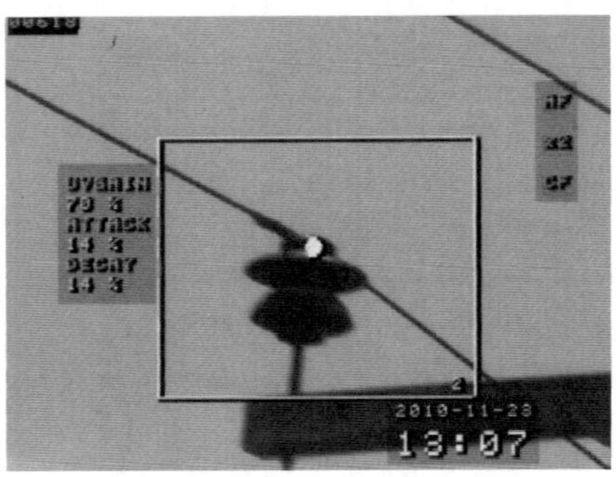

Figure 3. Scanned image on SEB overhead lines extracted from CoroCAM 504

Figure 4. Scanned image on SEB overhead lines extracted from CoroCAM 504

4. Experimental Methods

A preliminary study through field data collection and laboratory experiments shall be carried out to determine the corona count acceptance criteria to ease the field crews when evaluating their findings.

4.1 Laboratory Set Up

The testing shall be carried out at Sarawak Energy Berhad high voltage laboratory located at Jalan Belian, Kuching, Sarawak.

4.2 *Equipment*

Throughout the laboratory testing, CoroCAM 504 shall be used as the tools for carrying out the scanning work. The high voltage equipment used is Hipotronics high voltage test set.

4.3 *Test Sample*

IOP Conf. Series: Materials Science and Engineering **217** (2017) 012030 doi:10.1088/1757-899X/217/1/012030

For this study, the pin insulators of the same brand rated at 36KV shall be used for the testing. The total number of pin insulators required are as follow:

- **New pin insulators:** One (1) number (The new pin insulator shall be used as the control specimen) and shall be numbered ***0001***.

- **Old/crack/used pin insulators:** ten (10) numbers with flash over sign but without visible crack on the pin insulator skirt and shall be numbered ***0002 to 0011***.

4.4 *Test Method*
All the pin insulators shall be installed based on SEB Overhead Lines Construction Standards. This is mainly to simulate the actual condition on site. It shall be tested under dry condition. All tests are carried out

***Test 1:**
Sample: New pin insulator 0001
Test voltage: **19KV and up 36KV or when the test set trips

***Test 2:**
Sample: Used pin insulator 0002 – 0011
Test voltage: **19KV and up to 36KV or when the test set trips

***Test 3:**
Sample: Used pin insulator 0002 – 0011*
Test voltage: **19KV and up to 36KV or when the test set trips

*The first layer of all the used pin insulators were intentionally broken
**19KV is normal phase to ground system operating voltage

All the tests started with injection voltage of 6.5KV and the voltage was raised at 0.5KV every one (1) minute. Shall the test equipment trips within the one (1) minute, the test will stop.

5. Experimental Results and Discussion
The results of the tests were tabulated as in table 4:

Test	Injected voltage (KV)	Photon Count
Test 1: 0001	19KV	Nil
Test 1: 0001	36KV	Nil
Test 2: 0002 – 0010	19KV	Nil
Test 2: 0002 – 0010	36KV	30 – 47
Test 3: 0002 – 0010	19KV	35 – 60
Test 3: 0002 – 0010	*31KV to 33.5KV	475 – 533

*the high voltage test set tripped at the max voltage of 31 to 33.5KV

Table 4. Experimental Test Results of various scenarios.

Figure 5. Pin insulator chosen with flash over sign (To be used for Test 2)

Figure 6. The pin insulator installed according to SEB overhead lines construction standard

Figure 7. The used pin insulator with broken skirt (To be used for Test 3)

Figure 8. Scanned image of corona discharge count from one of the specimen under test for Test 3

6. Discussion and Conclusion

Based on the findings of the above result, it is found that:

- Corona discharges are hardly visible for system voltage below 19KV for normal pin insulator and those pin insulators that has flash over sign.
- With all the normal conditions remained unchanged, any photon count above 30 may indicate insulation failure on pin insulator such as hair line crack or broken skirts and should be replaced immediately before it gets worse, as this may cause outages.

Preliminary results found from this study triggered for a more detailed study with the intention to set up a table of corona count acceptance table for ease of troubleshooting. Some of the possible work that can be carried are:

- To conduct the test with more specimens with various defects
- To conduct the test by simulating different weather condition (Raining, lightning days)
- To conduct the test with lower voltage level pin insulator (11KV)

Reference

[1] D. Barrett, "Wildlife Induced Outages And Protection of Overhead Lines And Substations," *Wildl. Asset Prot. White Pap.*, 2015.
[2] S. E. Berhad, "About Us," 2013. [Online]. Available: www.sarawakenergy.com.my. [Accessed: 28-Jun-2013].
[3] M. & O. Standards, "Distribution Asset Management Report 2016," 2016.
[4] A. D. & Performance, "Breakdown by Category (2014 - 2016)," 2017.
[5] F. X. Li, "Distribution Network Online Condition-Based Maintenance Techniques of exploration," *New Technol. New Prod.*, vol. 1, no. 3, pp. 10–11, 2012.
[6] Y. C. C.M. Ji, "Condition-based maintenance management of distribution network equipment," *Hebei Electr. Power*, vol. 4, no. 5, pp. 6–7, 2005.
[7] Infraspection Institute, "Standard for Infrared Inspection of Building Envelopes," *Infraspection Inst. Stand.*, 2008.

Bearing faults identification and resonant band demodulation based on wavelet de-noising methods and envelope analysis

Ahmed M Abdelrhman[1,*], Yong Sei Kien[1], M Salman Leong[2], Lim Meng Hee[3], Salah M Ali Al-Obaidi[2]

[1]Curtin University Malaysia, CDT 250 98009, Miri, Sarawak, Malaysia
[2]Institute of Noise and Vibration, University Technology Malaysia, Malaysia
[3]UTM Razak School of Engineering and Advanced Technology, Malaysia

E-mail: ahmed.mohammed@curtin.edu.my

Abstract. The vibration signals produced by rotating machinery contain useful information for condition monitoring and fault diagnosis. Fault severities assessment is a challenging task. Wavelet Transform (WT) as a multivariate analysis tool is able to compromise between the time and frequency information in the signals and served as a de-noising method. The CWT scaling function gives different resolutions to the discretely signals such as very fine resolution at lower scale but coarser resolution at a higher scale. However, the computational cost increased as it needs to produce different signal resolutions. DWT has better low computation cost as the dilation function allowed the signals to be decomposed through a tree of low and high pass filters and no further analysing the high-frequency components. In this paper, a method for bearing faults identification is presented by combing Continuous Wavelet Transform (CWT) and Discrete Wavelet Transform (DWT) with envelope analysis for bearing fault diagnosis. The experimental data was sampled by Case Western Reserve University. The analysis result showed that the proposed method is effective in bearing faults detection, identify the exact fault's location and severity assessment especially for the inner race and outer race faults.

1. Introduction

Rolling element bearings are classified as the most important and most common component in rotating machinery [1]. It is reported that failure in machinery is commonly caused by the faulty bearings. Figures show that faults in bearing causes approximately 45% in machine breakdown and 40% of problems in rotating machinery [2]. In today's industries, the faults diagnosis of the rotating machinery is becoming a challenging problem and it has gained the attention from worldwide as it is realized that the early faults detection is essential for proper maintenance and prevent catastrophic failures. Therefore, research on the fault diagnosis and fault detection are still ongoing and they are gaining wide attention to prevent sudden failures.

The vibration signals and acoustics emission (AE) is the most commonly used techniques in condition monitoring and diagnosis of rotating machinery. The possibility of machinery failure such as mass unbalance, shaft misalignment, gear failures and bearing defects could be detected using vibration signals of a machine operating either in good or faulty condition [3]. The vibration signals contain rich information and it has been widely used for condition monitoring without interfering the machine operations. So, the vibrations signal are favoured to become the common approach for rotating machinery condition monitoring and fault diagnosis [4].

Vibration signals of the bearing could be classified into stationary and non-stationary signals. In reality, the signal of the bearing is always non-stationary due to the dynamic as the load and speed condition varies over time. The non-stationary signals are somehow treated at stationary signals for short time window method for the ease of computation. The non-stationary signals are complex and associated with a lot of noises. This has caused difficulties in determining the bearing faults by conventional time domain and frequency domain methods which assume that the signal is strictly periodic for analysis [5]. The time-domain features are good for fault detection but weaker fault isolation as compared to the frequency-domain features [6]. The needs for effective detection of the non-stationary signal is getting more demanding to prevent the sudden machine failures. Therefore, various signal processing techniques have been developed to overcome the problem.

Time domain features represent the vibration signal in the function of time where the amplitude of the waveform is plotted over time. It represents the proximity, velocity, and acceleration of the vibration signal. It is favourable to be used for fault diagnosis due to the advantages of a minimum data loss prior the inspection [7]. The time domain features mainly focus on the statistical characteristic of the vibration signal such as the peak level, standard deviation, skewness etc. The disadvantage of the time domain is the possibility of using too much data for easy and clear fault diagnosis [8].

In frequency domain analysis, the Fast Fourier Transform (FFT) transform the signals very quickly and gives the same result as the slower Discrete Fourier Transform (DFT) [9]. The limitation of this approach is that it is not able to analyse the non-stationary signals [10]. The Short Time Fourier Transform is then developed to overcome the limitation of FFT. The STFT is only applicable to analyse the quasi-stationary signals that are stationary at the short time window scale but not applicable to analyse highly transient phenomena in signals, such as impulse [11]. SFTF is inherently not able to determine time-dependent variation in the structure of the window at various scale [12]. For time-frequency analysis, The Empirical Mode Decomposition (EMD) is gaining the attention of the researchers due to its ability of decomposing non-linear or non-stationary signal into a series of zero-mean amplitude-modulation frequency-modulation (AM-FM) [13]. However, some researchers stated that The EDM is mainly to qualitative diagnosis that focusing on the type of severity of the faults but not the quantitative aspect such as the defect size [14]. Wavelet Transform (WT) has been introduced as an effective tool for machinery fault diagnosis. Wavelet itself is not a new theory as it is made up by combining many existing, well-known and independent signal processing techniques such as sub-band coding, quadrature mirror filter, etc. Due to the wide application, wavelet theory is becoming more active in the research fields as it is capable of further improvement in both the mathematical understanding and the wide range of applications in science and engineering [15-17]. The major merits of wavelet are that it is suitable for wideband signals that are not periodic and may contain both sinusoidal and impulse components. It has the ability to analyse the singularities and irregular signal structure and provides an excellent frequency resolution of the spectra, the variance of estimated power spectra and complexity [18].

In this paper, a method for bearing faults detection is formulated by combing CWT and DWT with envelope analysis. The WT was employed as a de-noising process to filter out the other frequencies than the resonance frequency band. This served as a standard reference to centre the band-pass filter for envelope analysis. So, by knowing the exact location of filtering, this would be a better approach to predict the fault severities using the frequency analysis.

2. Background Theory

An antifriction bearing consists of different components such as outer race, inner race, rolling element, and the cage was considered in this study. The inner race is strongly mounted to the shaft to rotate in the shaft speed with the outer race stay stationary during operation while the rolling elements and cage are rotating or spinning at their own speed. In fault diagnosis, the rolling elements play a big role in the frequency recognition where unique frequency will be generated by the faulty component when the rolling elements rolled pass through the fault location. The derivation of the equations measuring the frequency of defective component is based on the relationship between the rotating elements of a bearing and the relative rotating speed. A defective bearing in a running machine can generate at least five frequencies [19], these frequencies are as following:

a) FTF - Fundamental Train Frequency. It is also known as the frequency of the defected cage:

$$f(Hz) = \frac{1}{2}s\left[\frac{BD}{PD}cos\beta\right] \tag{1}$$

b) BPFI - Bass Pass Frequency of the Inner Race. This is the frequency that produced during the rolling elements roll across the defected inner race:

$$f(Hz) = \frac{n}{2}s\left[1 + \frac{BD}{PD}cos\beta\right] \tag{2}$$

c) BPFO – Bass Pass Frequency of Outer Race. This is the frequency that produced during the rolling elements roll across the defected outer race.

$$f(Hz) = \frac{n}{2}s\left[1 - \frac{BD}{PD}cos\beta\right] \tag{3}$$

d) BSF – Ball Spin Frequency. This is the circular frequency of each rolling elements.

$$f(Hz) = \frac{PD}{2 \times BD}s\left[1 - \left(\frac{BD}{PD}cos\beta\right)^2\right] \tag{4}$$

e) Rolling Element Defect Frequency. This frequency is twice the BSF.

$$f(Hz) = \frac{PD}{BD}s\left[1 - \left(\frac{BD}{PD}cos\beta\right)^2\right] \tag{5}$$

where s is speed (revolutions per minute), n is the number of the rolling elements, β is contact angle (degrees), BD is ball or roller diameter and PD is Pitch diameter. In this paper, the contact angle is zero degree.

3. Signals Analysis Techniques

3.1. Continuous Wavelet Transform

Wavelet Transform is a revolution from the Fast Fourier Transform (FFT). The multi-resolution of time-frequency analysis has resolved the limitation of FFT. Consider a function ψ having complex-valued satisfying the following conditions:

$$\int_{-\infty}^{\infty} |\psi(t)|^2 dt < \infty \tag{6}$$

$$C_\psi = 2\pi \int_{-\infty}^{\infty} \frac{|\Psi(\omega)|^2}{|\omega|} d\omega < \infty \tag{7}$$

where Ψ is defined as the Fourier transform of ψ. The first condition represents the finite energy of the function ψ. The second condition is the admissibility condition, it represents that if $\Psi(\omega)$ is smooth then the $\Psi(0) = 0$. The function ψ is the mother wavelet. The theory could be explained as the following: Let $f(x)$ be the signal and the CWT of $f(x)$ is then defined as:

$$W_f(a,b) = \int_{-\infty}^{+\infty} f(x)\psi_{ab}^{*}(x)dx \tag{8}$$

The $(*)$ represent the complex conjugate

$$\psi_{ab}(x) = \frac{1}{\sqrt{|a|}}\psi\left(\frac{x-b}{a}\right) \qquad a,b \in R, a \neq 0 \tag{9}$$

The admissibility condition is described as

$$C_\psi = \int_0^\infty \frac{|\psi(\omega)|^2}{\omega} d\omega < \infty \tag{10}$$

And it yields to:

$$\int_{-\infty}^{\infty} \psi(x)dx = 0 \tag{11}$$

The $\psi(\omega)$ is Fourier transform of the$\psi(x)$. The admissibility condition describes that the Fourier transform of $\psi(x)$ will be cancelled off at zero frequency. From this, it is clearly to present that ψ is described as a wave or the mother wavelet. It has two characteristic parameters which is (a) dilation and (b) translation which varies continuously. The function of the translation is to control the position of wavelet in time. The high-frequency information could be obtained by narrowing down the wavelet while the dilated wavelet accesses the low-frequency information. This could be said that the parameter a varies with different frequencies [20].

3.2. Discrete Wavelet Transform
DWT is derived from the discretization of CWT to reduce the computation cost when operating at every scale. It is opposite to the CWT, the scaling parameter a and the time localization parameter b are defined as

$$a = 2^{\mathrm{j}}, \mathrm{j}\epsilon\mathrm{Z} \tag{12}$$

$$b = \mathrm{k}2^{\mathrm{j}}, \mathrm{j}, \epsilon\mathrm{kZ} \tag{13}$$

where $Z = 0, \pm 1, \pm 2, \dots \dots \dots$ The discretization scaling and time parameter is the property of DWT such that:

$$DWT\ j,k = 2^{-j/2}\ x\ t\ \psi^{*}\ 2^{-j}t - k\ dt \tag{14}$$

In order for the discretization to occur, the CWT becomes DWT function, the scaling function is given by:

$$\psi_{jk} = 2^{-j/2}\psi 2^{-j}t - k \tag{15}$$

$$\phi_{jk} = 2^{-j/2}\phi 2^{-j}t - k \tag{16}$$

Then, the wavelet coefficient of a signal $x\ t$ can be defined as

$$a_{2j}k = x\ t\ \phi_{j.k}\ t\ dt \tag{17}$$

$$a_{2j}k = x\ t\ \phi_{j.k}\ t\ dt \tag{18}$$

Consequently, the discretization property can decompose the signal into a hierarchical structure with wavelet details and approximations at various scale [21].

3.3. Envelope Analysis

The envelope analysis demodulates the amplitude modulating (AM) signal to extract the signal signature using three steps. Firstly, the signals are filtered using the band-pass filter cantered at the resonant frequency band. The filtering effects will eliminate the unwanted side band's high amplitude contents and suppress the high-frequency noise. Therefore, a signal with a good signal-to-noise ratio (SNR) can be obtained.

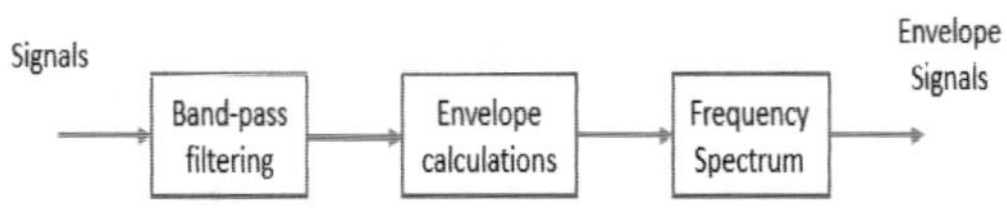

Figure 1. The signals flow in envelope analysis

The Hilbert Transformation (HT) is acted as the processor that perform the calculation or the envelope signal. Consider the input signal is FFT, the HT process signal and create analytical signal using the following algorithm:

$$Signal = X_{in} = fft(x_{in}) \tag{19}$$

$$X_a(n) = \begin{cases} X_{in}(n), & if\ n = \frac{N}{2} \\ 2 \times X_{in}(n), & if\ 0 < n <= \frac{N}{2} \\ 0, & if\ \frac{N}{2} < n < N \end{cases} \tag{20}$$

$$x_a = ifft(X_a) \tag{21}$$

The created signal is a complex signal in nature. The real part is the original signal and the imaginary part is the Hilbert transform of the original signal. The envelope signal is computed using

$$x_{env} = \sqrt{x_a \times conj(x_a)} \tag{22}$$

The signal is then converted to the frequency spectrum using the next equation where the $\overline{x}_{env}$ is the mean value of the envelope which is first substrate to remove the offset component. The Hanning window is employed to reduce the tendency of spectral leakage which is given by the following [22].

$$x_{env} = |fft((x_{env} - \overline{x_{env}}) \times hann(N))| \tag{23}$$

4. Data Acquisition

The vibration signals used in this study is obtained from the Case Western Reserve University. The apparatus used consists of 2 horsepower motor (left), a torque transducer/encoder (centre), a dynamometer (right), and control electronics (not shown) in the following figures. The vibration signals data sets for the different type of bearing condition were obtained, namely, inner race, outer race, rolling element fault and healthy bearing. The test bearings support the motor shaft so that the bearings can rotate as the motor rotate. Single point faults were introduced to the test bearing using the electro-discharge machining with fault diameter of 7 mils, 14 mils, 21 mils, 28 mils, and 40 mils (1mil=0.001 inches).

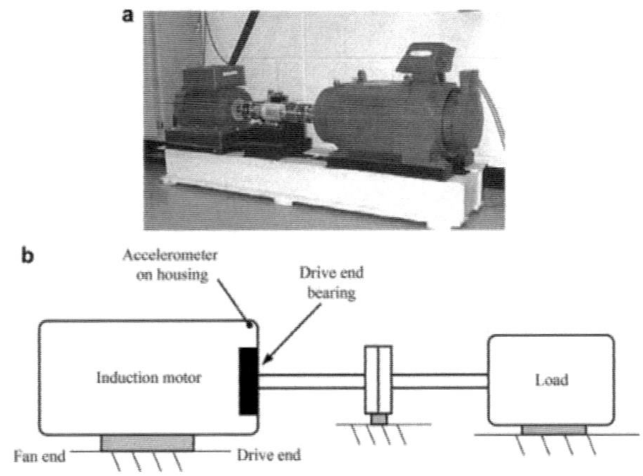

Figure 2. (a) Bearing test rig; (b) Its schematic representation

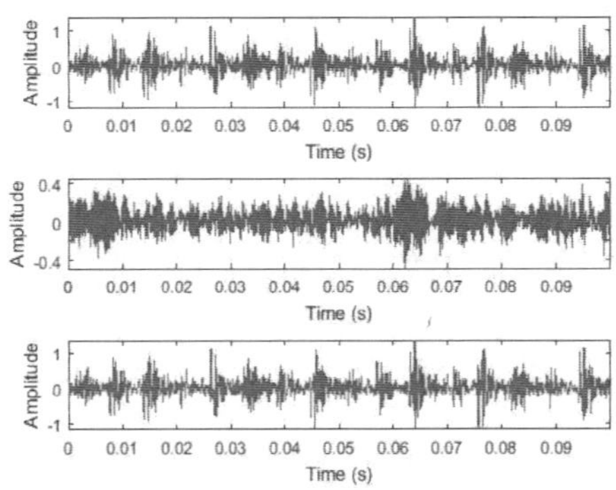

Figure 3. Time-wave of bearing raw signals: (upper) healthy (middle) inner race fault (bottom) outer race fault

The accelerometers were attached to the different positions of the housing with magnetic based to collect the vibration data. The positions include the 12 o'clock position at both the drive end and fan end of the motor housing where one of it is attached to the motor supporting base plate for some experiment. The vibration signals were collected using a 16 channel DAT recorder and were post processed using the MATLAB. The motor runs at different speeds typically from 1730 -1797 rpm and the bearing was run under different load condition 0, 1, 2, 3, 4 motor horsepower. The data were sampled at 12 kHz sampling frequency. To evaluate the purpose of this paper, the signals with zero loading, 0 horsepower obtained at the drive end location were used. The bearing was test run at 1797 rpm and the data used contained 120,000 data points with 10-second time span. The model of the bearing is 6205-2RS JEM SKF, deep groove ball bearing and the defect frequencies are summarized as the following [23].

Table 1. 2RS JEM SKF, deep groove ball bearing defect frequencies

Ball Passing Frequency	multiple of running speed in Hz **Defect Frequency (Hz)**
BPFI	5.4152
BPFO	3.5848
FTF	0.39828
$2 \times$ BPF	4.7135

5. Results and Discussion

5.1. Continuous Wavelet Transform and Envelope Analysis

In this study, Morlet wavelet mother function is employed for all the continuous wavelet transform process to correlate the bearing signals impulse and remove the noise present in the signals. In the process of CWT, the entire input signal was 'dismantled' and re-arranged according to the energy level content in the wavelet coefficient. The high-frequency components have a finer resolution and the lower have a courser resolution. So, the signal with higher energy content was located at the smaller scale while the signal with lower energy content was located at the higher scale. The energy content of the signal was identified by viewing the amplitude of the signal that corresponded to its scale in the wavelet coefficient. The higher amplitude component in the signal has brighter color at the coefficient map. According to the Figure 4, defective bearings often has lower scale than the normal bearing which is scale 3 and scale 9 respectively which indicated that the detective bearings have higher energy content than the normal bearing. The rolling element rolled pass the defective part created large impulse and produced high frequency during the operation. The defective bearing generated frequency as high as 2.6 kHz to 5.0 kHz which was known as the resonant frequency band of the defective bearing. The CWT was able to eliminate the other frequency components and remain only resonance frequency in the entire signals. As the resonance frequency band was identified, the band-pass filter was centered in such a way that it covered the entire resonance band in the envelope analysis. This is to ensure the band-pass filter could be accurately applied to the resonance band for high-frequency demodulation when assessing the increasing fault severities.

Figure 4(e), (f), (g) and (h) present the frequency spectrum of healthy bearing, inner race, outer race and rolling elements faults. The resonant band was treated as 2.5 kHz – 5 kHz. By centering the band-pass filter on the frequency band of 2.5 kHz – 5 kHz would yield frequency spectrum that reviewed the faulty characteristic frequency of the bearing in envelope analysis. To effectively assess the fault severities of the bearing, the exactly resonant frequency range have to be determined. This could be done using the CWT scale function to de-noising the signal and retain only the resonant band in the entire signal. For inner race fault, it is clearly seen that the peak generated by the characteristic frequency, BPFI, 162.18 Hz is prominent. The frequency spectrum did not show exactly the 162.18 Hz at the characteristic frequency as the bearings were not always in the no slip condition which the theoretical formula assumed that the bearing is in no slip condition. For the outer race fault, all the frequency spectrum showed that the BPFO 107.6Hz is prominent for 0.007" fault size and BPFO for the 0.021" fault size are evidenced but the opposite signals contained more other frequency component compared to the others. The frequency spectrum of rolling element fault contained the characteristic frequency of inner race, outer fault, and rolling element fault. In close inspection to the frequency spectrum of outer race fault, it was observed that the fundamental train frequency, FTF is presented along with second and third harmonics. Other than that, it was also observed that the unit speed and its harmonics are found in the spectrum that was labeled in 'x' multiple that indicated residual imbalance, loading, and looseness. There are many other frequencies and their harmonics contained in frequency are mainly due to many other frequency modulations presented in the resonance frequency band. In this analysis, results showed the harmonic of BPFI and BPFO indicated the fault location contained fragment denting or frosting [19]. The amplitude of the characteristic frequency and its harmonic are shown to be increased when the fault size increase.

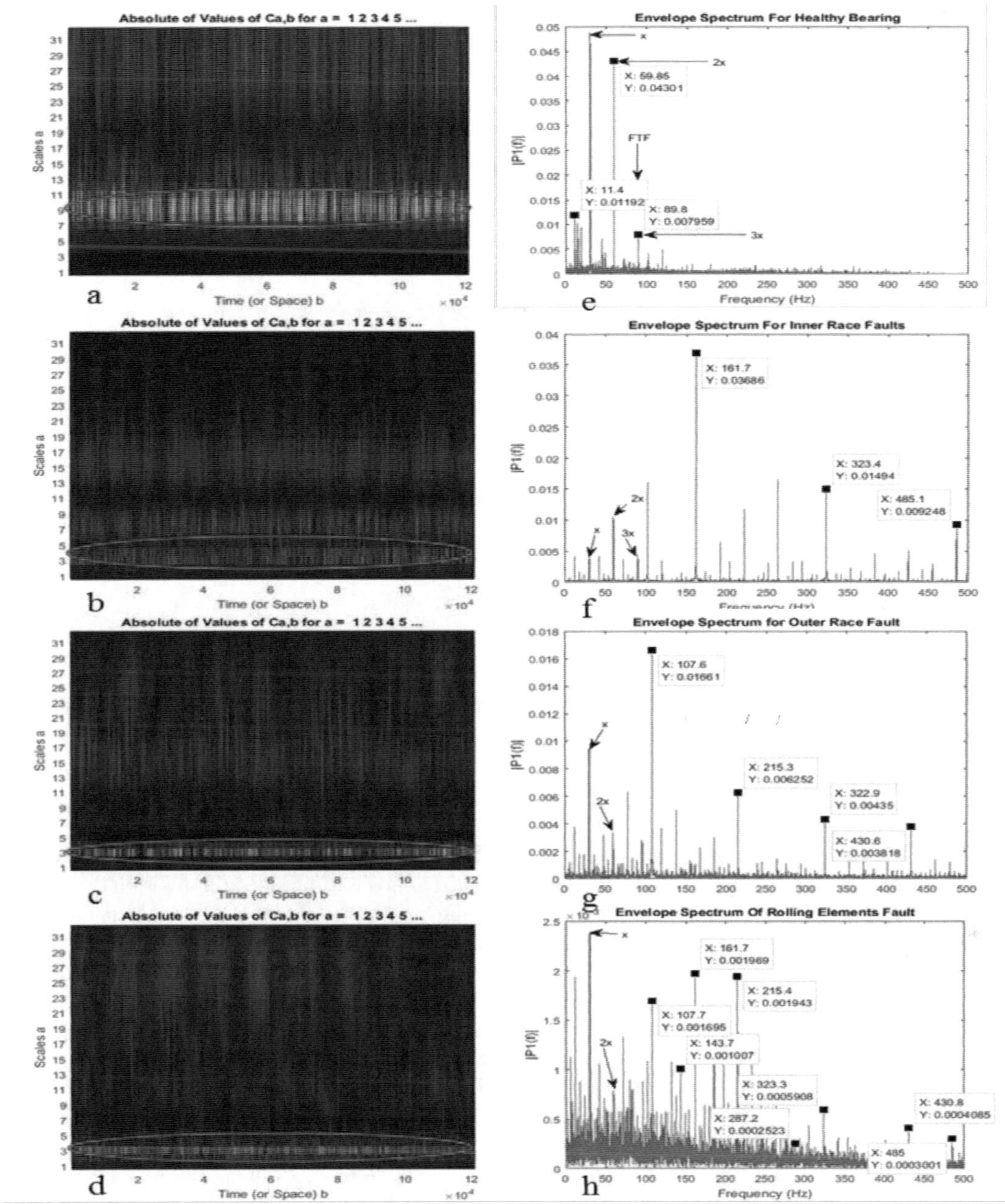

Figure 4. Wavelet coefficients map of different bearing faults, (a) healthy (b) inner race (c) outer race (d) Rolling elements; frequency spectrum for (e) healthy (f) inner race fault (g) outer race fault (h) rolling elements fault.

5.2. Discrete Wavelet Transform and Envelope Analysis

DWT is another important part of the wavelet transform that could perform very well in de-noising using the built-in high pass and low pass filters. The main function of the DWT is to decompose the signal into smaller frequency band to detect the pattern that is not visible in the signals. On the other hand, DWT also reassembles back the decomposed signal to its original signal without any loss of information. In this section, detection of bearing fault using the signal reconstruction in DWT and envelope analysis is presented. The mother wavelet employed for this work is 'db4' wavelet. The 'db4' function is similar to the bearing impulse but when compared to the Morlet function, a distinct difference was found between them in the terms of the time and amplitude due to the orthogonality.

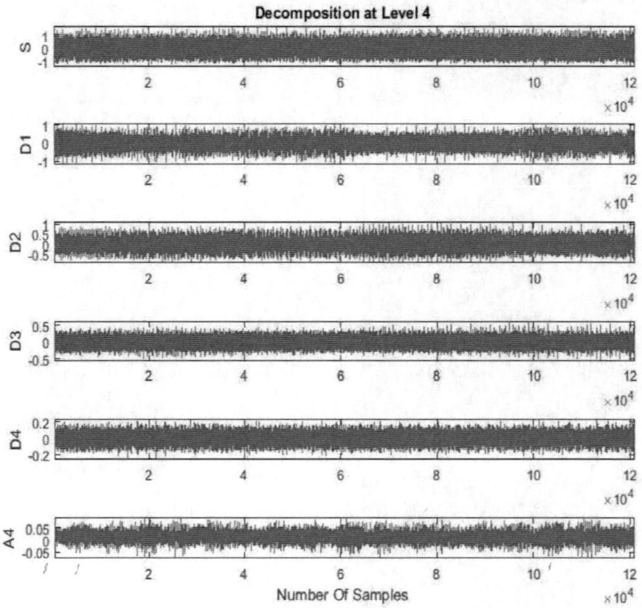

Figure 5: Signal Reconstruction Using DWT

The DWT decomposed the signal into the detail part and approximate part. The detail part of the decomposition contains the energy distribution of the signal which it is the key point for fault diagnosis. The benefit of the signal reconstruction is that the number of the samples are retained after the decomposition so that the frequency resolution can be maintained at the higher decomposition level. Using DWT to search for the high- frequency carrier is generally easier and has less computation time compared to the CWT. DWT processed the signals by decomposing the signals through the high-pass and the low-pass filters in the down-sampling of 2. In a close look into this process, the high-pass filtering process stopped after the higher frequency signals components were identified which only the lower frequency component would be decomposed in the next subsequent low-pass and high-pass filtering.

In this study, all the signals for the different bearing condition are decomposed to four level of decomposition as shown in Figure 5. According to Nyquist, the 12 kHz sampling rate signal would have the frequency halved the sampling rate which is 6 kHz. The frequency bands which reduced half at each decomposition level, are sequentially 3 kHz,1.5 kHz, 750 Hz and 325 Hz. The faulty frequencies are in the range of 325 Hz. Therefore, further decomposition is not necessarily. The energy distribution in each of the decomposed signals is examined. Then, the frequency band which has the highest distributed energy among the detail part were extracted for envelope analysis. The resonance frequency band retained by the DWT is different from the one in CWT. DWT retained a wider frequency band due to the different wavelet function used. The 'db4' function is similar to the bearing impulse but when compared to the Morlet function, a distinct difference was found between them in the terms of the time and amplitude and resulted in the differences in the frequency spectrum in terms of the upper and lower frequency compared to the CWT.

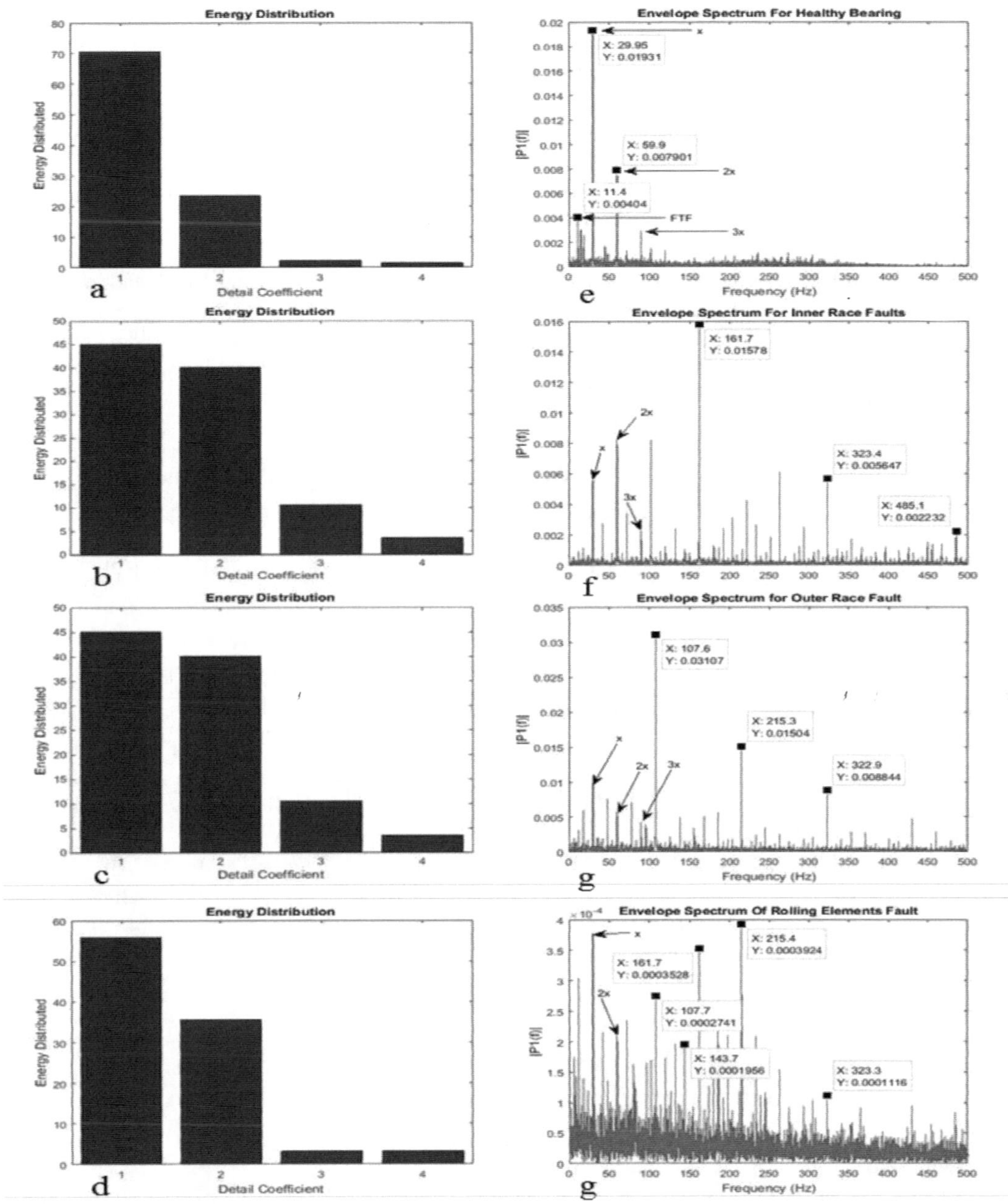

Figure 6. DWT energy distribution of different bearing faults (a) healthy (b) inner race (c) outer race (d) rolling elements; frequency spectrum for (e) healthy (f) inner race fault (g) outer race fault (h) rolling elements fault

Inner race fault was showed to be always diagnosable in envelope analysis regardless of which type of filter was used. A close look at the energy distribution of the decomposed signals in Figure 6, it was found that it has overall higher distributed energy in different frequency bands. Selecting the higher energy distribution frequency band in the wavelet based de-noising is just like the CWT de-noising. However, DWT decomposes the signal into resonance frequency bands with higher energy in the entire signal and quantify the energy content in the signal component.

Prior to envelope analysis, the SNR of both raw signal and wavelet de-noised signals were examined. The SNR has improved at least 2.5 dB after pre-processed using DWT. It was found that the 'db4' mother wavelet function with signal reconstruction reduce lesser noise compare to the CWT Morlet function. On other hand, the width of the resonance frequency would affect the overall pattern of the frequency spectrum. In this study, the resonance frequency band retained by the DWT is found to be different from the one in CWT. CWT have a cleaner spectrum compared to the DWT pre-processed opposite signal, however, some frequency components may lost and hence, other faults information also may be missed, because CWT filters eliminate the other frequency components and remain only resonant frequencies in the entire signal. Unlike, DWT which retained a wider frequency band due to the different wavelet function used which yield in wider fault's coverage. In this instance, the resonance frequency retained by DWT found to be contained the information of the misalignment and looseness as shown in Figure 6. The harmonic of the unit speed labeled by 1x, 2x and 3x indicated the looseness. Other than that, the 'db4' function in DWT produced similar pattern to the bearing impulse, however, when compared to the Morlet function in CWT, a distinct difference was found between them in the terms of the time and amplitude which resulted in differences in the frequency spectrum in terms of the upper and lower frequencies.

Moreover, it is also observed that the amplitude of the envelope spectrum for both bases are not the same. Signals pre-processed by DWT yielded more logical amplitude for the frequency spectrum in which the amplitude for BPFO was lower than BPFI. This fulfilled the theory that the vibration energy produced by outer race fault transmitted in a shorter distance to reach the sensor. Thus, the BPFO should have greater amplitude than BPFI.

Table 2. The summary of the amplitude of characteristic frequency for increased fault size.

Component	Fault Sizes	Characteristic frequency on spectrum				Frequency Spectrum Pattern
		CWT	Amplitude	DWT	Amplitude	
Inner race	0.007"	Prominent Peak	0.03686	Prominent Peak	0.0157800	Similar
	0.014"	Evidenced Peak	0.05390	Evidenced Peak	0.0390700	Similar
	0.021"	Evidenced Peak	0.34890	Evidenced Peak	0.1361000	Similar
Outer race	0.007"	Prominent Peak	0.01661	Prominent Peak	0.0310700	Similar
	0.021" Opposite	Evidenced Peak	0.02377	Evidenced Peak	0.0251900	Not Similar
	0.021" Orthogonal	Evidenced Peak	0.33230	Evidenced Peak	0.0527900	Not Similar

6. Conclusion

This paper presented the use of CWT and DWT and envelope analysis for resonance band demodulation. The CWT and DWT have proven to be an excellent filtering technique and resonant band demodulation methods to remove the unwanted frequency components from the signals. The proposed technique has shown to easier, faster and accurate to centre the band-pass filter for envelope analysis. The characteristic frequencies are evident in the frequency spectrum. Both CWT and DWT produced similar frequency spectrum pattern but different amplitude values, the distinguish similarity was the prominent characteristic frequency amplitude in the frequency spectrum. However, DWT has shown to be more superior as it is easier, less computational time, less noise and retained a wider frequency band which contained more faults diagnostic information. The frequency spectrum of rolling elements fault, found to be contained the characteristic frequency of inner race, outer fault and rolling element fault. It is also observed that the amplitudes of the characteristic frequencies increased when the fault occur and when the fault size increased. This feature could be used to distinguish the different bearing fault types, identify the fault locations and also assess the fault serveries.

7. References

[1] El-Thalji I and Jantunen E. 2015. A summary of fault modelling and predictive health monitoring of rolling element bearings. *Mechanical systems and signal processing*.**60**. 252-72

[2] Yadav M and Wadhwani S. 2011. Automatic fault classification of rolling element bearing using wavelet packet decomposition and artificial neural network. *International Journal of engineering and technology*.**3**. 270-6

[3] Samanta B and Al-Balushi K. 2003. Artificial neural network based fault diagnostics of rolling element bearings using time-domain features. *Mechanical systems and signal processing*.**17**. 317-28

[4] Ak Chandel R P. 2013. Bearing Fault Classification Based on Wavelet Transform and Artificial Neural Network. *IETE Journal of Reseach*.**59**. 219-25

[5] Bhavaraju K M, Kankar P, Sharma S C and Harsha S. 2010. A comparative study on bearings faults classification by artificial neural networks and self-organizing maps using wavelets. *International Journal of Engineering Science and Technology*.**2**. 1001-8

[6] Yan W, Qiu H and Iyer N. 2008 Feature extraction for bearing prognostics and health management (phm)-a survey (preprint). ed Secondary. Yan W, Qiu H and Iyer N: DTIC Document)

[7] Stalin S. 2014. Fault Diagnosis and Automatic Classification of Roller Bearings Using Time-Domain Features and Artificial Neural Network. *International Journal of Science and Research (IJSR)*.**3**. 824-51

[8] Patel V. 2014. Recent Advanced In AI Based Intelligent Fault Techniques for Rolling Element Bearing–A Review.

[9] Oulmane A, Lakis A and Mureithi N. 2014. Application of Fourier Descriptors & Artificial Neural Network to Bearing Vibration Signals for Fault Detection & Classification. *Universal Journal of Aeroneutical & Aerospace Science*.**2**. 37-54

[10] Gligorijevic J, Gajic D, Brkovic A, Savic-Gajic I, Georgieva O and Di Gennaro S. 2016. Online Condition Monitoring of Bearings to Support Total Productive Maintenance in the Packaging Materials Industry. *Sensors*.**16**. 316

[11] Feng Z, Liang M and Chu F. 2013. Recent advances in time–frequency analysis methods for machinery fault diagnosis: a review with application examples. *Mechanical systems and signal processing*.**38**. 165-205

[12] Yiakopoulos C and Antoniadis I. 2002. Wavelet based demodulation of vibration signals generated by defects in rolling element bearings. *Shock and Vibration*.**9**. 293-306

[13] Yao D, Jia L, Qin Y and Yang J. 2014. Faults diagnosis of railway rolling bearing by using time-frequency feature parameters and genetic algorithm neural network. *Pulse*.**10**. 4

[14] Rai A and Upadhyay S. 2016. A review on signal processing techniques utilized in the fault diagnosis of rolling element bearings. *Tribology International*.**96**. 289-306

[15] Abdelrhman A M, Leong M S, Hee L M and Al-Obaidi S M A.2015 *Book.Segregation of Close Frequency Components Based on Reassigned Wavelet Analysis for Machinery Fault Diagnosis*: Springer). pp.581-90

[16] Abdelrhman A M, Leong M S, Hee L M and Ngui W K. 2015 A Comparative Study of Reassigned Conventional Wavelet Transform for Machinery Faults Detection. In: *Applied Mechanics and Materials,* pp. 90-4

[17] Abdelrhman A M, Leong M S, Hee L M and Ngui W K. 2013 Application of wavelet analysis in blade faults diagnosis for multi-stages rotor system. In: *Applied Mechanics and Materials*. pp. 959-64

[18] Jayakumar K and Thangavel S. 2015. Industrial drive fault diagnosis through vibration analysis using wavelet transform. *Journal of Vibration and Control*. 1077546315606602

[19] Taylor J I and Vibration Consultants I. 1994. *The vibration analysis handbook*.Vibration Consultants)

[20] Jawadekar A, Paraskar S, Jadhav S and Dhole G. 2014. Artificial neural network-based induction motor fault classifier using continuous wavelet transform. *Systems Science & Control Engineering: An Open Access Journal*.**2**. 684-90

[21] Vishwash B, Pai P S, Sriram N, Ahmed R, Kumar H and Vijay G. 2014. Multiscale slope feature extraction for gear and bearing fault diagnosis using wavelet transform. *Procedia Materials Science*.**5**. 1650-9

[22] Feng G-J, Gu J, Zhen D, Aliwan M, Gu F-S and Ball A D. 2015. Implementation of envelope analysis on a wireless condition monitoring system for bearing fault diagnosis. *International Journal of Automation and Computing*.**12**. 14-24

[23] Univesity C W R. 2016 Bearing Data Center. ed Secondary. Univesity C W R (United State: Case Western Reverse Univesity)

IOP Conf. Series: Materials Science and Engineering **217** (2017) 012032 doi:10.1088/1757-899X/217/1/012032

Experimental study on vertical static stiffnesses of polycal wire rope isolators

P.S. Balaji[1], Leblouba Moussa[2], Noman Khandoker[3], Ting Yuk Shyh[3], M.E. Rahman[3*], Lau Hieng Ho[3]

[1]Department of Mechanical Eng., National institute of technology Warangal, India.
[2]Department of Civil & Environmental Engineering, University of Sharjah, UAE
[3]Faculty of Engineering and Science, Curtin University, Malaysia.

*E-mail: merahman@curtin.edu.my

Abstract. Wire rope isolator is one of the most effective isolation system that can be used to attenuate the vibration disturbances and shocks during the operation of machineries. This paper presents the results of investigation on static elastic stiffnesses (both in tension and in compression) of Polycal Wire Rope Isolator (PWRI) under quasi-static monotonic loading conditions. It also studied effect of variations in height and width of PWRI on its static stiffnesses. Suitable experimental setup was designed and manufactured to meet the test conditions. The results show that their elastic stiffnesses for both tension and compression loading conditions are highly influenced by their geometric dimensions. It is found that their compressive stiffness reduced by 55% for an increment of 20% in their height to width ratio. Therefore, the stiffness of PWRI can be fine-tuned by controlling their dimensions according to the requirements of the application.

1. Introduction

Vibration is one of the most destructive means for industrial equipment or structures. The sources of vibration are either by nature such as an earthquake, wind and ocean waves or man-made such as the operation of heavy machineries and piling constructions[1, 2]. These vibrational disturbances can gradually reduce the life of machines or structural elements in the industry [3]. Hence, vibration isolation system is used to minimize the harmful effects of vibration disturbance. Polycal Wire rope isolator (PWRI) is effective in isolating the system from vibrational environment. Wire Rope Isolators (WRI) possess high damping capabilities which can help in vibrational energy dissipation [4]. Additionally, it can provide protection by limiting the amount of displacements occurred. PWRI is assembled by stranded wire ropes held with rugged metal retaining discs as shown in Figure 1. This design provides flexibility and energy absorption characteristics together with large displacement capacity to obtain its damping effect from the sliding friction between the twisted wire ropes [5]. WRI has benefits such as high damping effectiveness, low maintenance, thermal insensitivity, durability, and low cost which are the reasons for its popularity in the isolation field. In addition, the performance of WRI is highly dependent on its geometric dimensions [6]. Therefore, it is practically essential to gain better understanding of the behaviour of the WRI.

WRI is mainly governed by its stiffness (K) and damping coefficient (C) whereby stiffness represents the load-supporting and damping coefficient represents the energy-dissipating capabilities of the isolators respectively [3, 7]. There are plenty of different geometries and stiffness values of wire rope isolator available in the market. The selection of suitable wire rope isolator for any particular application generally relies on the static stiffness. Besides, the knowledge on static stiffness is also applicable in the appropriate design of WRI [8]. In other words, it indicates the load-carrying capacity of the WRI. The static stiffness in wire rope isolation system can be estimated from the monotonic loading test.

Figure 1. Polycal Wire Rope Isolator (PWRI)

Monotonic loading can be attained by applying load in vertical direction and can be presented in a load-displacement graph [3]. The slope of the load-displacement plot indicates the vertical stiffness. In the previous studies, WRI appeared to be non-linear at high displacement amplitudes but at small displacement amplitudes the load-displacement relationship can be assumed to be linear [5, 9]. Therefore, the detailed study on the static stiffness WRI is quite limited in the literatures as most literatures are available for cyclic loading [3]. Demetriades et al. [5, 10] performed the study on the stiffness and damping capacity of helical wire rope isolators and suggested that the stiffness properties of the WRI is a function of its geometric dimensions and wire rope material properties. Balaji et al. [8, 11] has developed the analytical models for the stiffness of helical WRI and these models can be used to design the helical WRI. Massa et al.[12] designed an ball transfer unit to improve the vertical stiffness of the WRI. The studies [5, 10, 13, 14] available in the literature are mainly available for helical type WRI however the studies on PWRI are very limited. Hence, in this work, the vertical static stiffness and its influence of height-width ratio is studied experimentally using monotonic loading tests. This study can enhance the understanding of stiffness characteristics of PWRI.

2. Methodology

The experimental work was conducted to understand the behaviour of polycal wire rope isolator with different geometric dimensions under vertical loading (Figure 2(a)). Vertical loading was applied in both tension and compression directions. All the polycal wire rope isolators were tested under monotonic quasi-static loading condition to determine its static elastic stiffness. The three isolators (Figure 2(b)) are tested in this experimental work. Each isolator has its own individual geometric dimensions varying with wire rope diameter, height and width as given in Table 1.

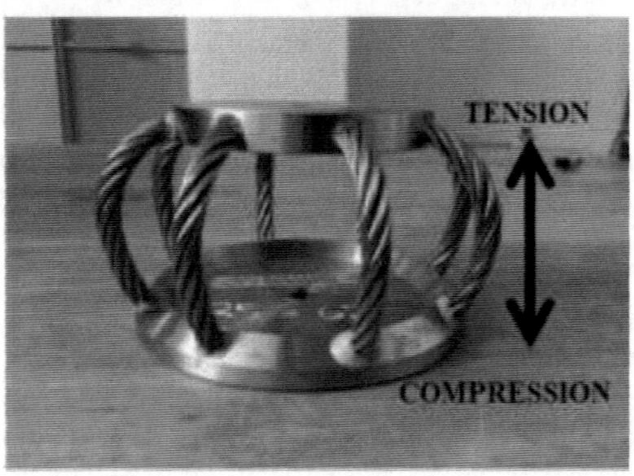

(a)

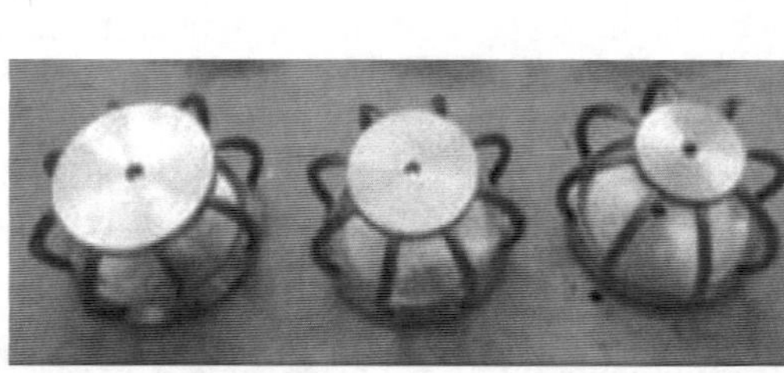

(b)

Figure 2. (a) Loading directions of PWRI for Experiment Work (b) Polycal Wire Rope Isolator for Experimental Work

Table 1. Specifications of WRI

Isolators No.	Wire Rope Diameter (mm)	W (mm)	H (mm)	D1 (mm)	D2 (mm)	d1 (mm)	d2 (mm)
1	6.4	124	62	102	60	M8	M6
2	6.4	120	71	102	60	M8	M6
3	6.4	120	76	102	60	M8	M6

ØW = Width Dia., H = Height, D_1 = Outer Dia. at bottom, D_2 = Inner Dia. at top, d_1 =Bolt Dia. at top, d_2 = Bolt dia. at bottom. The dimensions are illustrated in Figure 3.

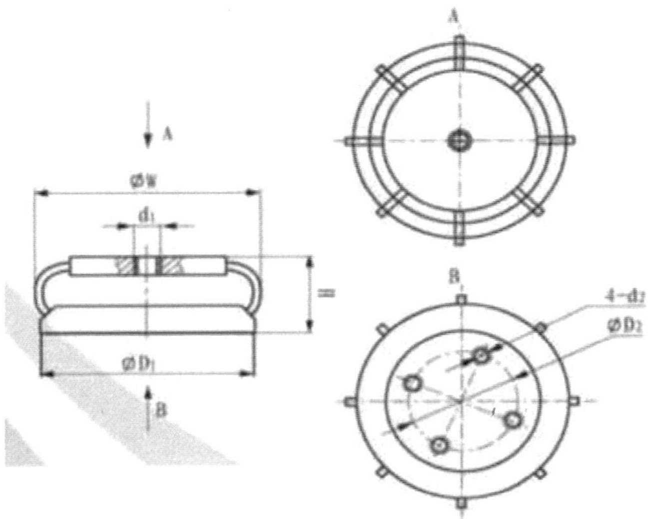

Figure 3. Dimensions of WRI (WU XI HONGYUAN)

3. Experimental Set Up

All experimental tests were conducted in Curtin University, Malaysia, using the INSTRON Machine 5982 available in the laboratory as shown in Figure 4. It can produce up to 100kN of loading which is sufficient for the test. It provides high accuracy data acquisition facilitated by Bluehill Software. By using it, the load-displacement graph is displayed in the computer during the test. In order to apply monotonic loading in vertical direction, suitable fixtures were designed. The loading rate is 2 mm/min both in tension and in compression. The test was done in the displacement control mode. The INSTRON top and bottom grip is controlled by the hydraulic power for effective griping. Furthermore, the top and bottom plate of PWRI are not compatible for directly gripped by the loading machine. Hence, there is a need of a fixture for of the PWRI. The drawing of the fixture used in this work is shown in the Figure 5-6. These fixtures enable the INSTRON machine to apply the desired loading. Figure 5-6 showed the designed fixtures for both tension and compression loading conditions.

IOP Conf. Series: Materials Science and Engineering **217** (2017) 012032 doi:10.1088/1757-899X/217/1/012032

Figure 4. Loading Test of WRI in Vertical Directions

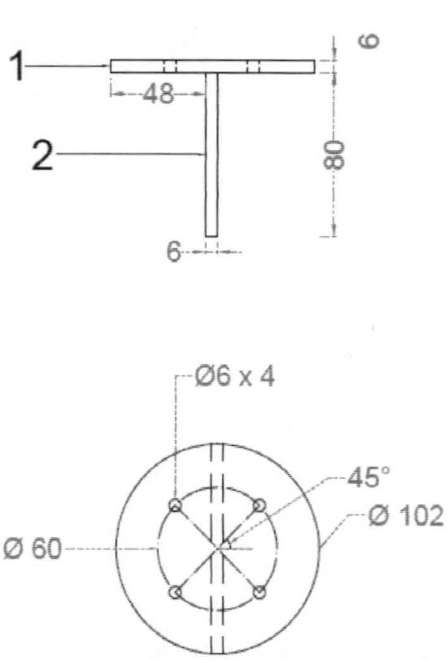

Figure 5. Bottom fixture for gripping the bottom metal retainer of PWRI

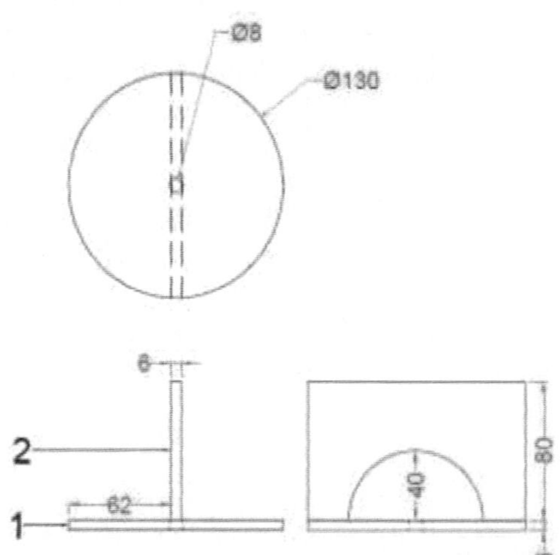

Figure 6. Top fixture for applying tension/compression loading

4. Results and discussions

This section presents the result of the monotonic loading tests. The static elastic stiffness of WRI was determined from the behaviour in monotonic loading condition. The slope of the load-displacement graph represents the static stiffness of WRI and it is obtained by best-fitted linear curve from the experimental data points.

4.1. Static Vertical Stiffness

Load-displacement plot in the vertical direction are presented in Figure 7-8 and the vertical stiffnesses data are presented in Table 2. It can be seen from the Table 2 that the vertical stiffness value under the compressive loading condition is less than the stiffness value in tensile loading condition. This can be explained by the hardening and softening phenomenon in PWRI material. Hardening occurs under tensile loading, which gives more frictions between the wire strands, while softening occurs under compression loading which gives lower frictions between the wire strands. Hence, the isolators will have higher stiffness under tension and lower stiffness under compression loading conditions.

Apart from that, isolators with different geometric dimensions have different static stiffnesses. It can be observed in Figure 7, Figure 8 that isolator 1 is the stiffest among all other isolators, and isolator 3 is the least stiff. This is because isolator 1 has the lowest height and the width among all the isolators. Similarly, the isolator 3 has the highest height and lowest width among all the isolators. Hence, it can be concluded that stiffness of WRI is greatly influenced by the height and width. These findings are in support of the published data in current literature [8].

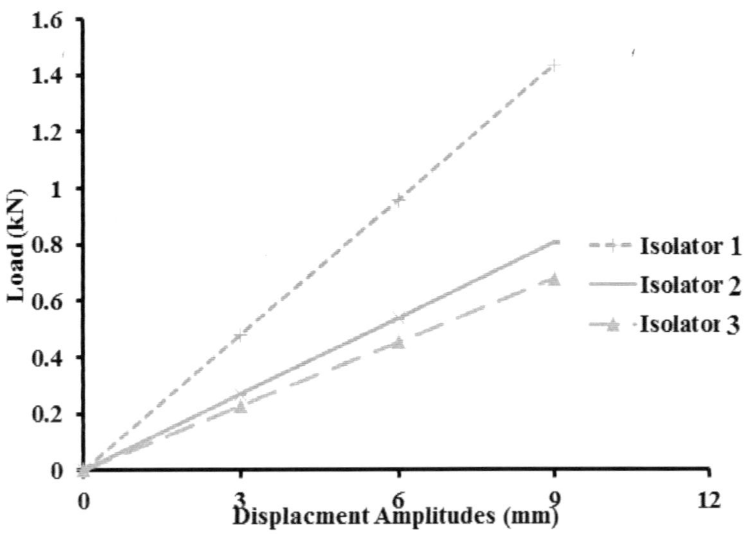

Figure 7. Load-Displacement plots in Compression

IOP Conf. Series: Materials Science and Engineering **217** (2017) 012032 doi:10.1088/1757-899X/217/1/012032

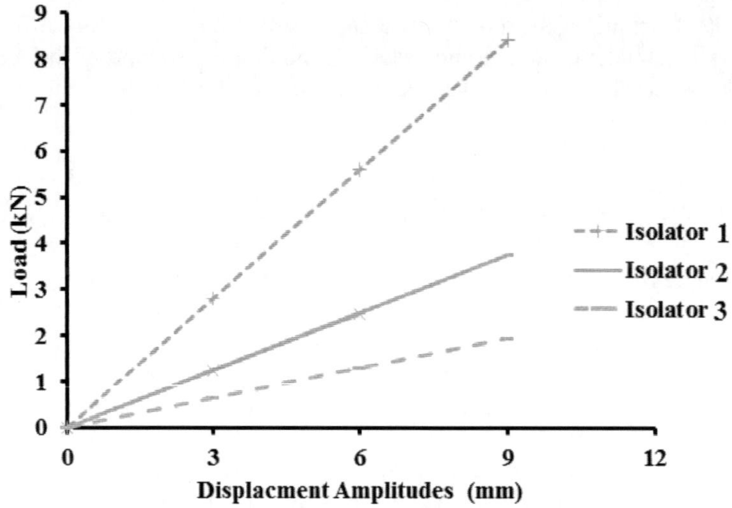

Figure 8. Load-Displacement plots in Tension

Table 2. Static Stiffness for WRI in compression

Isolators	Static Stiffness (N/mm)	
	In compressive loading	In tensile loading
1	160	933
2	89	416
3	75	216

4.2 Influence of Height to width ratio

This section discusses the influence of height to width ratio of wire rope isolator diameter on their compressive static stiffness. The geometry of PWRI depends on its height and width. Therefore, the changing of height and width of PWRI affects the static stiffness value. The comparison among Isolator 1, 2 and 3 was made to highlight the influence of height to width ratio on the static compressive stiffness since they differ only in their height to width ratio. Figure 9 shows that the static stiffness decreases when the height to width ratio increases. An increase in height to width ratio by 20% decrease the compressive static stiffness by 55%. This decrease in static stiffness can be explained by slenderness aspect of the PWRI. Larger ratio means a slender PWRI and it may cause instability for specific application. Therefore, the compressive static stiffness of a PWRI is important to know for the selection of suitable dimensions of PWRI for any particular application.

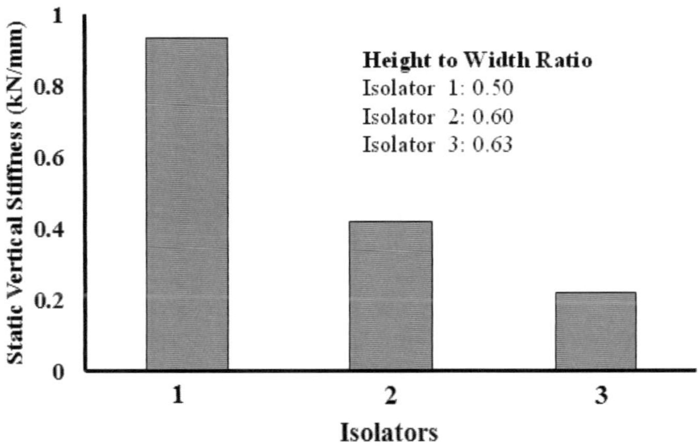

Figure 9. Comparison of Static Vertical Compressive Stiffness for Isolator 1, 2 and 3

5. Conclusions

In this work, monotonic loading tests were performed to study the static stiffnesses of the polycal wire rope isolators. The stiffness of PWRI was observed to be less stiff due to compressive loading as compared to the tensile loading, which can be attributed to the hardening and softening behavior of the wire rope material. It was also observed that the static stiffnesses of PWRI were greatly influenced by their geometric dimensions. It was clearly understood that their stiffness could be controlled by their height to width ratio depending on the application requirements. High stiffness can be obtained by lowering the height to width ratio and vice versa is applicable.

Acknowledgement
This paper is based upon work supported by the Ministry of Higher Education (MOHE), Malaysia under the FRGS grant scheme (FRGS/1/2015/TK01/CURTIN/02/2).

References

1. D Karnopp 1995 Active and Semi-Active Vibration Isolation *Journal of Mechanical Design,* **117**(B) 177-185.
2. E I Rivin 2003 *Passive Vibration Isolation. ASME Press.*
3. PS Balaji, ME Rahman, Leblouba Moussa, HH Lau Wire rope isolators for vibration isolation of equipment and structures–A review *IOP Conference Series: Materials Science and Engineering*. 2015. IOP Publishing.
4. ML Tinker, MA Cutchins 1992 Damping phenomena in a wire rope vibration isolation system. *Journal of Sound and Vibration* **157** 7-18.
5. GF Demetriades, MC Constantinou, AM Reinhorn 1993 Study of wire rope systems for seismic protection of equipment in buildings. *Engineering Structures* **15** 321-334.
6. GC Foss 2006 Modal Damping Estimates from Static Load-Deflection Curves, *24th Conference and Exposition on Structural Dynamics (IMAC - XXIV)* Missouri 54-61.
7. PS Balaji, Leblouba Moussa,ME Rahman, Loo Tshun Vuia 2015 Experimental investigation on the hysteresis behavior of the wire rope isolators Journal of Mechanical Science and Technology **29** 1527-1536.

8. PS Balaji, Leblouba Moussa, ME Rahman, HH Lau HH 2016 Static lateral stiffness of wire rope isolators *Mechanics Based Design of Structures and Machines, An International Journal* **44** 462-475.
9. P Paolacci, R Giannini 2008 Study of The Effectiveness of Steel Cable Dampers For The Seismic Protection of Electrical Equipment *The 14th World Conference on Earthquake Engineering* Beijing, China.
10. GF Demetriades, MC Constantinou, AM Reinhorn 1992 *Study of wire rope systems for seismic protection of equipment in buildings (NCEER-92-0012)*, in *Department of Civil Engineering*. 1992, State Univeristy of New York: New Yotk.
11. PS Balaji, Leblouba Moussa, ME Rahman ,HH Lau 2016 An analytical study on the static vertical stiffness of wire rope isolators *Journal of Mechanical Science and Technology* **30** 287-295.
12. G Massa, Giandomenico Pagano, Stefano Rocca, Ernesto Strano, Salvatore 2013 *Sensitive equipments on WRS-BTU isolators Meccanica* **48** 1777-1790.
13. S.Alessandri, Giannini R, Paolacci F, Malena M 2015 Seismic retrofitting of an HV circuit breaker using base isolation with wire ropes. Part 1: Preliminary tests and analyses *Engineering Structures* **98** 251-262.
14. Alessandri, S.Giannini, R.Paolacci, F. Malena 2015 Seismic retrofitting of an HV circuit breaker using base isolation with wire ropes. Part 2: Shaking-table test validation *Engineering Structures* 2015 **98** 263-274.

IOP Conf. Series: Materials Science and Engineering **217** (2017) 012033 doi:10.1088/1757-899X/217/1/012033

Multi Response Optimization of Laser Micro Marking Process:A Grey- Fuzzy Approach

[1]I.Shivakoti, [1]P.P Das, [2]G.Kibria, ,[1]B.B. Pradhan, [1]Z.Mustafa,[1]R.K. Ghadai

[1] Mechanical Engineering Department, Sikkim Manipal Institute of Technology, Sikkim, India

[2] Mechanical Engineering Department, Aliah University, Kolkata, India.

E-mail: ishwer.s@smit.smu.edu.in

Abstract. The selection of optimal parametric combination for efficient machining has always become a challenging issue for the manufacturing researcher. The optimal parametric combination always provides a better machining which improves the productivity, product quality and subsequently reduces the production cost and time. The paper presents the hybrid approach of Grey relational analysis and Fuzzy logic to obtain the optimal parametric combination for better laser beam micro marking on the Gallium Nitride (GaN) work material. The response surface methodology has been implemented for design of experiment considering three parameters with their five levels. The parameter such as current, frequency and scanning speed has been considered and the mark width, mark depth and mark intensity has been considered as the process response.

1. Introduction

The laser marking is the method of producing a mark on the surface of the work material. It is considered to be an efficient marking process as compared to traditional marking process. It produces clean and rapid marking onto the work material [1]. The demand of laser marking is increasing rapidly due to its diversified application . The product traceability and its identification is one of the important applications of laser marking [2].The fuzzy logic was first introduced in 1965 by Lotfi A. Zadeh[3].Fuzzy logic technique deals with the information which is uncertain. The hybrid approach of Grey based fuzzy shows the great improvement in the manufacturing process [4-5].

The improper selection of process parameters always shows the negative effect on the machining process[6]. The Grey based fuzzy logic when implemented for process parameter optimization of CNC milling shows a significant improvement in the process [7]. The Grey based Fuzzy logic technique is suitable for all the machining process. The hybrid aspect of Grey-Fuzzy provides good and understandable results [8]. The hybrid approach of Grey-Fuzzy logic technique combined with RSM design of experiment can be effectively used for process parameter optimization in EDM to achieve the minimum aspect of surface integrity [9].

The Grey-Fuzzy decision system can be used to solve the discrete multi-attributes decision problem in the manufacturing firms for process improvement [10]. The Grey- Fuzzy logic technique improves the grey-fuzzy reasoning grade and provides fewer uncertainties in the output as compared to grey relational analysis technique [11]. The paper presents the selection of optimal process parametric combination for laser beam micro marking process in order to obtain the effective marking on the Gallium Nitride (GaN) work material.

IOP Conf. Series: Materials Science and Engineering **217** (2017) 012033 doi:10.1088/1757-899X/217/1/012033

2. Experimental Planning and Methodology

The experiment was conducted on Laser Machine (Nd-YAG, Diode pumped). The parameter such as current, frequency and scanning speed with their five levels are considered for designing the experiment using Response surface methodology. As per the design the experiment has been conducted and responses such as mark width, mark depth and mark intensity are calculated.

Table 1. Parameters and their levels

Parameters	Unit	Levels				
		L1	L2	L3	L4	L5
Current	A	18	18.5	19	19.5	20
Pulse frequency	Hz	4000	4200	4400	4600	4800
Scanning speed	mm/s	5	7	9	11	13

The mark width and mark depth are evaluated using precision microscope and the image taken from the microscope has been processed through image processing in Matlab to calculate the mark intensity. As per the design of experiment twenty set of experiments has been conducted. The Gallium Nitride which is considered to be a difficult-to-cut material due to high hardness has been used as the work material. Table 1 shows the parameter and their levels.

3. Optimization using Grey-Fuzzy logic Technique

The Grey relational analysis soft-tool can be used to handle those data which is having a multiple response characteristics. The Grey system forms the basis for the Grey relational analysis where some part of information is provided and some part of information is unknown. Usually this situation is termed as grey i.e. neither white nor black. The Grey relational analysis involves following steps [12].

3.1. Normalization

The Normalization is the process of converting the experimental data set into the range between 0 to 1. The process response such as mark width, mark depth and mark intensity has been normalized. The response which is higher the better type was normalized using the equation 1 and the response of lower the better type was normalized using equation 2. Table 2 shows the experimental results and its normalized value.

$$m_{ij}=(x_{ij}-\min(x_{ij}))/(\max(x_{ij})-\min(x_{ij})) \quad (1)$$

$$m_{ij}=(\max(x_{ij})-(x_{ij}))/(\max(x_{ij})-\min(x_{ij})) \quad (2)$$

where m_{ij} is normalized value of x_{ij} for response j (j = 1,2,3,…., n) of experiment i (i = 1,2,3,…, m).

3.2. Grey relational coefficient

The grey relational coefficient determines the closeness between x_{ij} to x_{0j}. Higher the grey relational coefficient depicts x_{ij} is closer to x_{0j}. Table 3 shows the Grey relational coefficient values. The Grey relational coefficient is calculated as follows

$$\gamma\left(x_{0j}, x_{ij}\right)=\left(\Delta_{\min}+\xi\Delta_{\max}\right)/\left(\Delta_{ij}+\xi\Delta_{\max}\right) \quad (3)$$

The distinguish coefficient is taken as 0.5. Further the Grey relational coefficient has been used in Fuzzy logic in order to evaluate Grey-fuzzy relational grade (GFRG).

3.3. Fuzzy logic

In this research work Fuzzy logic is coupled with the Grey relational analysis. The experimental data are normalized in order to avoid unit influence for mark width, mark depth and mark analysis. Fuzzy logic consists of fuzzification and defuzzification. The fuzzification is the method for converting the input data into linguistic variable. The linguistic variable is assigned as high, medium and low. The fuzzy interference system is based on IF-THEN rules. The defuzzification converts the output data in terms of numerical form. Fig 1, 2& 3 shows the membership function for input and output and Fuzzy rules respectively.

Table 2. Experimental data and its normalized values

	Parametric combination			Experimental results			Normalized value		
Expt. No	Current (A)	Frequency (Hz)	Scanning Speed (mm/sec)	Mark width (μm)	Mark Depth (μm)	Mark Intensity	Mark width (μm)	Mark depth (μm)	Mark Intensity (μm)
1	18.5	4200	7	143.90	55.94	0.255	0.1916	0.3499	0.3117
2	19.0	4400	9	141.97	62.93	0.325	0.242	0.4403	0.5951
3	19.0	4000	9	149.14	89.11	0.425	0.0554	0.7789	1
4	19.0	4400	9	134.40	69.58	0.377	0.4388	0.5263	0.8057
5	19.0	4800	9	134.01	103.92	0.391	0.4489	0.9704	0.8623
6	20.0	4400	9	149.53	106.21	0.311	0.0454	1	0.5385
7	19.0	4400	5	130.91	99.29	0.321	0.5295	0.9105	0.5789
8	19.5	4200	11	148.56	102.33	0.411	0.0705	0.9498	0.9433
9	18.5	4600	7	115.78	73.04	0.301	0.923	0.5711	0.498
10	18.5	4600	11	117.91	56.86	0.281	0.8676	0.3618	0.417
11	18.0	4400	9	112.82	28.88	0.178	1	0	0
12	19.5	4600	11	151.27	74.98	0.303	0	0.5961	0.5061
13	19.0	4400	9	143.66	63.73	0.326	0.1979	0.4507	0.5992
14	19.0	4400	13	136.73	63.30	0.343	0.3783	0.4451	0.668
15	19.0	4400	9	144.88	57.94	0.355	0.1663	0.3758	0.7166
16	19.0	4400	9	140.41	53.63	0.353	0.2824	0.3201	0.7085
17	19.0	4400	9	138.86	71.25	0.343	0.3227	0.5479	0.668
18	19.5	4600	7	127.38	98.76	0.312	0.6212	0.9037	0.5425
19	19.5	4200	7	149.72	104.02	0.324	0.0403	0.9717	0.5911
20	18.5	4200	11	131.86	61.59	0.306	0.5048	0.423	0.5182

Table 3. Grey relational coefficient

Sl No	Mark width (μm)	Mark depth (μm)	Mark Intensity
1	0.4347	0.4208	0.3821
2	0.4718	0.5525	0.3975
3	0.6934	1	0.3461
4	0.5135	0.7201	0.4712
5	0.9441	0.7841	0.4757
6	1	0.52	0.3437
7	0.8482	0.5428	0.5152
8	0.9088	0.8981	0.3498
9	0.5383	0.499	0.8666
10	0.4393	0.4617	0.7906

11	0.3333	0.3333	1
12	0.5532	0.5031	0.3333
13	0.4765	0.5551	0.384
14	0.474	0.601	0.4458
15	0.4448	0.6382	0.3749
16	0.4238	0.6317	0.4106
17	0.5252	0.601	0.4247
18	0.8385	0.5222	0.569
19	0.9464	0.5501	0.3425
20	0.4643	0.5093	0.5024

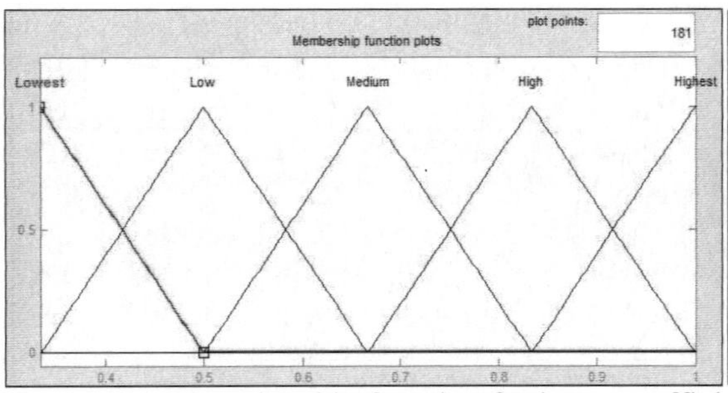

Fig. 1 Membership function for input coefficient

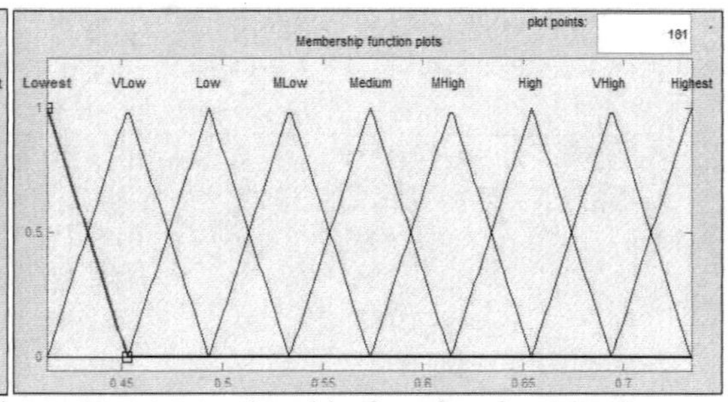

Fig. 2 Membership function for GFRG

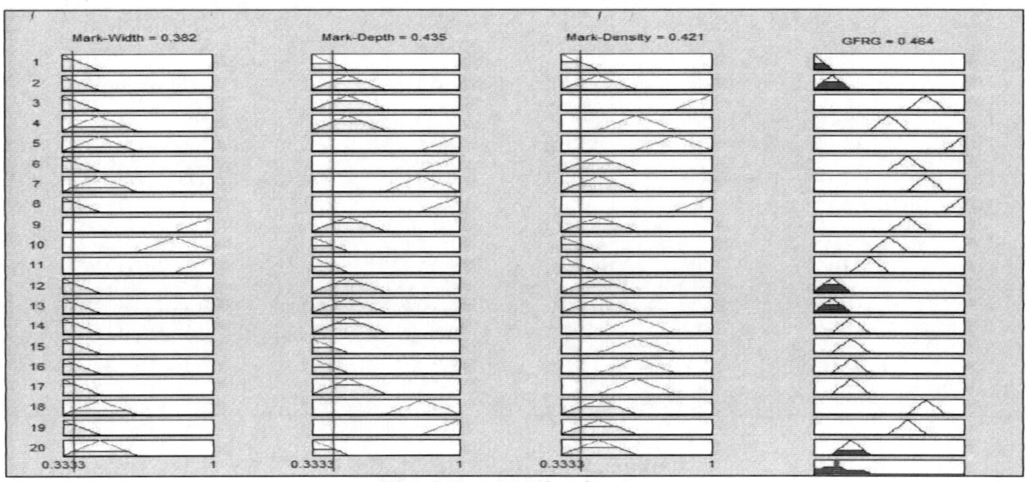

Fig. 3 Fuzzy rule viewer

Table 4. Grey- Fuzzy relational grade

Sl No	Mark width (µm)	Mark depth (µm)	Mark Intensity (µm)	Grey-Fuzzy relational grade (GFRG)	Rank
1	0.4347	0.4208	0.3821	0.464	19
2	0.4718	0.5525	0.3975	0.501	17
3	0.6934	1	0.3461	0.574	9
4	0.5135	0.7201	0.4712	0.553	10
5	0.9441	0.7841	0.4757	0.721	1
6	1	0.52	0.3437	0.614	6
7	0.8482	0.5428	0.5152	0.654	3

8	0.9088	0.8981	0.3498	0.719	2
9	0.5383	0.499	0.8666	0.614	7
10	0.4393	0.4617	0.7906	0.574	8
11	0.3333	0.3333	1	0.533	11
12	0.5532	0.5031	0.3333	0.454	20
13	0.4765	0.5551	0.384	0.499	18
14	0.474	0.601	0.4458	0.523	12
15	0.4448	0.6382	0.3749	0.51	16
16	0.4238	0.6317	0.4106	0.52	13
17	0.5252	0.601	0.4247	0.516	14
18	0.8385	0.5222	0.569	0.654	4
19	0.9464	0.5501	0.3425	0.617	5
20	0.4643	0.5093	0.5024	0.511	15

Table 5. Response table for GFRG

Levels	Current (Amp)	Frequency (Hz)	Scanning Speed (mm/sec)
1	0.533	0.574	**0.654**
2	0.540	0.577	0.587
3	0.557	0.543	0.554
4	0.608	0.574	0.564
5	**0.614**	**0.721**	0.523

Table 6. ANOVA for GFRG

Source	DF	SS	MS	F	P	%
Regression	3	0.1060	0.0353	58.16	0.000	19.46
Linear	3	0.1060	0.0353	58.16	0.000	19.46
Current (Amp)	1	0.0759	0.0714	117.51	0.000	39.31
Frequency (Hz)	1	0.0014	0.0108	17.91	0.001	5.99
Scanning speed (mm/sec)	1	0.0286	0.0286	47.20	0.000	15.79
Residual error	16	0.0097	0.0006			
Total	19	0.1158				

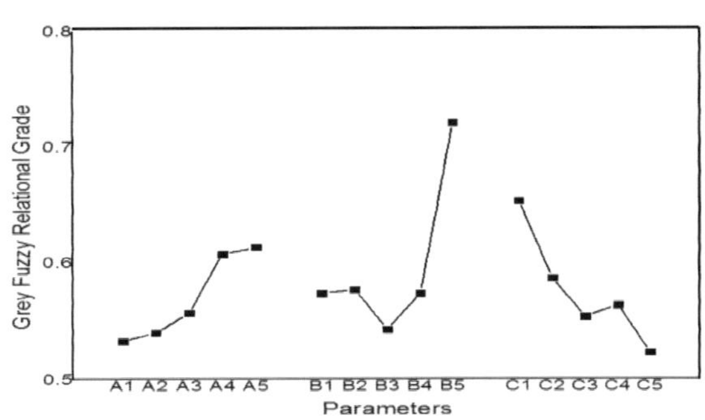

Fig 4. Response plot for GFRG

IOP Conf. Series: Materials Science and Engineering **217** (2017) 012033 doi:10.1088/1757-899X/217/1/012033

The Grey relational coefficient value is processed through the Fuzzy logic technique and the Grey-Fuzzy relational grade has been calculated. The Fuzzy IF-THEN rule has been utilized for the same and the defuzzification has been done in order to convert the linguistic variable into numerical value. Table 4 shows the Grey-Fuzzy relational grade (GFRG).

4. Results and Discussion

The Grey relational coefficient is used to determine the GFRG. The ranking of GFRG has been done.The parametric combination having the highest value of GFRG represents the strong correlation with the reference sequence. Table 5 sows the response table for GFRG.The response table has been made considering the different levels at which the marking is performed. The maximum value of each level has been used to select the optimal values. Fig 4 shows the response plot for GFRG. The table revels that the optimal parametric combination for minimum mark width, maximum mark depth and maximum mark intensity are 20Amp/ 4800 Hz/ 5 mm/sec.The ANOVA for GFRG is shown in table 6 in order to validate the optimum levels obtained. The p value in the ANOVA table is less than 0.05 which depicts that the obtained GFRG is significant. Further the percentage contribution of each parameters on the response has been calculated in order to observe the most significant parameter. The current is the most contributing factor having the highest percentage followed by scanning speed and frequency.

5. Conclusion

The paper discuss the Fuzzy based Grey relational optimization for optimizing multi-criteria performance. RSM based design of experiment has been adopted and the micro marking has been performed in the Gallium Nitride (GaN) work material. The grey relational coefficient is used for evaluating Grey-Fuzzy relational grade and the ranking of GFRG has been done. The optimal parametric combination obtained is 20Amp/ 4800 Hz/ 5 mm/sec. Further the ANOVA for GFRG has been performed in order to validate the GFRG values. The percentage contribution of each parameter on the response has been obtained. The obtained results may be utilised for laser beam micro marking of Gallium Nitride (GaN).

References

[1] T.W.Ng, S.C.Yeo, Journal of Material Processing Technology, Vol 104, 2000, pp. 280-283.

[2] Chen, T.C., Darling, R.B, Journal of Materials Processing Technology, Vol 169, 2005, pp. 214-218.

[3] Zadeh LA, Fuzzy sets. Information and control, Vol 8, 1965, pp 338-353.

[4] Lin C.L, Lin J.L, Ko.T, International Journal of Advance Manufacturing Technology, Vol 19, 2002, pp. 271-277.

[5] Lin J.L, Lin C.L, Journal of Material Processing Technology, Vol 160, 2004, pp. 9-14.

[6] I.Shivakoti, G.Kibria, S.Diyaley, B.B. Pradhan, International Journal of Scientific & Engineering research, Vol 3, 2012, pp. 1-6.

[7] Biswajit Das, S.Roy, R.N. Rai, S.C. Saha, Engineering Science and Technology an International Journal, Vol 19, 2016, pp. 857-865.

[8] N. Tamiloli, J. Venkatesan, B. Vijaya Ramnath, Measurement, Vol 84, 2016, pp. 68-82.

[9] Shailesh Dewangan, Soumya Gangopadhyay, Chandan Kumar Biswas, Engineering Science and Technology, an International Journal, Vol 18, 2015, pp. 361-368.

[10] S.M. Mousavi, S.mirdamadi, A. Siadat, J. Dantan, R. Tavakkoli-Moghaddam, Engineering Applications of Artificial Intelligence, Vol 39, 2015, pp. 157-167.

[11] P.Suresh, K. marimuthu, S. Ranganathan, T. Rajmohan, Transactions of Nonferrous Metals Society of China, Vol 24, 2014, pp. 2805-2814.

[12] I.Shivakoti, G.Kibria, S.Diyaley, B.B.Pradhan, Journal of Computational and Applied research in Mechanical Engineering, Vol 3, 2013, pp. 27-39.

Grey fuzzy logic approach for the optimization of DLC thin film coating process parameters using PACVD technique

[1]R.K. Ghadai*, [1]P.P. Das, [1]I.Shivakoti, [2]S. C. Mondal, [3]B. P. Swain

[1]Mechanical Engineering Department, Sikkim Manipal Institute of Technology, Sikkim, India
[2]Mechanical Engineering Department, Indian Institute of Engineering science and Technology, Shibpur, India.
[3]Centre for Materials Science and Nanotechnology, Sikkim Manipal Institute of Technology, Sikkim, India

E-mail: ranjankumarbls@gmail.com

Abstract. Diamond-like carbon (DLC) coatings are widely used in medical, manufacturing and aerospace industries due to their excellent mechanical, biological, optical and tribological properties. The selection of optimal process parameters for efficient characteristics of DLC film is always a challenging issue for the materials science researchers. The optimal combination of the process parameters involved in the deposition of DLC films provide a better result, which subsequently help other researchers to choose the process parameters. In the present work Grey Relation Analysis (GRA) and Fuzzy-logic are being used for the optimization of process parameters in DLC film coating by using plasma assist chemical vapour deposition (PACVD) technique. The bias voltage, bias frequency, deposition pressure, gas composition are considered as input process parameters and hardness (GPa), Young's modulus (GPa), ratio between diamond to graphic fraction, (I_D/I_G) ratio are considered as response parameters. The input parameters are optimized by grey fuzzy analysis. The contribution of individual input parameter is done by ANOVA. In this analysis found that bias voltage having the least influence and gas composition has highest influence in the PACVD deposited DLC films. The grey fuzzy analysis results indicated that optimum results for bias voltage, bias frequency, deposition pressure, gas composition for the DLC thin films are -50 V, 6 kHz, 4 μbar and 60:40 % respectively.

1. Introduction

Diamonds like carbon (DLC) films have attracted to many researchers due to their excellent mechanical, optical, biological and tribological properties [1]. DLC is an amorphous carbon that contains sp^3 (diamond-like) bonding where sp^2 (graphitic) as the defective carbon fraction, depending on deposition techniques. Due to its excellent properties DLC films have been used in many industrial sectors such as medical, manufacturing and aerospace industries.[1]. For the deposition of DLC thin film various deposition techniques are used and many people have studied the influence of different deposition parameters in a systematic manner [2]. During the deposition of DLC film various input parameters like bias voltage, bias frequency, deposition pressure, gas composition and electrode separation are involved in various techniques. The problem is to determine the optimal response parameters like coating thickness, surface roughness, coating material, hardness, elasticity, residual stress, and fracture toughness of the coating, bond layer, and substrate [3]. Taguchi technique is used to optimize the deposition parameters like bias voltage (V), bias frequency (kHz), deposition pressure (μbar), gas composition (%) of DLC Coatings with the IC-PECVD technique to get optimum response parameters i.e. hardness and roughness (Ra) [4].

The multi-response optimization problems could be solved by using different methods such as genetic algorithm (GA), artificial neural network (ANN) grey relational analysis (GRA), response surface methodology (RSM) and fuzzy logic [5]. GRA and fuzzy logic are used as the effective tool for the multi objective optimization. The grey relational grade (GRG) is calculated by doing average of the grey relational coefficient of each response to convert the optimization of the complex performance characteristics into optimization of a single GRG [6]. In the present work GRA and fuzzy logic are being used for the optimization of process parameters in DLC film coating by using PACVD technique.

2. Experimental Description and

Jatti et al. [7] deposited DLC thin film by using plasma enhanced chemical vapor deposition (PECVD) technique. In the experiment the reaction chamber was powered by a 13.56 MHz RF power (50 W) source. For the deposition of the DLC film a p-type silicon wafer (100) of 1.5 cm x 1.5 cm square area of 500 μm thickness was taken as substrate. The total flow rate of methane (CH_4) and hydrogen gases were maintained as 10 sccm by varying the composition of the precursor gases. In the present case bias voltage, bias frequency, deposition pressure and gas composition were considered as input process parameters [Table 1]. An L9 Taguchi orthogonal array was used for the experimental design and it is indicated in table 1.

Table 1. Parameters and their levels [7]

Parameters	Unit	Levels		
		L1	L2	L3
Bias voltage	V	-50	-100	-150
Bias frequency	kHz	0.2	6	40
Deposition pressure	μbar	2	4	6
Gas composition	%	60:40	80:20	90:10

In the present work hardness (GPa), Young's modulus(GPa), I_D/I_G Ratio are considered as response parameters. The mechanical properties like hardness and Young's modulus were found out by nano indentation. From Raman spectra the I_D/I_G ratio was obtained by computing the areas of D-peak and G-peak and taking their ratio.

3. Methodology

3.1 Grey relation analysis(GRA)

Unlike Taguchi the Grey relational analysis (GRA) is used to optimize the multi objective process outcome which make it complicated [8]. The following three steps are used to optimize the process parameter in GRA.

In the first step, the measurement values of centre line average I_D/I_G ratio, hardness (GPa) and Young's modulus (GPa) are to normalize in the range of zero to one which is called grey relational normalization. Such normalization is required since the range and the unit in one response may vary from the others. For the case of hardness and Young's modulus the response characteristics will be 'higher-the-better' and the normalizing equation for them is as follows in equation (1)

$$z_{ij} = (x_{ij} - Min(x_{ij}))/(Max(x_{ij}) - Min(x_{ij})) \quad (1)$$

But in case of I_D/I_G ratio the response characteristics will be 'lower-the-better' and the normalizing equation is stated in equation 2.

$$z_{ij} = (Max(x_{ij}) - x_{ij})/(Max(x_{ij}) - Min(x_{ij})) \quad (2)$$

Where z_{ij}= Normalized value for i[th] experiment and for j[th] response, x_{ij}) = Data obtained for the experiments. Max(x_{ij})) and Min(x_{ij}) are the maximum and minimum value of data sequence of a particular experimental outcome.

$$\varphi_{ij} = (\Delta_{min} - \delta\Delta_{max})/(\Delta_{ij} + \delta\Delta_{max}) \quad (3)$$

Where deviation sequence is denoted by Δ_{ij}, that is the difference between x_{0j} and x_{ij}, x_{oj} is the ideal normalized value or the target value of the jth response, Δ_{max} is the highest value of Δ, $\Delta_{min}$is the lowest value of Δ, δ is the distinguishing coefficient usually ranges from 0 to 1, in this case it is taken as 0.5.

$$\gamma_i = \frac{1}{n}\sum_{j=1}^{n}\varphi_{ij} \quad (4)$$

Where $\gamma_i$the grey relational grade and n is the number of process response.

3.2. Fuzzy based modeling

In grey relational analysis, the use of lower-the-better, higher-the-better performance characteristics shows that some kind of uncertainty presents in the results. This problem can be efficiently checked by using fuzzy logic [9].

In fuzzy-logic, the fuzzifier uses different membership function to fuzzify the input grey relational coefficient of the input parameters. Here triangular membership function is taken with five levels for each of the three input value and nine levels for output GFRG In total nine fuzzy rules are developed from the nine different experiments by correlating the grey relational coefficient with GRG. Mamdani inference engine is use to defuzzify the input values. This inference engine performs fuzzy reasoning on fuzzy rules by taking max–min inference for generating a fuzzy value.

4. Results and Discussion

The experimental results are given in table 2. Here the results obtained by using grey relational analysis, grey fuzzy reasoning grey and ANOVA are presented. From the experimental results the normalized data can be calculated by using equation (1) and (2). The grey relational coefficients and the overall grey relational grade for each of the combination of the parameters is given in Table 3. For the response parameter I_D/I_G ratio 'lower-the-better' criterion is preferred and for hardness and Young's modulus 'higher-the-better' criterion is preferred. In order to obtain a better quality in the performances and to decrease the problem in the data, grey-fuzzy logic method is additionally used for computing the grey fuzzy reasoning grey (GFRG).

Table 2. Input and response parameters [7]

	Input parameters				Response parameters		
Expt. No	Bias voltage(V)	Bias Frequency(kHz)	Deposition Pressure(μbar)	Gas composition (%)	I_D/I_G ratio	Hardness (Gpa)	Young's Modulus (Gpa)
1	-50	0.25	2	60:40	0.11	17.73	186.74
2	-100	6	4	80:20	0.1	16.79	189.94
3	-150	40	6	90:10	0.3	15.63	173.86
4	-50	0.25	2	60:40	0.44	16.57	175.66
5	-100	6	4	80:20	0.32	17.96	191.22
6	-150	40	6	90:10	0.33	16.65	172.81
7	-50	0.25	2	60:40	0.74	16.09	185.46
8	-100	6	4	80:20	0.77	16.88	175
9	-150	40	6	90:10	0.17	17.91	188.59

To generate the grey fuzzy output, MATLAB (2012a) toolbox is used. The grey coefficients of I_D/I_G, hardness and Young's modulus are considered to be the inputs for which triangular membership function having five fuzzy subsets described by lowest, low, medium, high and highest as shown in Fig.1. Similarly for the grey fuzzy reasoning grey (GFRG) nine fuzzy subsets denoted by lowest, very low, low, medium low, medium, medium high, high, very high and highest as shown in Fig.2. Nine fuzzy rules are generated based on the nine experiments for activating the fuzzy inference system using the MATLAB toolbox as shown in Fig.2. In Fig. 3 the location of triangle indicates the determined fuzzy sets for each input and output. The height of the darkened area in each triangle corresponds to the fuzzy membership value for that fuzzy set. The yellow colour shows the input membership function and the blue colour shows the output membership function. It is seen that when the input for I_D/I_G is 0.827, hardness is 0.959 and Young's modulus is 0.778, than the GFRG is 0.848. Similarly, the GFRG values for all the nine experiments are calculated and shown in Table 3. From the Table 3 it is seen that the highest GFRG value of experiment number 5 and 9 are the optimal combination of given parameters.

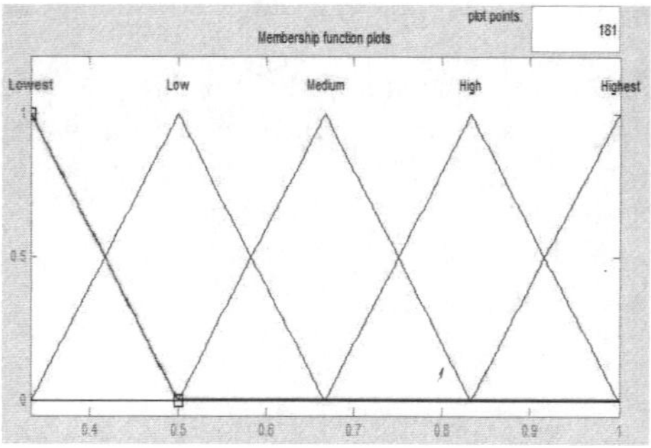

Fig. 1 Membership function for input

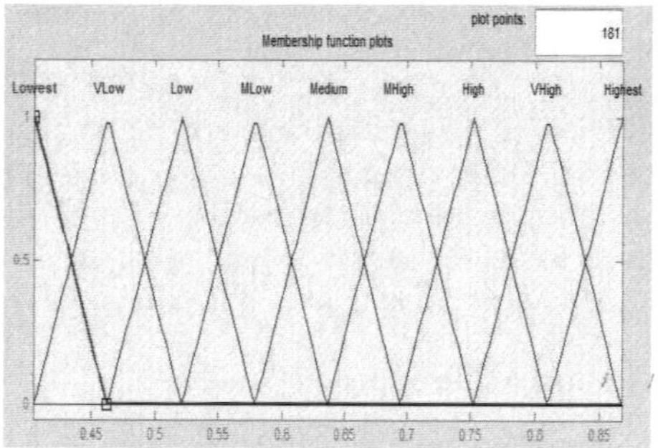

Fig. 2 Membership function for GFRG

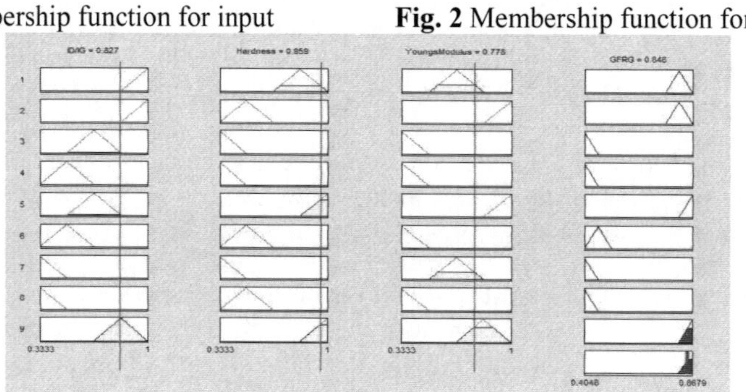

Fig. 3 Fuzzy rule viewer

Table 3: Normalized value and Grey relational coefficient

Sl no	I_D/I_G ratio	Hardness (Gpa)	Young's Modulus (Gpa)	I_D/I_G ratio	Hardness (Gpa)	Young's Modulus (Gpa)	GRG	GFRG
	Normalized data			Grey relational coefficient				
1	0.9851	0.9013	0.8263	0.81	0.8351	0.6727	0.8263	0.81
2	1	0.4979	0.7923	0.81	0.499	0.878	0.7923	0.81
3	0.7015	0	0.4353	0.424	0.3333	0.3465	0.4353	0.424
4	0.4925	0.4034	0.4413	0.46	0.456	0.3717	0.4413	0.46
5	0.6716	1	0.8679	0.848	1	1	0.8679	0.848
6	0.6567	0.4378	0.4656	0.461	0.4707	0.3333	0.4656	0.461
7	0.0448	0.1974	0.4475	0.424	0.3838	0.6151	0.4475	0.424

8	0	0.5365	0.4048	0.423	0.5189	0.3621	0.4048	0.423
9	0.8955	0.9785	0.8545	0.848	0.9588	0.7777	0.8545	0.848

Table 4: Response table for GFRG

Parameters	Levels		
	L1	L2	L3
Bias voltage(V)	**0.6813**	0.589	0.565
Bias frequency(kHz)	0.5647	**0.693**	0.5777
Deposition pressure(μbar)	0.5647	**0.706**	0.5653
Gas composition (%)	**0.8353**	0.565	0.4357

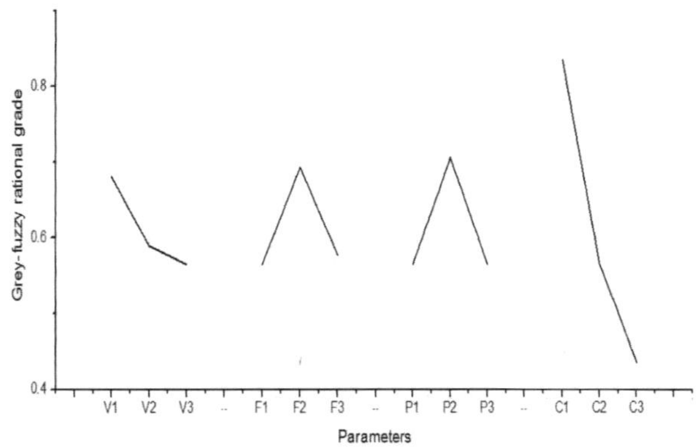

Fig. 4 GFRG response plot

The average values of grey fuzzy reasoning grade values are tabulated in Table 4. The average values are plotted in the graph which is shown in Fig. 4. From the graph it is found that the optimum results for bias voltage, bias frequency, deposition pressure, and gas composition for the DLC thin films are -50 V (V1), 6 kHz (F2), 4 μbar (P2) and 60:40 % (C1) respectively.

4.1 ANOVA for GFRG

To find the significance of different input parameters the grey-fuzzy reasoning grade obtained is subjected to ANOVA which is shown in Table 5. As the degrees of freedom for residual error is zero, it does not provide enough data's. When four input parameters with three level values are considered and an L9 Orthogonal Array is chosen for analysis, this kind of things happens. Due to that pooling is performed and is defined as a common practice of revising and re-estimating ANOVA results. For two reasons it is recommended. First one is when a number of factors are included in an experiment, the laws of nature make it probable that half of them would be more influential than the rest and the second one is in statistical predictions, which encounters two types of mistakes: alpha and beta mistakes. When something is important but actually is not, that is called alpha mistake and the beta mistake is just the reverse of the alpha mistake. Unfortunately, the test of significance can be done only when the error term has non zero DoF. Pooling is started with the factor which has the least influence. In this analysis, Bias voltage is having the least influence; hence it is pooled as shown in Table 6.

Table 5: ANOVA table before pooling

Source	DOF	SS	MSS	F value	P value
Bias voltage (V)	2	0.022545	0.011272	**	
Bias frequency (kHz)	2	0.030266	0.015133	**	
Deposition pressure (μbar)	2	0.039763	0.019881	**	
Gas composition (%)	2	0.249541	0.124770	**	
Error	0	*	*		
Total	8	0.342114			

Table 6: ANOVA table after pooling

Source	DOF	SS	MSS	F value	P value
Bias frequency (kHz)	2	0.03027	0.01513	1.34	0.427
Deposition pressure (μbar)	2	0.03976	0.01988	1.76	0.362
Gas composition (%)	2	0.24954	0.12477	11.07	0.083
Error	2	0.02254	0.01127		
Total	8	0.03211			

The Grey relational coefficient value is processed through the Fuzzy logic technique and the Grey-Fuzzy relational grade has been calculated. The Fuzzy IF-THEN rule has been utilized for the same and the defuzzification has been done in order to convert the linguistic variable into numerical value. Table 3 shows the Grey-Fuzzy relational grade (GFRG).

5. Conclusion

In this paper, the deposition of DLC film is carried out with the input parameters considered as bias voltage, bias frequency, deposition pressure and gas composition, and the response obtained are hardness, Young's Modulus and I_D/I_G ratio. The input parameters optimized by grey fuzzy analysis. The contribution of individual input parameter is done by ANOVA. In this analysis it is found that bias voltage is having the least influence and gas composition has highest influence. From grey fuzzy analysis it is found that for -50 V bias voltage, 6 kHz bias frequency, 4 μbar deposition pressure and 60:40 % gas composition, the optimum result will obtained.

References

[1] K. Bewilogua, D. Hofmann, Surf. Coat. Tech., Vol 242, 2014, pp. 214- 225.
[2] S.V. Hainsworth and N.J. Uhure, Int. Mater. Rev., Vol 52, 2007, pp. 153-174.
[3] O. Keles, Y. Taptik, O. L. Eryilmaz, M. Urgen, A. F. Cakir, Quality Engg., Vol 12, 1999, pp.29–36.
[4] T. P. Singh,Vijaykumar S. Jatti, Particulate Sci. Tech., Vol 33, 2015, pp. 119–123.
[5] K. Krishnaiah, P. Shahabudeen, PHI Learning Private Ltd, New Delhi, India, 2012.
[6] Y.F. Hsiao, Y.S. Tarng and W.J. Huang, Mater. Mfg Processes, Vol 23, 2008, pp. 51-58.
[7] Vijay Kumar S. Jatti, Meena Laad and T.P. Singh, Procedia Mater. Sci., Vol 6, 2014, pp. 1017 – 1023.
[8] P.N. Singh, K. Raghukandan and B.C. Pai, J. Mater. Proc. Tech., Vol. 155-156, 2004, pp. 1658–1661.
[9] J. Deng, The Journal of Grey System, Vol 1, 1989, pp. 1–24.

IOP Conf. Series: Materials Science and Engineering **217** (2017) 012035 doi:10.1088/1757-899X/217/1/012035

Modelling batch microwave heating of water

S P Yeong[1], M C Law[1] C C Vincent Lee[1] and Y S Chan[2]

[1] Mechanical Engineering Department, Curtin University Malaysia, CDT 250, 98009 Miri, Sarawak, Malaysia.

[2] Chemical Engineering Department, Curtin University Malaysia, CDT 250, 98009 Miri, Sarawak, Malaysia.

E-mail: yeong.siew.ping@postgrad.curtin.edu.my

Abstract. A numerical model of the microwave heating of distilled water is developed using COMSOL Multiphysics software to investigate the microwave effects on the heating rate. Three frequencies (0.915GHz, 2GHz and 2.45 GHz) have been applied in the model in order to study their influences on the water temperature. It is found that the water heats up at 2GHz and 2.45GHz, however, there is no sign of heating at 915MHz. This is supported with the figures of the electric field distribution in the microwave cavity. The results shown in the developed model is validated with the experimental results obtained at 2.45 GHz.

1. Introduction

Microwave is an alternative heat source which has attracted higher demands in industrial and household applications over the years. In contrast to conventional heating where heat is transferred from the surface to the inner volume of the object, microwave penetrates into materials and quickly converted into thermal energy within the volume. This instantaneous volumetric heating can heat up an object much faster and reduce the process time required.

However, one major disadvantage which prohibits the advancement in microwave technology is the uneven heating distribution within a material due to non-uniform electric field patterns. In order to reduce the uneven heating pattern, rotating glass turntable plate are used in the domestic microwave ovens to rotate the foods during heating. However, this method might not be suitable for industrial applications where the equipment are not designed to rotate. Hence, there is a need to investigate the natural heat convection in the sample to better understand the microwave dielectric heating phenomena.

As compared to microwave heating solid specimen, liquid samples are considerably more challenging to perform analysis due to the presence of fluid motion. This would lead to the complex interactions of flow fields, microwave fields, and local distributions of temperature within the liquid.

The numerical models developed are mostly solving Maxwell's equations, heat transfer and fluid flow equations [1-3]. Some uses Lambert's Law [4] to solve the electromagnetic field when the samples are thicker and have a larger depth than the penetration depth of microwaves, causing the microwave intensity to reduce exponentially with the depth. For samples with smaller, electromagnetic field inside the sample are usually solved using Maxwell's equation as the heat transfer rate is faster.

Zhang et al. have developed a three dimensional model to predict the distribution of electromagnetic fields, temperatures, and velocities of liquids enclosed in a container which is located within a microwave cavity. They have reported that convection plays an important role on the temperature distribution within the liquid [5]. Thus, fluid flow during a heating process should be investigated. On the other hand, Sabliov et al. numerically investigated the heating of a continuous flow liquid in a microwave system using ANSYS Multiphysics [6]. It was assumed that the dielectric properties of the liquid are temperature independent as their goal is to predict the effect of flow rate on the temperature distribution at 915MHz. The dielectric properties of the water they used (a constant value) is calculated from the empirical formula determined by Komarov and Tang [7] for tap water.

Ratanadecho et al. have investigated the microwave heating of a liquid layer both numerically and experimentally. Their work showed that the electric conductivity value of a liquid could strongly affect the degree of microwave penetration and heating rate within the liquid layer [1]. This conductivity rises as the concentration of ions present in the liquid increases. Cha-um et al. have numerically and experimentally studied the microwave heating of water and oil which possess different dielectric properties, using a rectangular waveguide. The temperature profiles and velocity field within sample are determined to examine the effects of microwave power level, position of the waveguide, and liquid layer size and thickness on the liquid heating [2]. Nonetheless, a research conducted by Rattanadecho [8] on wood reveals that microwave frequency could also affect the temperature distribution within a sample.

Although most of the previous investigations considered simulations of microwave heating in liquid, a little effort has been reported on performing investigation on the influence of frequencies in heating liquid. This work serves as a preliminary study on the microwave heating of liquids and focus on predicting the effect of microwave frequencies on the heating rate as well as the flow field through a numerical model. Distilled water is selected as the test fluid.

2. Model Description

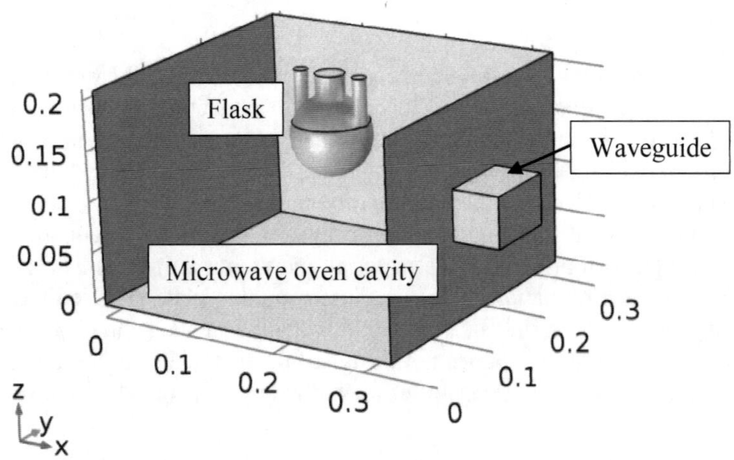

Figure 1 COMSOL model

A model as depicted in Figure 1 is developed. It consists of a waveguide, microwave resonant cavity and a three-neck flask which is filled with water. The dimension of the microwave oven and flask are listed in Table 1.

Table 1 Geometrical parameters of the COMSOL model

Parameters	Value
Oven cavity dimension (mm)	330 (w) × 309 (d) × 211 (h)
Waveguide (m)	0.05 (w) × 0.08 (d) × 0.05 (h)
Flask diameter (m)	0.08
Initial temperature, T_0 (K)	298.15
Flask height from the bottom of oven, h_f (m)	0.151

In this simulation, the relative permittivity of distilled water is assumed to vary with frequency but temperature independent. The values used are listed in Table 2 as follows:

Table 2 Relative permittivity of distilled water at room temperature (25°C)

Frequency, f (GHz)	ε'	ε''	Reference
0.915	78.9	3.60	[9]
2.00	77.9	7.57	[10]
2.45	78.0	12.50	[11]

Some assumptions that are applied in this model are as follows [2]:
1. The absorption of microwave by the air in a microwave cavity is negligible
2. The walls of the microwave oven are perfect electric conductors
3. The flask is a heat insulator and has no effect on the electromagnetic and temperature field
4. Fluid is incompressible
5. Gravity effect is considered

The numerical modelling of the microwave heating of water is carried out using COMSOL Multiphysics v5.2a software. The physic models that have been coupled in the COMSOL model are electromagnetic waves (frequency domain) under radio frequency module, heat transfer in fluids under heat transfer module, and laminar flow under fluid flow module. Radio frequency module is employed to solve Maxwell's equations, while heat transfer and fluid flow modules are used to solve Fourier and Navier Stokes equations. The radio frequency module is coupled with the fluid flow and heat transfer modules in such a way that the heat generation terms calculated using radio frequency module are then used to calculate the temperature field. Simulations are run on Intel Xeon CPU E5-2695 @ 2.3 GHz processor with 256GB RAM.

3. Governing Equations

Electromagnetic waves physic model is applied in the entire domain of the microwave oven whereas the latter two models are limited to the liquid region enclosed by the flask. The details related to each physics model are further described in the following subsections.

3.1. Electromagnetic waves, frequency domain

A rectangular port with Transverse electric (TE_{10}) mode is employed in this physics model to represent the microwave generator (magnetron). The port power is set at 300W. The modelling of microwave heating is governed by Maxwell's equation as follows:

$$\nabla\times\left(\frac{1}{\mu'}\nabla\times\vec{\mathbf{E}}\right)-\frac{\omega^2}{c^2}\left(\varepsilon'-j\varepsilon''\right)\vec{\mathbf{E}}=0 \tag{1}$$

IOP Conf. Series: Materials Science and Engineering **217** (2017) 012035 doi:10.1088/1757-899X/217/1/012035

where μ' is the relative permeability, $\vec{\mathbf{E}}$ is the electric field intensity inside the microwave cavity (V/m), ω is the angular frequency (rad/s), c is the speed of light in free space (3×10^8 m/s), ε' is the relative permittivity or dielectric constant, and ε'' is the relative dielectric loss of a material. Dielectric constant measures the ability of a material to store electrical energy whereas dielectric loss indicates the ability of a material to loss the electrical energy as heat energy. After obtaining the electric field intensity from Equation 1 and the material properties, the volumetric power P_V dissipated from the electric component of microwaves can be determined as follows (Salvi et al., 2011):

$$P_V = \sigma\left|\vec{\mathbf{E}}\right|^2 = 2\pi\varepsilon_0\varepsilon'' f\left|\vec{\mathbf{E}}\right|^2 \qquad (2)$$

where σ is the electrical conductivity (S/m) and ε_0 represents the free space permittivity (8.854×10^{-12} F/m).

Walls of the microwave cavity are assumed to be made of copper which approximates perfect electrical conductors. This indicates that the microwave generated from the magnetron could not penetrate through these boundaries and it will only exist inside the microwave cavity to heat up liquid molecules. On the other hand, the three opening ends of the three-neck flask are also defined as impedance boundaries, but with the properties of air.

3.2. Heat transfer in fluids

The heat transfer in the liquid domain is governed by Fourier energy equation as follows:

$$\rho c_p \frac{\partial \mathrm{T}}{\partial t} + \rho c_p \mathbf{u}\cdot\nabla\mathrm{T} = \nabla\cdot\left(k\nabla\mathrm{T}\right) + P_V \qquad (3)$$

where ρ is the density (kg/m^3), T is the temperature (K), c_p is the specific heat capacity (J/(kg.K)), k is the thermal conductivity (W/(m.K)), and P_V is the thermal power received from the electromagnetic wave (W). It is assumed that phase change does not take place.

Since the electromagnetic waves physics model is in frequency domain, the electric field generated in the COMSOL model would not change in a time dependent study. To compensate that, the multiphysics coupling between the electromagnetic waves and heat transfer in fluids models will be disabled and replaced with a user-defined heat source. The heat source term, retrieved from the electromagnetic waves model, is multiplied with a piecewise function to imitate the ON/OFF cycle in a microwave oven (model: Samsung ME711K), which is approximately 13s ON and 18s OFF per cycle. In addition, the flask which is made from glass is assumed to be a heat insulator. Hence, there is no heat flux from the liquid domain to the ambient air inside the cavity which remains at 20°C. The liquid domain is illustrated in Figure 2.

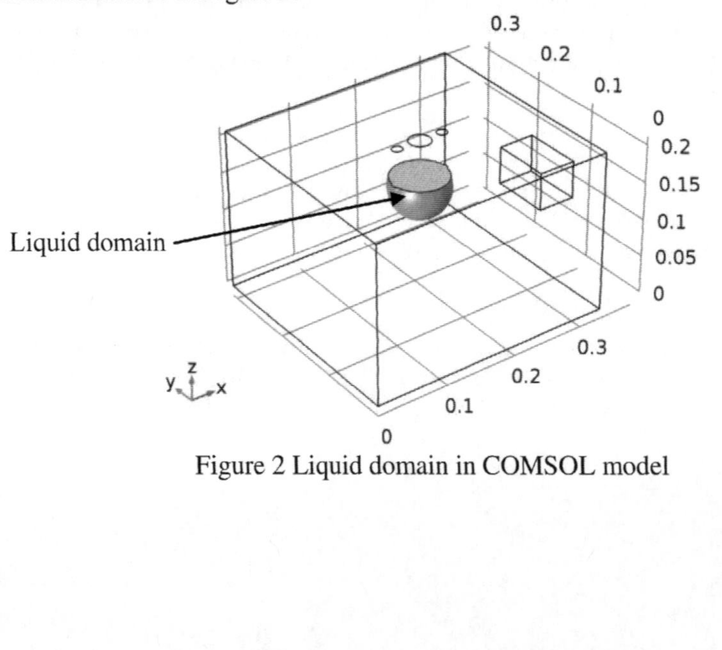

Figure 2 Liquid domain in COMSOL model

3.3. Laminar flow

The flow field of the liquid can be described by using single phase Navier-Stokes equations:

$$\frac{\partial \rho}{\partial t} + \nabla \cdot (\rho \mathbf{u}) = 0 \tag{4}$$

$$\rho \frac{\partial \mathbf{u}}{\partial t} + \rho \mathbf{u} \cdot \nabla \mathbf{u} = -\nabla p + \left[\mu \left(\nabla \mathbf{u} + (\nabla \mathbf{u})^T \right) - \tfrac{2}{3} \left(\mu (\nabla \cdot \mathbf{u}) \right) \right] - \rho g \tag{5}$$

where μ is dynamic viscosity (Pa.s), $(\nabla \mathbf{u})^T$ is transpose of $\nabla \mathbf{u}$ and g is gravitational force (kg.m/s2).

A pressure outlet with suppressed backflow is applied at the liquid-air interface. This allows the pressure build up in the liquid during heating to be released through the interface.

4. Model Validation

4.1. Mesh Independence Analysis

The type of elements used in this model is free tetrahedral. The mesh size in fluid domain is set to be smaller compare to the remaining domain as it is the domain of interest. In order to examine the effect of mesh size refinement on the model solution, the temperature difference of using different mesh size is studied as shown in Figure 3.

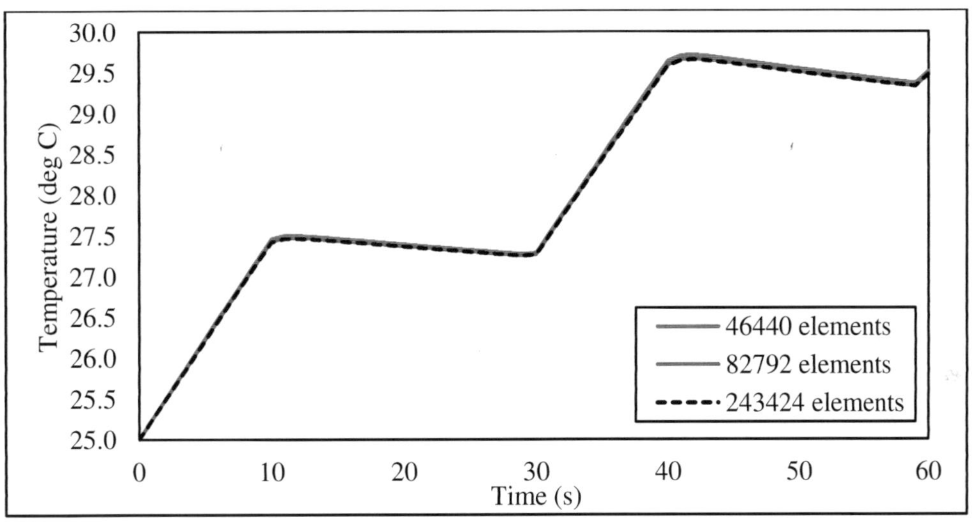

Figure 3 Effect of mesh size refinement on temperature change

As the mesh size refined, it is observed that the temperature change is insignificant. This shows that the mesh size do not need to be further refined and using 46440 elements is sufficient to obtain satisfying results. The simulation time required is approximately 12 minutes for each run.

4.2. Experimental Validation

In order to validate the model, an experiment has been conducted to measure the temperature rise of distilled water at 2.45GHz in a domestic microwave oven. The microwave power level is fixed at 300W. The temperature reading of the water is taken at a point inside the flask as shown in Figure 4 using the fibre optic thermometer for 2 minutes. The temperature measured in simulation model would be taken at this point as well.

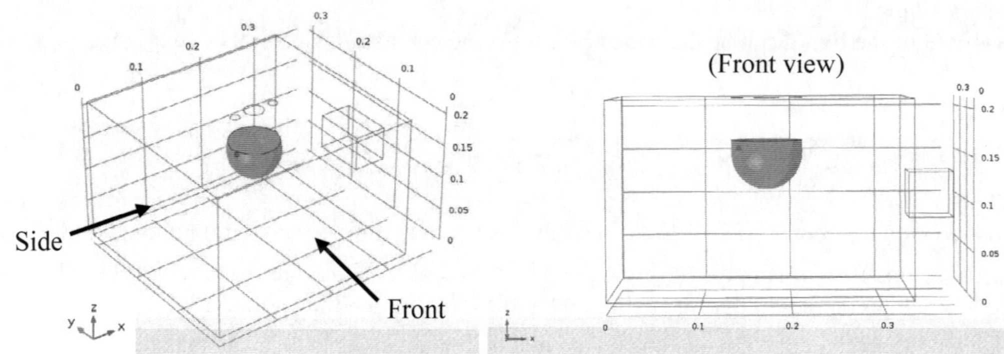

Figure 4 Temperature measuring point

Nine sets of readings has been taken in the experiment to plot the average temperature profile. The comparison between the simulated and experimental results is plotted in Figure 5.

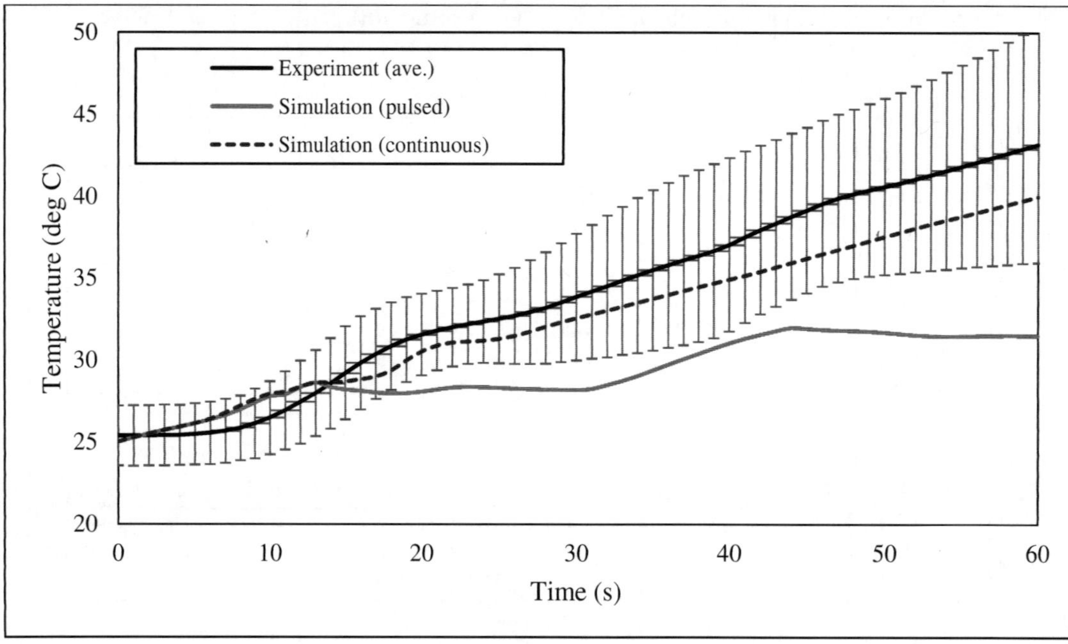

Figure 5 Temperature Profiles of distilled water at 2.45GHz microwave irradiation

As shown by the error bars of the experimental results, the standard deviation of temperature increases as time increases. It is found that the average temperature rise is roughly 17.8°C after heating for 60 seconds. However, the predicted temperature rise of distilled water using the simulation model is only around 6.5°C after a minute. The percentage error increases from 0.4% to 37% as time passes. Nonetheless, the prediction is within the deviation range of the experimental average temperature profile for the first 18 seconds. After that, the simulation result falls out of the range for the remaining time, although it is still slowly increasing.

The possible sources of error include the purity of the distilled water and the waveguide dimension. The electrical conductivity of the distilled water was not tested due to equipment unavailability. This is of concern as the presence of ions would affect both the dielectric constant and dielectric loss of water. These ions will be excited by the microwave and creates additional heat energy which consequently increases the temperature.

During the model development, the dimension of the waveguide is actually a rough estimation based on the actual microwave oven. The equipment was not disassembled to obtain the exact measurement of the waveguide. This could have partially contribute to the discrepancies in simulation results.

Besides, after averaging nine sets of temperature profile, the effect of microwave ON and OFF cycle on the temperature change of experimental results has been dampen. The temperature continues to rise between 18th seconds and 31st seconds although theoretically there is no heat generation. However, more work need to be done here to prove that during the OFF cycle, there is no extra heat generated in the microwave oven. This could be done by using thermal imaging infrared camera to capture the temperature differences of the liquid during the heating process. If it is proved that the heat is continuously generated during the heating process in the experiment, then this model is able to predict the temperature difference as well under continuous heating condition as shown in Figure 5. The predicted temperature results would be slightly below the experimental results but still within the deviation range.

Nonetheless, the author strongly believes that there is no heat generated during the OFF cycle of microwave unless there is further confirmation on the experimental results using the thermal imaging technique. Hence, the pulsed setting will still be used in present work to investigate the effect of frequencies on microwave heating although the error might attain as high as 37%.

5. Effect of Frequencies

The temperature change of the water at three different frequencies is shown in Figure 6. It is found that the temperature increases as time goes by. After 13 seconds, the temperature will drop for 18 seconds as the pulsed microwave is turned off temporarily. As the microwave turns on again, the temperature will then continue to increase. Surprisingly, there is no significant temperature rise in the case of 915 MHz even after 60s. The water does not seem to be heated up after a minute of microwave irradiation. For the 2 GHz frequency, the temperature profile is almost identical to that of 2.45GHz. The slight temperature difference between the two frequencies is suspected to be caused by the convective flow happens within the water.

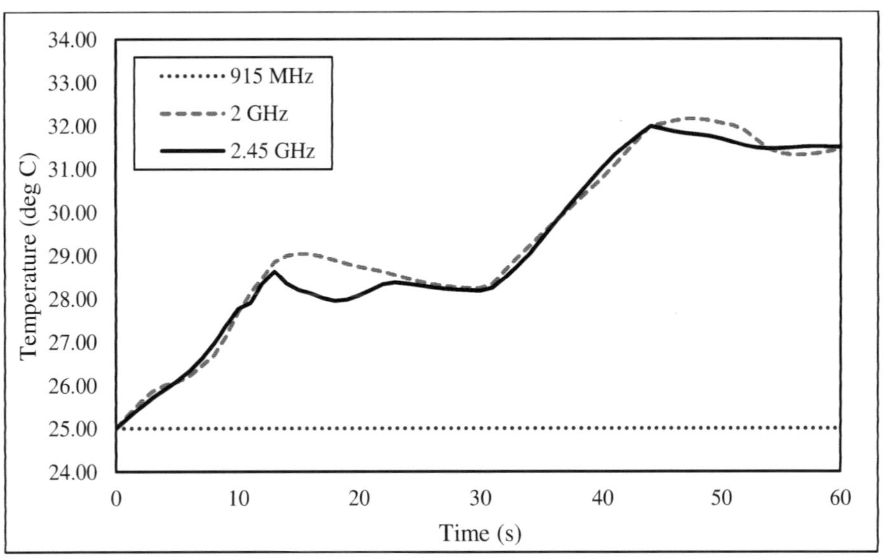

Figure 6 Temperature of distilled water at several frequencies

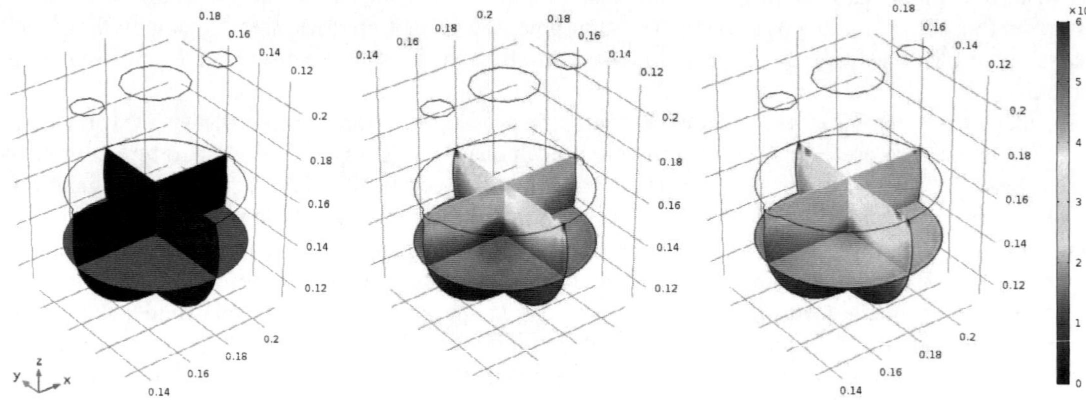

Figure 7 Fluid flow velocity (m/s) in distilled water at t=60s
(left: 0.915GHz, middle: 2GHz, right: 2.45 GHz)

As microwave is switched on, a fraction of the water absorbs the electromagnetic energy and heats up. This results in temperature variance within the water which promotes convection. As illustrated in Figure 7, the fluid flow mostly occurs near the water surface, except for the case of 0.915GHz, where there is no change in the velocity. The maximum velocity observed at 2 GHz and 2.45GHz is around 6×10^{-6} m/s, occurs at the water surface near the flask wall, which can be imagined to be in a ring shape, as indicated by the four red dots on the two planes. The velocity in 2.4GHz is slightly faster compared to 2GHz, and the heat distributes faster to the bottom section of the flask. This might have caused the sudden drop in temperature after 13th seconds and 44th seconds (as shown in Figure 6) for 2.45GHz as cooler water flows upwards from the bottom faster and absorbs heat.

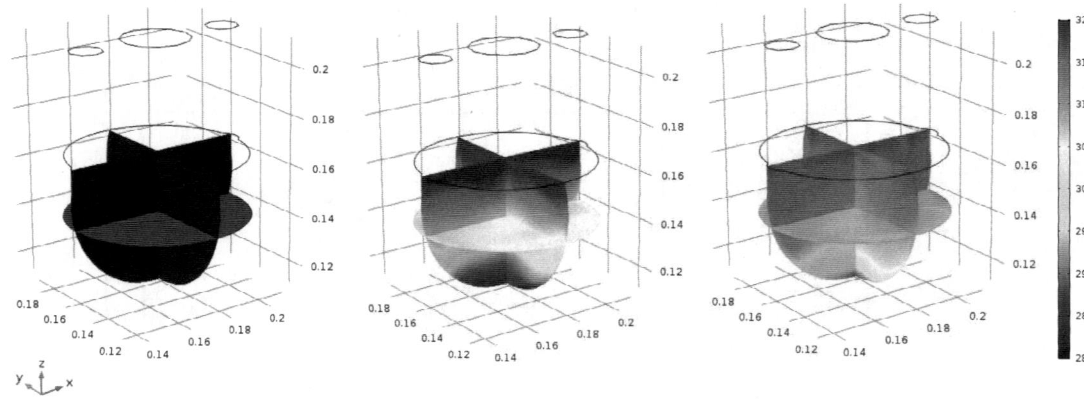

Figure 8 Temperature distribution (°C) in distilled water at t=60s
(left: 0.915GHz, middle: 2GHz, right: 2.45GHz)

Figure 8 is plotted to better visualize the temperature variance of the water after 60 seconds. There is no change in temperature for 0.915GHz, as indicated in Figure 6. For 2GHz, the water temperature is obviously higher near the water surface while the temperature distribution for 2.45GHz is more thorough to the entire volume. Since the heat absorbed by the water is mainly from the electric field (microwave) generated, the electric fields at the three frequencies are plotted in Figure 9.

The electric field in the microwave cavity is very weak for 0.915 GHz. The highest electric field is only 4000 V/m which is found at the waveguide. As a result, the water has not absorbed enough amount of energy to raise its temperature.

For 2GHz, the electric field has becomes stronger and reaches 1.5×10^4 V/m, nearly fourfold of 0.915GHz. This is the same for 2.45 GHz as well. However, it is observed that the electric field is stronger in 2 GHz compared to 2.45GHz. The strong electric field as indicated by the red region occurs near the flask at 2GHz as shown from the front and side view in Figure 9. This explains the high temperature observed near the water surface for 2 GHz as depicted in Figure 8. Electric field around 6000 V/m could also be observed at the center of the liquid domain.

For 2.45 GHz, the strong electric fields locate about the water surface near the flask wall and at the bottom section of the flask. The total area of strong electric field for 2.45GHz are relatively smaller than 2 GHz when viewing from all three planes. This indicates that the water receives less electromagnetic energy at 2.45 GHz frequency.

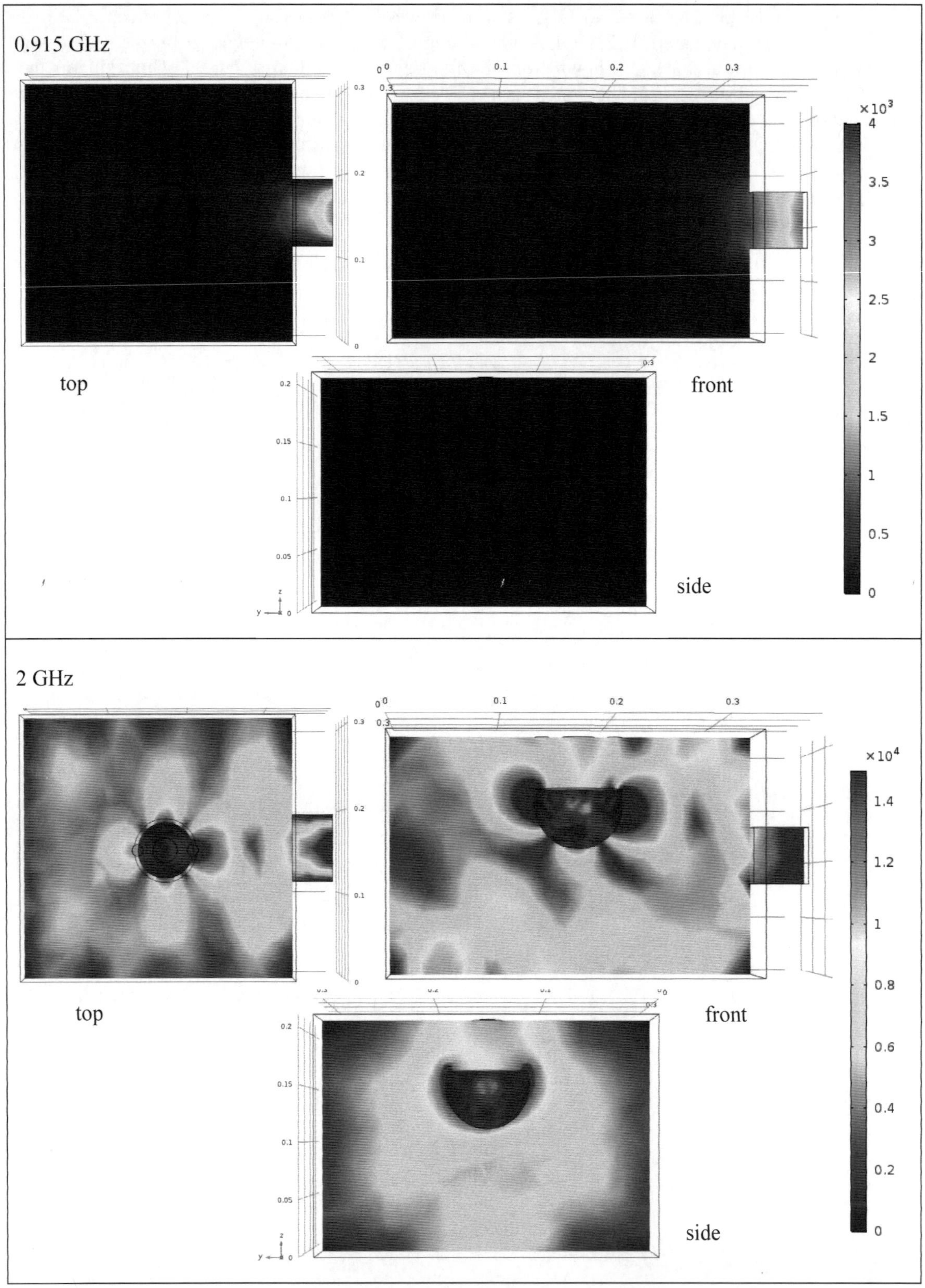
0.915 GHz
top
front
side
×10^3
4
3.5
3
2.5
2
1.5
1
0.5
0
2 GHz
top
front
side
×10^4
1.4
1.2
1
0.8
0.6
0.4
0.2
0

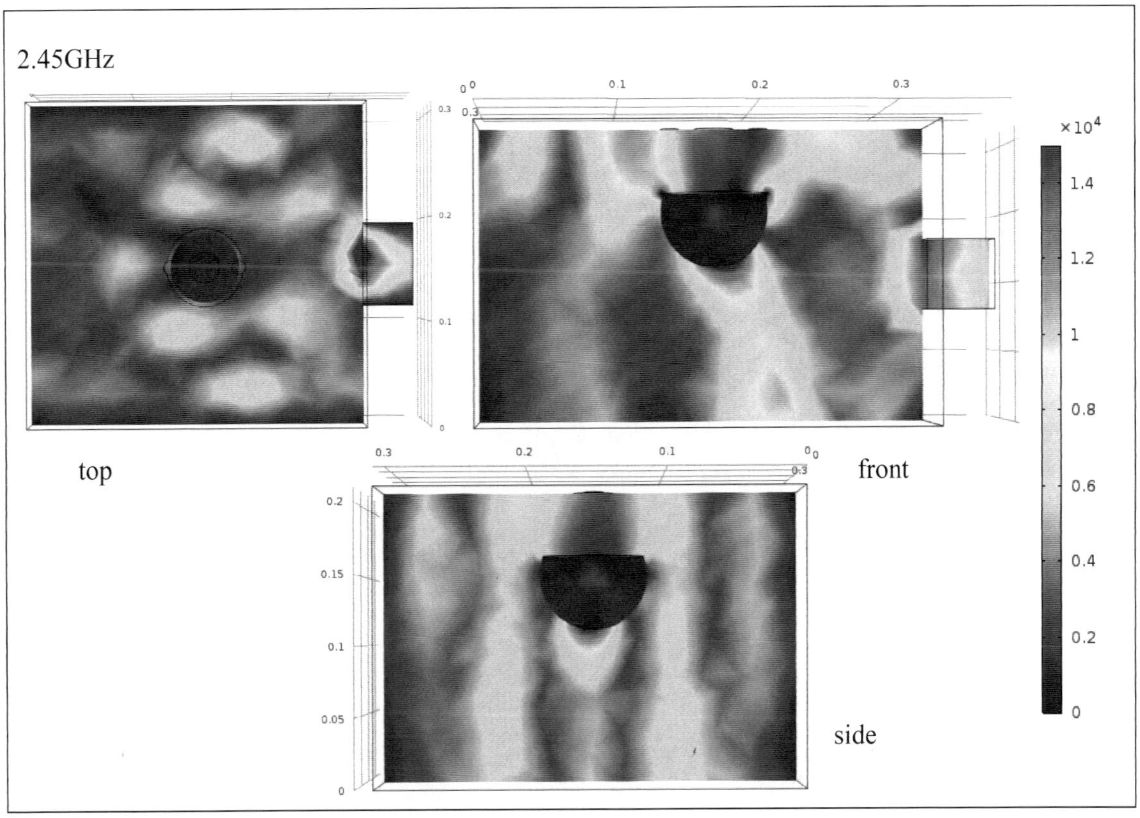

Figure 9 Electric field distribution (V/m) (top: 0.915GHz, middle: 2GHz, bottom: 2.45GHz)

6. Conclusion

A simulation model of the batch microwave heating of water is developed using COMSOL Multiphysics software. It is observed that the water is heated up at similar rate at 2 GHz and 2.45 GHz but the heating rate at 2GHz outperforms 2.45GHz. However, the water does not heat up at 915MHz due to weak electric field distribution in the microwave cavity. More experimental work should be carried out to validate the simulation model.

References

[1] Ratanadecho, P., K. Aoki, and M. Akahori, *A numerical and experimental investigation of the modeling of microwave heating for liquid layers using a rectangular wave guide (effects of natural convection and dielectric properties).* Applied Mathematical Modelling, 2002. **26**(3): p. 449-472.

[2] Cha-um, W., P. Rattanadecho, and W. Pakdee, *Experimental and Numerical Analysis of Microwave Heating of Water and Oil Using a Rectangular Wave Guide: Influence of Sample Sizes, Positions, and Microwave Power.* Food and Bioprocess Technology, 2011. **4**(4): p. 544-558.

[3] Yousefi, T., et al., *An investigation on the microwave heating of flowing water: A numerical study.* International Journal of Thermal Sciences, 2013. **71**: p. 118-127.

[4] Datta, A., H. Prosetya, and W. Hu, *Mathematical Modeling of Batch Heating of Liquids in a Microwave Cavity.* Journal of Microwave Power and Electromagnetic Energy, 1992. **27**(1): p. 38-48.
[5] Zhang, Q., T.H. Jackson, and A. Ungan, *Numerical modeling of microwave induced natural convection.* International Journal of Heat and Mass Transfer, 2000. **43**(12): p. 2141-2154.
[6] M. Sabliov, C., D.A. Salvi, and D. Boldor, *High Frequency Electromagnetism, Heat Transfer and Fluid Flow Coupling in ANSYS Multiphysics.* Journal of Microwave Power and Electromagnetic Energy, 2006. **41**(4): p. 5-17.
[7] Komarov, V.V. and J. Tang, *Dielectric permittivity and loss factor of tap water at 915 MHz.* Microwave and Optical Technology Letters, 2004. **42**(5): p. 419-420.
[8] Rattanadecho, P., *The simulation of microwave heating of wood using a rectangular wave guide: Influence of frequency and sample size.* Chemical Engineering Science, 2006. **61**(14): p. 4798-4811.
[9] Ikediala, J.N., et al., *Development of a saline water immersion technique with RF energy as a postharvest treatment against codling moth in cherries.* Postharvest Biology and Technology, 2002. **24**(2): p. 209-221.
[10] Zahn, M., et al., *Dielectric properties of water and water/ethylene glycol mixtures for use in pulsed power system design.* Proceedings of the IEEE, 1986. **74**(9): p. 1182-1221.
[11] Komarov, V., S. Wang, and J. Tang, *Permittivity and measurements.* Encyclopedia of RF and microwave engineering, 2005.

Acknowledgement
This research has received funding support from FRGS (ref: FRGS/2/2014/TK06/CURTIN/02/1).

Institute of Physics
Dirac House, Temple Back
Bristol BS1 6BE UK

ISSN: 1757-8981
ISBN 978-1-5108-4418-6

9 781510 844186

7th Global Conference on Materials Science and Engineering (CMSE 2018)

IOP Conference Series: Materials Science and Engineering
Volume 474

Xi'an, China
1 - 4 November 2018

IOP
Institute of Physics